Y0-BTC-445

AMERICAN MATHEMATICAL SOCIETY
COLLOQUIUM PUBLICATIONS, VOLUME X

ALGEBRAIC GEOMETRY AND THETA FUNCTIONS

BY

ARTHUR B. COBLE

PROFESSOR OF MATHEMATICS IN THE
UNIVERSITY OF ILLINOIS

NEW YORK
PUBLISHED BY THE
AMERICAN MATHEMATICAL SOCIETY
501 WEST 116TH STREET
1929

LÜTCKE & WULFF, HAMBURG, GERMANY

PREFACE

The present volume is an amplification of the Colloquium Lectures delivered at Amherst in September, 1928, under the title The Determination Of The Tritangent Planes Of The Space Sextic Of Genus Four. In order to present clearly the current state of that problem a comparison with the better known cases of genus two and genus three seems desirable. Preliminary chapters on algebraic geometry and theta functions are incorporated in order to facilitate reading by recalling fundamental ideas of these two subjects in such fashion as will be most helpful in later applications.

An important object is the correlation of two series of memoirs, the one by F. Schottky and the other by the author. In the latter series the properties of discrete sets of points in projective spaces, congruent to each other under regular Cremona transformation, are developed. Such sets irrespective of the number of points and dimension of the space have associated groups which are isomorphic with theta modular groups. On the other hand Schottky, with the theta relations as a starting point, defines a few sets of points in terms of theta modular functions of genera two, three, and four. These two theories are unified by the theorem proved herein that the sets thus defined are transformed under period transformation of the moduli into sets congruent to the original set under Cremona transformation.

The extension of the highly developed theory of the bitangents of a plane quartic curve to the tritangent planes of the space sextic is a matter of obvious interest. The extension of Aronhold's algebraic exposition was proposed in 1915 (and later withdrawn) by the Berlin Academy as the Steiner prize problem. Wirtinger, Roth, and Milne have discussed a single quadratic system of contact quadrics which contains 28 pairs of tritangent planes. In the particular case when the sextic is on a quadric cone a planar set of eight points serves the same purpose as an Aronhold set of seven bitangents of a quartic. The present state of this problem in the general case is discussed in the last section.

Algebraic curves, surfaces, and correspondences have been given in the Clebsch-Aronhold symbolic notation. As a rule this is merely a shorthand device but occasionally symbolic calculations are necessary. In a number of cases geometric configurations have been defined by algebraic forms with unrestricted coefficients and with variables drawn from different domains. The few concepts of group theory which constantly recur are of the simplest type.

The historical side of the subject is quite well covered by the Encyklopädie article of Krazer-Wirtinger and by the Report On Special Topics In Algebraic Geometry recently issued by the National Research Council. The references given are mainly for informational purposes. The content of sections 13, 48, 49, 55, 56 is novel. Other new results appear in sections 28, 32, 33, 38, 39, 41, 50, 51, 54.

Much of the author's own work has been done as a research assistant of the Carnegie Institution of Washington, D. C. The value of this connection as a stimulus to consecutive research, and also the participation of the Institution in the support of this publication, are gratefully acknowledged.

January, 1929.

ARTHUR B. COBLE.

CONTENTS

CHAPTER I

TOPICS IN ALGEBRAIC GEOMETRY

CHAPTER II

TOPICS IN THETA FUNCTIONS

CHAPTER III

GEOMETRIC APPLICATIONS OF THE FUNCTIONS OF GENUS TWO

CHAPTER IV

GEOMETRIC APPLICATIONS OF THE FUNCTIONS OF GENUS THREE

CHAPTER V

GEOMETRIC ASPECTS OF THE ABELIAN MODULAR FUNCTIONS OF GENUS FOUR

CHAPTER VI

THETA RELATIONS OF GENUS FOUR

CHAPTER I

TOPICS IN ALGEBRAIC GEOMETRY

1. Linear systems of plane curves. With $x_0 : x_1 : x_2$ as the homogeneous coördinates of a variable point x in the plane, a curve of order n is the locus defined by the equation,

$$(\alpha x)^n = (\alpha_0 x_0 + \alpha_1 x_1 + \alpha_2 x_2)^n$$

$$= \sum \binom{n}{i_0\, i_1\, i_2} \alpha_{i_0 i_1 i_2} x_0^{i_0} x_1^{i_1} x_2^{i_2} = 0 \qquad (i_0 + i_1 + i_2 = n), \tag{1}$$

where $\binom{n}{i_0\, i_1\, i_2}$ is a polynomial coefficient and the actual coefficient, $\alpha_{i_0 i_1 i_2}$, is expressed in terms of the symbols α_0, α_1, α_2 by

$$\alpha_{i_0 i_1 i_2} = \alpha_0^{i_0} \alpha_1^{i_1} \alpha_2^{i_2}. \tag{2}$$

The $r+1$ curves of order n

$$(\alpha_0 x)^n, (\alpha_1 x)^n, \cdots, (\alpha_r x)^n$$

are linearly independent if the matrix

$$\left\| \alpha^{(k)}_{i_0 i_1 i_2} \right\| \qquad (k = 0, \cdots, r) \tag{3}$$

of $r+1$ rows formed from their coefficients is not zero. In that case the aggregate of curves obtained from variable λ's in the equation,

$$\lambda_0 (\alpha_0 x)^n + \lambda_1 (\alpha_1 x)^n + \cdots + \lambda_r (\alpha_r x)^n = 0 \tag{4}$$

constitutes a *linear system of order* n and *dimension* r. For $r = 1, 2, 3$ we call the system a *pencil, net, web* respectively; or in general a system (∞^r).

The maximum dimension is $N-1$, $N = (n+1)(n+2)/2$, i. e., any linear system of order n is contained in the aggregate of all curves of order n. A linear system of dimension

$r<N-1$ may equally well be defined as the aggregate of curves of order n whose coefficients $\alpha_{i_0 i_1 i_2}$ satisfy $N-1-r$ independent linear relations of the form,

$$(5) \qquad \sum a^{(k)}_{i_0 i_1 i_2} \alpha_{i_0 i_1 i_2} = 0 \qquad (k = 0, 1, \cdots, N-2-r),$$

where again the matrix

$$(6) \qquad \left\| a^{(k)}_{i_0 i_1 i_2} \right\|$$

is not zero. An invariantive statement of the relation (5) is that the curve, $(\alpha x)^n = 0$ of order n is *apolar* to the curve

$$(7) \quad (a\xi)^n = (a_0 \xi_0 + a_1 \xi_1 + a_2 \xi_2)^n = \sum a_{i_0 i_1 i_2} \xi_0^{i_0} \xi_1^{i_1} \xi_2^{i_2} = 0$$

of class n, i. e., the simultaneous invariant (5), or $(\alpha a)^n$, vanishes. Thus with every linear system of dimension $r<N-1$ and order n there is associated a linear system of dimension $N-2-r$ and class n,

$$(8) \quad l_0 (a_0 \xi)^n + l_1 (a_1 \xi)^n + \cdots + l_{N-2-r} (a_{N-2-r} \xi)^n = 0,$$

such that each system contains all the curves apolar to every curve of the other. We say that the two linear systems are *apolar*. Their matrices, (3) and (6), are also apolar, i. e., the row-product (5), formed for any row of the one with any row of the other, vanishes. A theorem of Grassmann[32] (cf. [17] I, p. 158) states that the determinants with proper signs formed from the sets of $r+1$ columns of the one matrix are proportional to the determinants formed from the respective complementary sets of $N-r-1$ columns of the apolar matrix. It is the values of these determinants, values subject, for $0<r<N-2$, to certain quadratic relations, which characterize the linear systems rather than the particular curves selected for their expression in (4) and (8).

A particular type of apolarity relation serves to express the conditions that all the curves of a linear system may have the same behavior at a given point called a *base-point* of the system. If the apolar linear system contains a linear system whose curves consist of a point, $(b\xi) = 0$, repeated $n-k+1$ times together with an *arbitrary* curve of class $k-1$, say $(c\xi)^{k-1} = 0$, then $(b\xi)^{n-k+1}\cdot(c\xi)^{k-1}$ is apolar to every curve $(\alpha x)^n$ of (4), i. e. $(\alpha b)^{n-k+1}(\alpha c)^{k-1} = 0$ for every choice of c, or $(\alpha b)^{n-k+1}(\alpha x)^{k-1} \equiv 0$. This imposes, for given b, $k(k+1)/2$ linear conditions on the coefficients α which for the particular choice of $b = 0:0:1$ are obviously independent. The conditions express that the curves α have a k-fold point at b. The product of the k tangents is $(\alpha b)^{n-k}(\alpha x)^k = 0$. If in this we set $x = b', b'', \cdots$ we obtain further linear conditions on the α's which express that branches of the curves α at b have the fixed directions $bb', bb'', \cdots$.

A linear system which contains *all* the curves of order n which have a given *base*, $B(b_1^{k_1}, b_2^{k_2}, \cdots, b_j^{k_j})$, i. e. which have at the point b_i at least a k_i-fold point, is said to be *complete*; otherwise *incomplete*. The apolarity relations which define a complete system imply that certain intersections of two members of the system are fixed. The limiting cases for which one or more of the base points approach a given base point along a given curve are included. Thus if b_1^1 approaches $b_2^{k_2}$ along a given line, the curves have a k_2-fold point at b_2 with this given line as one of the k_2 tangents.

Two theorems of Bertini[6] are useful. The first states that if the general curve of the system has a k-fold point variable with the curve then all the curves of the system have a fixed $(k-1)$-fold part; the second that if the general curve is degenerate then either all the curves have a fixed part or each is made up of h variable members of a fixed pencil. If we assume that a common part has been removed from the members of the system and that the curves then are not composed of members of a pencil we may conclude that the general curve of the system is irreducible and has no multiple points outside the base points of the system. Unless definitely stated otherwise we will consider only systems of this latter character, and these will be termed *proper*.

Let then a complete linear system be defined by a given base, $B(b_1^{k_1}, \cdots, b_j^{k_j})$. The dimension r of the system is given by

1*

(9) $$r = s_n + [n(n+3) - \sum_j k_j(k_j+1)]/2$$

where s_n is the number of independent relations among the linear conditions imposed by given B on curves of order n. If $s_n = 0$ the system is *regular*, otherwise *irregular* with *irregularity* (superabundance) s_n. It may be proved that $s_n = 0$ for sufficiently large n ([51] p. 53). Let the *grade* D of the linear system be the number of variable intersections of two general curves of the system. Then allowing for the fixed intersections at B we have

(10) $$D \doteq n^2 - \sum_j k_j^2 .$$

If also p is the genus of a general curve of the system, then

(11) $$p = [(n-1)(n-2) - \sum_j k_j(k_j-1)]/2 .$$

Thus r, D, p are connected by the relation

(12) $$r = D - p + 1 + s_n .$$

The members of a linear system which pass through a point P of the plane in general position constitute a new linear system (∞^{r-1}). If the new system has acquired along with P the base points $P_1, P_2, \cdots, P_{t-1}$ then the points of the plane divide into sets each of t points such that any curve of the system which passes through one point of a set must also pass through the other points of the set. If $t = 1$ the system is termed *simple*; if $t > 1$ *not simple*. But simple or not there is but one curve of the system on r points in general position; and conversely if an algebraic system of dimension r is such that it contains but one curve on r points in general position then it is a linear system ([64] p. 12, pp. 15–24; [51] pp. 56–57).

2. Mappings determined by linear systems. Given a proper linear system $|C_n|$ of order n and of dimension $r > 2$ we select $r+1$ independent curves and set

(1) $$\varrho y_0 = (\alpha_0 x)^n, \ \varrho y_1 = (\alpha_1 x)^n, \cdots, \varrho y_r = (\alpha_r x)^n.$$

We interpret $y_0 : y_1 : \cdots : y_r$ as the homogeneous coördinates of a point y in a linear space S_r and have for each point x

of the plane which is not a base point of $|C_n|$ a point y of S_r. Since the curves of $|C_n|$ are not composed of members of a pencil, the point y will run over a manifold M_2 of dimension two in S_r. If $\eta_0 : \cdots : \eta_r$ are dual coördinates in S_r, i. e. if $(\eta y) = 0$ is the equation of an S_{r-1} in S_r, then y lies on the S_{r-1}, η, if x is on the curve, $\sum_{i=0}^{r} \eta_i (\alpha_i x)^n = 0$, of $|C_n|$. Thus the linear sections η of M_2 arise from the curves of the planar system. The order of M_2, the number of variable points in which it is met by an S_{r-2} defined by η, η', is the number D of variable intersections x of two curves provided $|C_n|$ is simple; otherwise is D/t. Castelnuovo [9] has proved that if $t > 1$ the sets of t points x may be put into one-to-one correspondence with the points x' of a plane. Thus in all cases the points y of M_2 in S_r are in one-to-one or birational correspondence with the points of a plane and M_2 is therefore a *rational surface* in S_r. If conversely such a rational surface is given with points y in rational correspondence with points x of a plane, its linear sections η determine a proper linear system in the plane.

We observe that the coördinates y in (1) are fixed by the choice of $r+1$ independent curves of $|C_n|$ and the choice of a factor of proportionality in each. If the given choice is altered, the point x determines a new point y' which arises from y by linear transformation or collineation. If on the other hand we carry out in the plane a collineation which carries x into x' then $(\alpha_i x)^n$ becomes $(\alpha'_i x')^n$ and the point y on M_2 determined by x is the same as that determined by x' since $\varrho y_i = (\alpha_i x)^n = (\alpha'_i x')^n$. Thus the relation between M_2 and the linear system $|C_n|$ is definite only to within projective transformation both in S_r and in the plane. We extend this later to Cremona transformation in the plane.

In the birational relation between the plane x and $M_2(y)$ there are two kinds of exceptional points. We consider only the case of simple linear systems. The first type is a k-fold base point b. Setting b in (1) the coördinates y vanish and the point y is indeterminate. Let c be a general point and let x approach b along the line bc i. e. let λ approach 0 in $b+\lambda c$. When we set $x=b+\lambda c$ in (1) the terms in λ vanish up to the degree k since $(\alpha_i b)^{n-k+1}(\alpha_i x)^{k-1}\equiv 0$. The terms in λ^k remain and higher powers of λ may be dropped. Thus $\varrho y_i=(\alpha_i b)^{n-k}(\alpha_i c)^k$. If now $(\eta y)=0$, the point c is on a tangent to the curve $\sum \eta_i(\alpha_i x)^n=0$ at its k-fold point b. Hence all sections η of M_2 on y correspond to curves of $|C_n|$ with a fixed tangent at b in the direction bc. As c varies along a line $c+\lambda c'$ not on b, the coördinates y_i are expressed as rational functions of degree k in the parameter λ and the points y which correspond to the various directions at b run over a rational curve on M_2 of order k. It may happen however that the directions at b divide into ∞^1 sets of t each such that any curve α which passes through b with one direction in a set of t must have branches in each direction of the set. Then the rational curve on M_2 which corresponds to the directions at b is a t-fold curve of order k_1, where $k=tk_1$. We call these base points the *fundamental* or *F-points* of the rational transformation (1); and the rational curves which correspond to directions about them, the corresponding *principal* or *P-curves*.

The second type of exceptional point lies on M_2 and occurs when the linear system $|C_n|$ has in the plane a P-curve i. e. a curve which meets the curves of the system only at the base points. If x is on the P-curve the system (∞^{r-1}) of $|C_n|$ on x contains the P-curve as a factor and the residual variable part $|C_m|$ will meet the P-curve in k variable points. Since $|C_m|$ is the same for all positions of x on the P-curve the surface M_2 has a k-fold point corresponding to the P-curve in such a way that the ∞^1 directions at the k-fold point correspond to the ∞^1 points on the P-curve.

The simplest example of M_2 in S_r is the quadric in S_3, the map of the plane by the complete system of conics $|C_2|$ on two base points, $B(b_1, b_2)$. The line of the pencil on b_1 with parameter t is cut by $|C_2|$ in one variable point and thus maps into a generator t of the quadric; similarly the lines on b_2 map into the cross generators τ of the quadric. If t_0, τ_0 are the parameters in these respective pencils of the P-curve, $b_1 b_2$, then directions at b_1 have a variable t but common τ_0. Thus the F-points b_1, b_2 map respectively into the generators τ_0, t_0 which are the P-curves on the quadric. Points on the P-curve $b_1 b_2$ have parameters t_0, τ_0 and map into the intersection P of the two P-curves on the quadric. The transformation is reversed by projection of the quadric upon the plane from P. The generator τ_0 on P cuts the plane at b_1; t_0 on P at b_2. The directions at P in the tangent plane project into points on $b_1 b_2$. A general plane section projects into a conic on b_1, b_2.

3. Linear systems ($r = 2$, $D = 1$); **Cremona transformations.** If $|C_n|$ is a proper net for which $D = 1$, if C is a general irreducible curve of the net, and C', C'' two other curves which with C define the net, then C is but by the pencil $C' + \lambda C''$ in a single variable point and therefore is rational. Moreover the net is a complete system. For if C'' were a further independent curve with the same behavior at the base points, a member of the net $C' + \lambda C'' + \mu C'''$ could be determined with two variable intersections, and therefore a part common, with C. From the relation $r = D - p + 1 + s_n$, valid for complete systems, in which $r = 2$, $D = 1$, $p = 0$, we find that $s_n = 0$. Such a net defined by the requirements that the general curve is irreducible, and that $r = 2$, $D = 1$ is called *homaloidal*.

Let the homaloidal net in S_x be defined by its base points $p_1, p_2, \cdots, p_\varrho$ of orders $r_1, r_2, \cdots, r_\varrho$. From the values of p and D as in I we find that

$$(1) \qquad \begin{aligned} r_1^2 + r_2^2 + \cdots + r_\varrho^2 &= n^2 - 1, \\ r_1 + r_2 + \cdots + r_\varrho &= 3(n-1). \end{aligned}$$

The mapping

$$(2) \qquad y_i = (\alpha_i x)^n \qquad (i = 0, 1, 2),$$

is now from point x of S_x to point y of the plane M_2^1 or S_y. The lines, $(\eta y) = 0$, of S_y are in projective correspondence with the curves, $\sum \eta_i (\alpha_i x)^n = 0$, of the homaloidal net. To the variable intersection y of two lines η, η' in S_y there corresponds the variable intersection x of the two corresponding curves of the net. The transformation thus established between the points of the two planes—valid throughout except for a finite number of points—is called a *Cremona transformation* T. For proper choice of integers n, r_i, ϱ the net in S_x is determined when the ϱ base points are selected. The projectivity with lines in S_y involves no new absolute constants. Thus T depends upon $2\varrho - 8$ absolute projective constants.

Lines in S_x are mapped into rational curves of order n in S_y which in general are irreducible. These curves constitute an algebraic system of dimension two such that on two generic points y there is a single curve of the system. Thus the system is a proper net for which $r = 2$, $D = 1$, i. e. a homaloidal net in projective correspondence with the lines of S_x. We may then solve (2) in the form

$$x_i = (\beta_i\, y)^n \qquad (i = 0, 1, 2). \tag{3}$$

If this net in S_y has base points $q_1, \cdots, q_\sigma$ of orders $s_1, \cdots, s_\sigma$ then T has $2\sigma - 8$ absolute projective constants whence $\sigma = \varrho$. As before

$$\begin{aligned} s_1^2 + s_2^2 + \cdots + s_\varrho^2 &= n^2 - 1, \\ s_1 + s_2 + \cdots + s_\varrho &= 3(n-1). \end{aligned} \tag{4}$$

We call $p_1, \cdots, p_\varrho$ the *fundamental points* or *F-points* of the transformation T from S_x to S_y; and similarly $q_1, \cdots, q_\varrho$ the F-points of the *inverse* transformation T^{-1} from S_y to S_x.

As in the more general mapping of **2**, directions at the F-point p_i $(i = 1, \cdots, \varrho)$ of order r_i in S_x correspond to the points on a P-curve, P_i, of order r_i in S_y. To the pencil of lines on p_i there corresponds in S_y a pencil of variable rational curves of order $n - r_i$, each taken with the fixed part P_i which is cut by the variable part in the point which corresponds to the direction at p_i on the variable line. Thus a P-curve in S_y is the fixed part of a degenerate pencil of the net in S_y. Conversely the points of such a fixed part must correspond to a single point — necessarily an F-point — in S_x. The orders also of such fixed parts determine the orders of the F-points in S_x. Thus either net once given determines the nature of the other net as well as (for $n > 2$ and $\therefore \varrho > 4$) the position of the F-points of the other net when four of these have been chosen.

Let the P-curve P_i in S_y which corresponds to the F-point p_i in S_x pass through the F-point q_j $(j = 1, \cdots, \varrho)$ in S_y with say α_{ij} branches. Then there are α_{ij} directions at q_j which correspond to directions at p_i. Hence the P-curve, Q_j, in S_x which corresponds to directions at q_j in S_y also passes α_{ij} times through p_i.

When the F-points of T and T^{-1} are in general position, the P-curves have the following properties: (a) they are rational; (b) they have multiple points only at the F-points; (c) they are completely determined by their behavior at the F-points; (d) they meet the general curve of the net only at the F-points; (e) they meet each other only at the F-points; and (f) as an aggregate they make up the jacobian of the net. The last of these follows from the fact that the jacobian is the locus of double points of curves of the net. The others are immediate consequences of properties of T or of the relations 3 (4).

4. Further relations on the integers n, r_i, s_j, α_{ij} associated with T. With reference to a given Cremona transformation T and a given set P_m^2 of m discrete points $p_1, \cdots, p_\varrho, \cdots, p_m$ in S_x of which the first ϱ are the F-points of T, any given curve C_x has a „singularity complex" (S. Kantor, [37] pp. 293–316) which consists of the set of integers (positive or zero),

$$\gamma_0, \gamma_1, \cdots, \gamma_\varrho, \gamma_{\varrho+1}, \cdots, \gamma_m$$

where γ_0 is the order of C_x; $\gamma_1, \cdots, \gamma_\varrho$ the multiplicities of C_x at the F-points $p_1, \cdots, p_\varrho$ of T; and $\gamma_{\varrho+1}, \cdots, \gamma_m$ the multiplicities of C_x at the $m-\varrho$ further points of P_m^2 which are ordinary points for T but which for some reason or other we may wish to notice.

The curve C_x is then transformed by T into a curve C'_y in S_y with the singularity complex

$$\gamma'_0,\ \gamma'_1,\ \cdots,\ \gamma'_\varrho,\quad \gamma'_{\varrho+1},\ \cdots,\ \gamma_m$$

with respect to a set of points Q^2_m of m points in S_y of which $q_1, \cdots, q_\varrho$ are the F-points of T^{-1} and $q_{\varrho+1}, \cdots, q_m$ are the correspondents or *images* under T of the ordinary points $p_{\varrho+1}, \cdots, p_m$. Again γ'_0 is the order of C'_y and γ'_i $(i = 1, \cdots, m)$ the multiplicity of C'_y at q_i. Two planar sets of points P^2_m, Q^2_m so related that ϱ points in each set will serve for F-points of T and T^{-1} respectively while the remaining $m-\varrho$ points of each set are corresponding pairs of T and T^{-1} will hereafter be said to be *congruent sets of points under* T.

A line is transformed by T into a curve of order n with a point of order s_j at q_j $(j = 1, \cdots, \varrho)$. These orders are multiplied by γ_0 for the transform of C_x of order γ_0. However, for each branch of C_x through p_i, the curve P_i separates from the transform. At ordinary corresponding pairs the behavior of C_x and its transform is the same. Hence the singularity complexes γ and γ' of C_x and C'_y are connected by the *linear transformation* S:

$$(1)\quad S:\quad \begin{aligned} \gamma'_0 &= n\gamma_0 - r_1\gamma_1 - \cdots - r_i\gamma_i - \cdots - r_\varrho\gamma_\varrho, \\ \gamma'_j &= s_j\gamma_0 - \alpha_{1j}\gamma_1 - \cdots - \alpha_{ij}\gamma_i - \cdots - \alpha_{\varrho j}\gamma_\varrho, \\ \gamma'_{\varrho+k} &= \gamma_{\varrho+k} \\ (j &= 1, \cdots, \varrho;\quad k = 1, \cdots, m-\varrho). \end{aligned}$$

It is clear that two curves C_x, D_x with complexes γ, δ will have the same number of intersections outside the set P^2_m in S_x as their transforms have outside the set Q^2_m in S_y, i. e.

$$\gamma'_0\,\delta'_0 - \gamma'_1\,\delta_1 - \cdots - \gamma'_m\,\delta'_m = \gamma_0\,\delta_0 - \gamma_1\,\delta_1 - \cdots - \gamma_m\,\delta_m.$$

Thus S has the invariant quadratic form

$$(2)\qquad Q = \gamma_0^2 - \gamma_1^2 - \gamma_2^2 - \cdots - \gamma_m^2 \qquad (m \geqq \varrho).$$

Moreover the genus of C_x is invariant under T whence $(\gamma_0 - 1)(\gamma_0 - 2) - \sum_{i=1}^{m} \gamma_i(\gamma_i - 1)$ is invariant. By combining this with (2) we find the invariant linear form

$$L = 3\gamma_0 - \gamma_1 - \gamma_2 - \cdots - \gamma_m \qquad (m \geqq \varrho). \tag{3}$$

As an easy algebraic consequence we have

(4) *The invariance under the linear transformation S of the forms Q, L furnishes the following relations on the integers (positive or zero) n, r_i, s_j, α_{ij} associated with the Cremona transformation T:*

$$\begin{aligned}
\textstyle\sum_i r_i^2 &= n^2 - 1, & \textstyle\sum_j s_j^2 &= n^2 - 1,\\
\textstyle\sum_i r_i &= 3(n-1), & \textstyle\sum_j s_j &= 3(n-1),\\
\textstyle\sum_i \alpha_{ij}^2 &= s_j^2 + 1, & \textstyle\sum_j \alpha_{ij}^2 &= r_i^2 + 1,\\
\textstyle\sum_i \alpha_{ij} &= 3s_j - 1, & \textstyle\sum_j \alpha_{ij} &= 3r_i - 1,\\
\textstyle\sum_i \alpha_{ij} r_i &= n s_j, & \textstyle\sum_j \alpha_{ij} s_j &= n r_i,\\
\textstyle\sum_i \alpha_{ij}\alpha_{ik} &= s_j s_k, & \textstyle\sum_j \alpha_{ij}\alpha_{kj} &= r_i r_k
\end{aligned}$$
$$(i, j = 1, \cdots, \varrho).$$

The integers n, r_i (or n, s_j) are subject also to a series of inequalities, which when the r's are arranged in order of magnitude,

$$r_1 \geqq r_2 \geqq r_3 \geqq \cdots \geqq r_\varrho,$$

take the form

$$\begin{aligned}
n &\geqq r_1 + r_2,\\
2n &\geqq r_1 + r_2 + \cdots + r_5,\\
3n &\geqq 2r_1 + r_2 + \cdots + r_7,\\
3n &\geqq r_1 + r_2 + \cdots + r_9,\\
&\cdots\cdots\cdots
\end{aligned} \tag{5}$$

These express respectively that the net does not necessarily contain as a factor the line $p_1 p_2$ or say the curve $(12)^1$, nor the curve $(12345)^2$, nor the curve $(1^2\,234567)^3$, nor the curve $(123456789)^3$, etc. The first case of numbers n, r_i which satisfy the two relations 3(1) and which is ruled out by (5) is $n = 5$; $r_1, \cdots, r_8 = 3, 3, 1, \cdots, 1$. Whether the equalities 3 (1) and the inequalities (5) are sufficient to ensure that further integers s_j, α_{ij} can be determined which satisfy (4), and whether the satisfaction of equations (4) is sufficient to ensure that a transformation T with the corresponding set of integers exists, are questions not yet answered.

A variety of properties of this set of integers associated with T have been given by Clebsch, Bertini, Montesano, and others ([52] Chap. IV, §§ 1, 2).

5. On the nature of the Cremona group and its invariants. If T_1 is a Cremona transformation from point x to point x', T_2 another from point x' to x'' (x, x', x'' all in the same plane), the product $T_1 T_2$ is a Cremona transformation from x to x''. The totality of all Cremona transformations in the plane (including as particular cases the projective transformations) constitute the Cremona group in the plane. S. Kantor[40] describes this group as one whose elements contain certain continuous parameters $a_1, a_2, \cdots$, and certain discontinuous parameters $\alpha_1, \alpha_2, \cdots$; each set of parameters being infinite in number. The parameters α vary with the *type* of T. We say that two transformations with integers n, r_i and n', r_i' are of different type if $n \neq n'$; or if when $n = n'$ the r_i are not merely a permutation of the r_i'. When the type is fixed the continuous parameters a of that type vary with the choice of the direct and inverse F-points subject to the $2\varrho - 8$ relations which connect them. We obtain in the next two sections a separation of these two types of parameters in so far at least as concerns certain sub-aggregates of the Cremona group.

The Cremona group in the plane is more amenable to general treatment than the similar group in higher spaces as a consequence of the theorem of Noether (cf. [52] Cap. IV, 7) that T can be expressed as a product of quadratic transformations. Perhaps a simpler statement is that the Cremona group can be generated by collineations and a single involutorial quadratic transformation A_{123} with F-points at p_1, p_2, p_3 and P-curves, $P_i = \overline{p_j p_k}$ $(i = 1, 2, 3)$.

There is no simple basis for an invariant theory of the Cremona group. It is evident that a linear system, $|C_{\gamma_0}|$, of curves is transformed into a linear system $|C'_{\gamma'_0}|$ with the same r, D, p. But these numbers are merely arithmetic invariants of the system. Two systems with the same invariants r, D, p are not usually conjugate under the Cremona group. But a complete linear system defined by a base B is transformed by T into a complete linear system defined by a base B' and the two sets of points B, B' (with the inclusion perhaps of base points of zero multiplicity) are *congruent under* T as defined in **4**. Thus the planar set of m points $p_1, \cdots, p_m$, or P_m^2, usually discrete, as the carrier, actual or potential, of an infinite variety of complete linear systems, seems a natural basis for an invariant theory. This notion will recur frequently.

A complete linear system, defined on a set P_m^2, consists of a continuous aggregate of curves. We shall have occasion also to discuss the discontinuous aggregate of principal curves or P-curves (cf. **9**, **56**) defined by P_m^2, an aggregate which is infinite for $m > 8$.

6. Types of Cremona transformations. The arithmetic linear group, $g_{m,2}$. If the set of points P_m^2 is congruent to the set of points Q_m^2 under T_1 and if Q_m^2 is congruent to R_m^2 under T_2, then P_m^2 and R_m^2 are congruent under the product $T_1 T_2$. A curve C_x with complex γ at P_m^2 passes under T_1 into a curve $C'_{x'}$ with complex γ' at Q_m^2, and this under T_2 into a curve $C''_{x''}$ with complex γ'' at R_m^2. Then γ' is expressed linearly in terms of γ by S_1 (**4** (1)), γ'' in terms of γ' by S_2, and γ'' in terms of γ by the product $S_1 S_2$. The class of all planar sets Q_m^2 congruent to a given set P_m^2 has the property that any two members of the class are congruent under a definite transformation T, localized by the two sets, with its $\varrho \leqq m$ direct and inverse F-points in the two sets. These transformations T regarded as planar transformations constitute a *groupoid* (cf. [66] p. 5) rather than a group, i.e., elements T defined by P_m^2, Q_m^2 and T' defined by $P_m^{2\prime}$, $Q_m^{2\prime}$ have a product TT' defined by P_m^2, $Q_m^{2\prime}$ only when $P_m^{2\prime}$ coincides with Q_m^2.

The elements S however are merely descriptive and are independent of the position of the sets of points. If m is given they constitute a group which expresses the various ways in which curves or linear systems of curves defined at one set of m points may be transformed into similar systems defined at another set of m points. We call this group of linear transformations with integer coefficients, the group $g_{m,2}$ and seek a system of generators.

It may happen that the sets P_m^2 and Q_m^2 are congruent under collineation ($\gamma_0' = \gamma_0$) but in such wise that the points p in natural order correspond to the points q in some permuted order. Thus $g_{m,2}$ contains the permutation group of $\gamma_1, \cdots, \gamma_m$ whose elements we shall indicate by writing their cycle form in terms of $1, \cdots, m$. Also P_m^2, Q_m^2 may be congruent under the quadratic transformation A_{123} so that $g_{m,2}$ contains the element

$$(1) \qquad A_{123}: \begin{array}{l} \gamma_0' = 2\gamma_0 - \gamma_1 - \gamma_2 - \gamma_3, \\ \gamma_1' = \gamma_0 \qquad\;\; - \gamma_2 - \gamma_3, \\ \gamma_2' = \gamma_0 - \gamma_1 \qquad\;\; - \gamma_3, \\ \gamma_3' = \gamma_0 - \gamma_1 - \gamma_2 \qquad\;\;, \\ \gamma_i' = \gamma_i \qquad\qquad\qquad (i = 4, \cdots, m). \end{array}$$

Since the ternary Cremona group is generated by collineations and A_{123} there follows:

(2) *The linear group, $g_{m,2}$ is generated by the permutation group Π of $\gamma_1, \cdots, \gamma_m$ and the element A_{123}.*

We mention some examples. The quadratic inversion with center p_1 and conic of fixed points on p_2, p_3 has $S = A_{123} \cdot (23)$. The quadratic transformation A_{124} has $S = (34) \cdot A_{123} \cdot (34)$. The cubic involution with double F-point at p_1 and simple F-points at p_i such that the P-curve of p_i is the line $p_1 p_i$ $(i = 2, \cdots, 5)$ has $S = A_{123} \cdot A_{145} \cdot (23)(45)$. Similarly the Jonquières involution of order n has

$$S = A_{123} \cdot A_{145} \cdot \cdots \cdot A_{1,2n-2,2n-1} \cdot (23) \cdots (2n-2, 2n-1).$$

Since T preceded or followed by a collineation is a transformation T' of the same type as T itself we have the theorem:

(3) *The types of Cremona transformations with* $\varrho \leqq m$ *F-points are in one-to-one correspondence with the double cosets of the group* $g_{m,2}$ *with respect to its subgroup* Π *of order* $m!$

We recall (cf. Miller, Blichfeldt, Dickson[44.1] pp. 25–26) that a resolution of $g_{m,2}$ into double cosets with respect to Π has the form

$$g_{m,2} = \Pi \cdot \Pi + \Pi S_1 \Pi + \Pi S_2 \Pi + \cdots$$

where S_i is so chosen as not to lie in a preceding double coset. No element of $g_{m,2}$ in one double coset is found in another but the elements within a coset may each be repeated k times. If T_i has α_0 F-points of zero order (ordinary points), α_1 simple F-points, $\cdots\cdots$, α_j F-points of order $j\,(\alpha_0 + \alpha_1 + \cdots + \alpha_j = m)$ then in the double coset $\Pi S_i \Pi$ each element occurs $k = \alpha_0!\ \alpha_1! \cdots \alpha_j!$ times.

The only sure method of finding actually existent new types is by constructing products of known types i. e. by finding products of elements of $g_{m,2}$ (cf.[52] Chap. IV, § 2).

In order to identify $g_{m,2}$ with certain known groups we shall frequently use a certain conjugate set of involutions in $g_{m,2}$ which generates the group. Any element of $g_{m,2}$, as a linear transformation, has a set of "multipliers". For an involution these are ± 1. The subgroup Π of $g_{m,2}$ is generated by transpositions of type (12). For $\gamma_1 - \gamma_2$ the transposition has one multiplier -1 and for $\gamma_1 + \gamma_2$ and every other γ a multiplier $+1$. Hence (12) is a central involution with space of fixed points $\gamma_1 - \gamma_2 = 0$ and center at the pole of this fixed space with respect to the invariant quadratic form Q [4 (2)]. Thus the involution is defined by its space of fixed points $\gamma_1 - \gamma_2 = 0$ and has the determinant -1. We determine the conjugate involutions by the fixed spaces conjugate to $\gamma_1 - \gamma_2 = 0$. Under A_{123}, $\gamma_3 - \gamma_4$ becomes $\gamma_0 - \gamma_1 - \gamma_2 - \gamma_4$ and under (34) this becomes $\gamma_0 - \gamma_1 - \gamma_2 - \gamma_3$. But $A_{123}\,(34)\,A_{123} = A_{124}$ and $A_{123} = (34)\,A_{124}\,(34)$. Hence A_{123} is in the conjugate set of involutions and is defined by $\gamma_0 - \gamma_1 - \gamma_2 - \gamma_3$. Similarly the quintic transformation $A_{123}\,A_{456}\,A_{123}$ is in the conjugate set and is defined by

$2\gamma_0-\gamma_1-\gamma_2-\cdots-\gamma_6$. Thus we find a sequence of conjugate fixed spaces

$$
\begin{aligned}
&\gamma_1-\gamma_2,\\
&\gamma_0-\gamma_1-\gamma_2-\gamma_3,\\
&2\gamma_0-\gamma_1-\gamma_2-\cdots\cdots-\gamma_6,\\
&3\gamma_0-2\gamma_1-\gamma_2-\cdots\cdots-\gamma_8,\\
&4\gamma_0-2\gamma_1-2\gamma_2-2\gamma_3-\gamma_4-\cdots-\gamma_9,\\
&\cdots\cdots\cdots\cdots\cdots\\
&c_0\gamma_0-c_1\gamma_1-\cdots\cdots\cdots\cdots\cdots-c_m\gamma_m,
\end{aligned}
\tag{4}
$$

together of course with all that arise from these by the operations of Π. We observe that the coefficients c of the general form satisfy, in the early cases, the relations

$$
\begin{aligned}
c_0^2-c_1^2-\cdots-c_m^2&=-2,\\
3c_0-c_1-\cdots-c_m&=0.
\end{aligned}
\tag{5}
$$

Since according to 4 the left members are invariant under $g_{m,2}$, the coefficients c satisfy these relations in all cases. If the general fixed space (4) is transformed by A_{123} it is reproduced along with the additive term

$$(c_0-c_1-c_2-c_3)(\gamma_0-\gamma_1-\gamma_2-\gamma_3). \tag{6}$$

With the aid of this simple rule we find that the entire set of conjugate generating involutions for $m=9$ is defined by the forms,

$$
\begin{aligned}
A(k)&=3k\gamma_0-(k+1)\gamma_1-k(\gamma_2+\cdots+\gamma_8)-(k-1)\gamma_9,\\
B(k)&=(3k+1)\gamma_0-(k+1)(\gamma_1+\gamma_2+\gamma_3)-k(\gamma_4+\cdots+\gamma_9),\\
C(k)&=(3k+2)\gamma_0-(k+1)(\gamma_1+\cdots+\gamma_6)-k(\gamma_7+\gamma_8+\gamma_9)\\
&\qquad (k=0,1,2,\cdots).
\end{aligned}
\tag{7}
$$

Indeed for $k=0$ we find respectively the first three forms in (4). If then these forms are at most permuted under typical quadratic transformation the conjugate set is complete. The form $A(k)$ is transformed by A_{123}, A_{129}, A_{234}, A_{239} into a type $C(k-1)$, $A(k)$, $A(k)$, $B(k)$ respectively; $B(k)$ is transformed by A_{123}, A_{124}, A_{145}, A_{456} into a type

$C(k-1)$, $A(k)$, $B(k)$, $C(k)$; and $C(k)$ is transformed by A_{123}, A_{127}, A_{178}, A_{789} into $B(k)$, $C(k)$, $A(k+1)$, $B(k+1)$ respectively. Hence the infinite set (7) together with those which arise from them by permutations of Π constitute a complete conjugate set.

The specific form of the element of $g_{m,2}$ defined by the linear form

$$c_0\gamma_0 - c_1\gamma_1 - \cdots - c_m\gamma_m \tag{8}$$

in the above conjugate set is

$$\begin{aligned} \gamma_0' &= (c_0^2+1)\gamma_0 - c_0 c_1 \gamma_1 - \cdots\cdots - c_0 c_m \gamma_m, \\ \gamma_i' &= c_i c_0 \gamma_0 - c_i c_1 \gamma_1 - \cdots - (c_i^2-1)\gamma_i - \cdots - c_i c_m \gamma_m, \\ & (i = 1, \cdots, m). \end{aligned} \tag{9}$$

This is evidently true of the transposition, the quadratic, and the quintic transformation. To prove it true in general we verify that the transformation formed as in (9) from the transform of (8) by A_{123} is the transform of (9) by A_{123}.

We observe that $g_{m,2}$, generated by involutions of determinant -1, must have an invariant subgroup of index 2 whose elements have a determinant $+1$.

Since for $m=9$ the conjugate set of generating involutions contain an infinite number of elements there must be an infinite number of types of Cremona transformations with 9 or more F-points. The number with 8 or fewer F-points is finite. They are given in the following table.

(10)

	A_2	B_3	C_4	C_5	D_4	D_5	D_6	D_7	D_8	E_6	E_6^{-1}	E_7	E_8	E_8^{-1}	E_8'	E_9	E_9^{-1}	E_9'
α_1	3	4	3	0	6	3	1	0	0	4	3	2	2	0	1	1	0	0
α_2		1	3	6	0	3	4	0	0	1	4	3	0	5	3	1	3	4
α_3					1	1	2	4	7	3	0	2	5	2	2	3	3	0
α_4											1	1	1	0	2	3	1	4
α_5														1			1	

	E_{10}	E_{10}^{-1}	E_{10}'	E_{10}''	E_{11}	E_{11}'	E_{12}	E_{12}^{-1}	E_{12}'	E_{13}	E_{13}'	E_{14}	E_{14}'	E_{15}	E_{16}	E_{17}
α_1	1	0	0	0	0	0	0	0	0	0	0	0	0	0	0	0
α_2	0	1	2	0	1	0	1	0	0	0	0	0	0	0	0	0
α_3	2	5	2	7	2	4	0	3	2	1	0	1	0	0	0	0
α_4	5	0	3	0	3	3	4	1	4	3	6	0	3	1	0	0
α_5		2	1	0	2	0	3	4	1	3	0	6	3	4	3	0
α_6				1		1			1	1	2	1	2	3	5	8

Here A, B, C, D, E refer to 3, 5, 6, 7, 8 F-points and the subscript is the order. For given type, α_i is the number of i-fold F-points.

7. The continuous aggregate of Cremona transformations of given type. The Cremona group $G_{m,2}$ in $\Sigma_{2(m-4)}$. If T has ϱ F-points, the projective conditions on the sets of points P_m^2, Q_m^2 that they may be congruent under T coincide, when $m = \varrho$, with the $2\varrho - 8$ projective conditions on the 2ϱ F-points of T, T^{-1}; and, when $m = \varrho + 1$ and the pair of images $p_{\varrho+1}$, $q_{\varrho+1}$ appears, with the $2\varrho - 6$ projective conditions which define the transformation. Hence given the nature of T and the distribution of its F-points in the given set P_m^2, given also four points of the set Q_m^2 we seek to determine the position of the remaining points of Q_m^2.

If we regard the planar collineation group C as known and are primarily interested in Cremona transformations only in so far as they disturb projective relations we then would be concerned only with those properties of T which it has in common with the entire aggregate CTC' i. e. *the double coset of the Cremona group with respect to the group of collineations.* If we utilize C to place the first three points of P_m^2 at the reference points, the fourth point p_4 at the unit point, and if further we reduce the individual factors of proportionality in the remaining points to one common to all by making the last coördinate of each point the same quantity u, then the set P_m^2 is defined by a value system:

$$(1) \qquad P\colon \quad x_i,\ y_i,\ u \qquad (i = 5, \cdots, m).$$

The set P_m^2 is then determined by the $2(m-4)$ ratios $x_i : y_i : x_j : y_j : \cdots : u$ i. e. by the homogeneous coördinates of a point P in a linear space $\Sigma_{2(m-4)}$. We have thus mapped the *ordered sets* P_m^2 upon the points P of $\Sigma_{2(m-4)}$ by means of projective invariants of P_m^2. If we denote by (ijk) the determinant of three points p_i, p_j, p_k and set

$$(2) \qquad \begin{aligned} \pi &= (125)\,(126) \cdots (12m), \\ \pi_i &= \pi/(12i) \qquad (i = 5, \cdots, m), \end{aligned}$$

the explicit values of the coördinates of P in $\Sigma_{2(m-4)}$ are

$$
(3) \qquad \begin{aligned}
x_i &= (134)\,(124)\,(23\,i)\,\pi_i, \\
y_i &= (234)\,(124)\,(13\,i)\,\pi_i, \\
u &= (234)\,(134)\,\pi \qquad\qquad (i = 5, \cdots, m).
\end{aligned}
$$

It is inevitable that such a mapping will itself have singular points and will present in particular cases the analogue of F-points and P-loci. For example all ordered planar sets for which p_1, p_2, p_4 are on a line will map into a single point P in $\Sigma_{2(m-4)}$.

We utilize C' to reduce the set Q_m^2 to a similar canonical form and obtain a point Q in Σ. Since Q_m^2 is projectively and rationally determined when P_m^2 and the type of T with reference to it are given, and vice versa for T^{-1} we have

(4) *If the sets P_m^2 and Q_m^2 are congruent under planar Cremona transformation their representative points P and Q in $\Sigma_{2(m-4)}$ are conjugate under a Cremona transformation τ in Σ. The elements τ constitute a Cremona group $G_{m,2}$ in Σ.*

The generators of $G_{m,2}$ are obtained in the same way as those of $g_{m,2}$. It may be first that the sets P_m^2 and Q_m^2 are projective but not in the identical order. If P_m^2 is reordered the representative point P in $\Sigma_{2(m-4)}$ shifts to P' and for the $m!$ possible orders we obtain $m!$ points P' (including P) which form a conjugate set under a Cremona group $G_{m!}$ in Σ of order $m!$ which is isomorphic with the subgroup Π of $g_{m,2}$. A set of generating transpositions of $G_{m!}$ for the more general case P_m^k of m points in S_k has been given by the author ([11] I § 7). A sample for P_6^2 is given below in (6).

When $p_1, p_2, p_3;\ q_1, q_2, q_3$ are the reference points and $p_4 = q_4$ the unit point, A_{123} is merely

$$x_i' = 1/x_i.$$

This inverts the coördinates of the remaining points so that the explicit form of the element A_{123} of $G_{m,2}$ is

$$(5) \quad A_{123}\colon\ x_i' = 1/x_i,\ y_i' = 1/y_i,\ u' = 1/u \quad (i = 5, \cdots, m).$$

The explicit form of these generators for P_6^2 (cf. [17] III § 1), when x_5, y_5, x_6, y_6 are replaced by x, y, z, t is:

(6)		(12)	(23)	(34)	(45)	(56)	A_{123}
	$x' =$	y	xt	$u-x$	uy	z	$1/x$
	$y' =$	x	ut	$u-y$	ux	t	$1/y$
	$z' =$	t	zy	$u-z$	zy	x	$1/z$
	$t' =$	z	uy	$u-t$	tx	y	$1/t$
	$u' =$	u	yt	u	xy	u	$1/u$

A conjugate aggregate of points under $G_{m,2}$ in $\Sigma_{2(m-4)}$ represents in the plane the aggregate of sets Q_m^2, projectively distinct and ordered, which are congruent to a member of the aggregate. Two members P_m^2, Q_m^2 define a ternary transformation T whose F-points and ordinary pairs are in a prescribed position in P_m^2, Q_m^2. For any other set $P_m^{2\prime}$ in the aggregate there is a similar T' with similarly placed points for which $P_m^{2\prime}$, $Q_m^{2\prime}$ are congruent. Then T, T' yield the same transformation τ in Σ for which both P, Q and P', Q' are conjugate pairs. As a conjugate set of points in Σ varies through Σ we find exemplars of all projectively distinct sets P_m^2. Thus the mapping of sets P_m^2 in the plane upon points of $\Sigma_{2(m-4)}$ converts the groupoid property (cf. 6) of congruent planar sets into the group property of points under $G_{m,2}$; and furthermore converts the independent variation in the plane of the $m-4$ projectively independent points of P_m^2 into the variation in Σ of the single point P.

From the formation of the generators above we find that

(7) *The Cremona group $G_{m,2}$ in $\Sigma_{2(m-4)}$ is isomorphic with the linear group $g_{m,2}$. This isomorphism is simple except in the particular cases $G_{5,2}$, $G_{7,2}$, $G_{8,2}$ for which it is respectively* $1:16$, $1:2$, $1:2$.

These cases of multiple isomorphism arise when *for a general set P_m^2* congruence under T implies projectivity. The author has proved ([17] II §§ 2, 3) that this occurs in only four cases: (1) under quadratic transformation, A_{123}, the points p_1, p_2,

p_3, p_4, p_5 are projective to $q_1, q_2, q_3; q_5, q_4$; (2) under cubic transformation $A_{123}\, A_{145}$ (23) (45), $p_1, \cdots, p_5$ are projective to $q_1, \cdots, q_5$; (3) under the octavic Geiser transformation (D_8 of **6** (10)), $p_1, \cdots, p_7$ are projective to $q_1, \cdots, q_7$; and (4) under the 17-ic Bertini transformation (E_{17} of **6** (10)), $p_1, \cdots, p_8$ are projective to $q_1, \cdots, q_8$. In each case the addition of another ordinary pair destroys the projective relation.

For sufficiently general sets P_m^2 there are no other cases where congruence implies projectivity. Nevertheless we shall have much to do later with special sets of points both in the plane and in space for which further types of congruence imply projectivity. If P_m^2 is such a set the point P in $\Sigma_{2(m-4)}$ belongs to a conjugate set of points which is smaller in number than in general; or more precisely P is a fixed point under a subgroup of $G_{m,2}$.

8. Cremona groups as Galois groups in certain problems. The simple case, preliminary to the above, of P_m^1, m points on a line, has important applications. We transform P_m^1 by collineation into the canonical form

$$(1) \qquad 1, 0;\ 0, 1;\ 1, 1;\ x_4, u;\ x_5, u;\ \cdots;\ x_m, u.$$

Thus P_m^1 is mapped into a point P in Σ_{m-3}

$$P: \qquad x_4, x_5, \cdots, x_m, u.$$

Since Cremona transformations do not occur on the line, the $G_{m,1}$ reduces to the subgroup $G_{m!}$ determined by reordering the points of P_m^1. The $m!$ points P obtained by transforming any permutation of (1) into the same form as (1) are a conjugate set under $G_{m!}$ in Σ_{m-3}. This $G_{5!}$ in Σ_2 was first observed by Autonne[1], the $G_{6!}$ in Σ_3 by S. Kantor[39], and finally the $G_{m!}$ in Σ_{m-3} by E. H. Moore[47] who recognized the ratios of the coördinates of P in the double ratios of P_m^1.

It is clear that $G_{m!}$ is a form of the Galois group of the equation of degree m, $(\alpha\, t)^m = 0$, whose roots determine the points of P_m^1. The author has used the group $G_{5!}$ to obtain a solution of the quintic equation[13], and $G_{6!}$ to obtain a solution of the sextic equation[15]. The general principle involved may be stated as follows:

(2) *If a given algebraic problem (or geometric problem when stated in algebraic form) has a finite number of solutions with a Galois group of order k in the original domain; if also by linear transformation of the original variables and parameters the original problem may be reduced to a canonical form C which involves r essential parameters $a_1, \cdots, a_r$ (not further reducible by such processes), and for which the solutions are all rationally known, then C may be obtained in k ways with parameters $a_1^{(i)}, \cdots, a_r^{(i)}$ $(i = 1, \cdots, k)$ which play the role of k points in S_r conjugate under a Cremona G_k which is isomorphic with the Galois group of the given problem.*

We apply this later to $G_{6,2}$ and the group of the 27 lines on a cubic surface, $G_{7,2}$ and the group of the 28 double tangents of a quartic curve, etc. Our final problem will be to connect the $G_{10,3}$ determined by a special set of 10 points in space with the group of the 120 tritangent planes of a space sextic of genus four.

9. P-curves and discriminant conditions of P_m^2. We call that aggregate of curves defined on P_m^2 such that each corresponds to the directions about a point in some set Q_m^2 congruent to P_m^2, *the P-curves of the set P_m^2*. They are in fact the P-curves of the transformations T^{-1}. Bertini[5] shows that they are rational curves whose multiple points may be arbitrarily assigned. If we denote a complex γ by its linear polar form as to Q, and ascribe to the directions about a point the multiplicity -1 at the point, the polar form for p_1 is γ_1 and the conjugates of this under $g_{m,2}$ determine the types of the P-curves. Thus for $m \leqq 8$ we have the types

$$
\begin{array}{ll}
 & \gamma_1 \\
 & \gamma_0-\gamma_1-\gamma_2, \\
 & 2\gamma_0-\gamma_1-\gamma_2-\cdots-\gamma_5, \\
(1) & 3\gamma_0-2\gamma_1-\gamma_2-\cdots-\gamma_7, \\
 & 4\gamma_0-2(\gamma_1+\gamma_2+\gamma_3)-(\gamma_4+\cdots+\gamma_8), \\
 & 5\gamma_0-2(\gamma_1+\cdots+\gamma_6)-(\gamma_7+\gamma_8), \\
 & 6\gamma_0-3\gamma_1-2(\gamma_2+\cdots+\gamma_8).
\end{array}
$$

For $m = 9$ an infinite number of types may be read off from the values of γ_i' in 6 (9) for the forms in 6 (7) but these will not exhaust the types. The aggregate of P-curves of P_m^2 defines a discontinuous division of the projective plane[7]. An immediate consequence of the definition is:

(2) *If P_m^2 is congruent to Q_m^2 under T the P-curves of P_m^2 are conjugate to those of Q_m^2 under T.*

If in P_m^2 itself or in any set Q_m^2 congruent to it two of the points are made to coincide in some direction then there is imposed on the set P_m^2 a projective condition which we call a *discriminant condition on P_m^2*. If P_m^2 and Q_m^2 are congruent under A_{123} and if in Q_m^2 q_4 concides with q_3 then in P_m^2 the points p_1, p_2, p_4 are on a line. Thus in P_m^2 there exists a complex whose polar form is $\gamma_0 - \gamma_1 - \gamma_2 - \gamma_4$ and this transforms under A_{123} into the form $\gamma_3 - \gamma_4$ for the coincidence in Q_m^2. Hence the types of discriminant conditions on P_m^1 are (cf. 6 (4)) the same as the types of conjugate generating involutions of $g_{m,2}$.

(3) *The discriminant conditions on the set P_m^2 are in one-to-one correspondence with the conjugate generating involutions of $g_{m,2}$. If such a condition is satisfied by P_m^2 the corresponding Cremona transformation T (whose S in $g_{m,2}$ is given in* 6 (9)) *degenerates into a collineation.*

Indeed if p_1, p_2 coincide the transposition (12) is effected by the identical collineation; and if p_1, p_2, p_3 are given on a line A_{123} is a collineation. Any other condition is the transform of $(p_1\, p_2\, p_3) = 0$ by some transformation T and the generating involution corresponding to it is the transform of A_{123} by T. But $T^{-1} A_{123}\, T$ is a collineation if A_{123} is a collineation, provided the product is formed for congruent sets.

We have characterized in the above a variety of linear systems, or particular curves such as P-curves, or conditions such as discriminant conditions, all attached to P_m^2; as well as the conjugates of this variety as they appear in the congruent sets Q_m^2, in each case by a system of linear forms conjugate under $g_{m,2}$. Thus $\gamma_0,\ 2\gamma_0 - \gamma_1 - \gamma_2 - \gamma_3, \cdots$ represent the types of homaloidal nets; $\gamma_1, \gamma_0 - \gamma_1 - \gamma_2, \cdots$

represent the P-curves; and $\gamma_1-\gamma_2, \gamma_0-\gamma_1-\gamma_2-\gamma_3, \cdots$ represent discriminant conditions. More generally if

$$(4)\qquad \begin{aligned} c_0^2-c_1^2-\cdots-c_m^2 &= r+p-1, \\ 3c_0-c_1-\cdots-c_m &= r-p+1 \end{aligned}$$

then $c_0\gamma_0-c_1\gamma_1-\cdots-c_m\gamma_m$ represents a linear system of dimension r and genus p. If the integer (positive or zero ordinarily) coefficients c are such that r and p are not negative, the system exists; if however r or p is negative the system is non-existent or virtual when P_m^2 is general. We may, however, as in the case of the discriminant conditions impose on P_m^2 the requirement that a virtual system be existent with an actual dimension r and genus p. Then the conjugates of the form under $g_{m,2}$ express the nature of the conjugate condition on the sets congruent to P_m^2.

10. Linear systems ($r=2, D>1$); birational transformations. If $|C_n|$ is a proper net for which $D>1$ the pencil of curves on x passes through $D-1$ further points x each of which determines the same pencil. Necessarily then the system is not simple and the points of the plane divide into sets of D points in a planar involution. The mapping,

$$(1)\qquad y_i=(\alpha_i x)^n \qquad (i=0,1,2),$$

furnishes for x an ordinary point in S_x a unique point y in S_y. For given y, determined by lines $(\eta y)=0$, $(\eta' y)=0$, (1) furnishes a group of points x which arise from the points on $\sum\eta_i(\alpha_i x)^n=0$, $\sum\eta_i'(\alpha_i x)^n=0$ which are variable with η, η'. We have thus a $(D, 1)$ correspondence from S_x to S_y (cf. [52] Chap. V).

Let $f(x)$ be a given curve in S_x which at one point p has the property that it does not contain any of the $D-1$ other points of the set to which p belongs. To the points x on $f(x)$ there correspond in (1) the points y on a curve $f'(y)$ in S_y; to a point y on $f'(y)$ and on η, η' the D points in S_x of which in general only one is on $f(x)$. This particular point x on the three curves $f(x)$, $\sum\eta(\alpha x)^n=0$, $\sum\eta'(\alpha x)^n=0$

can be obtained by rational processes in terms of η, η' which clearly can occur only in the combinations $(\eta\eta')_i = y_i$. Thus we find, for points x on the given curve $f(x)$ a unique solution of (1), namely

$$x_i = (\beta_i y)^{n'} \qquad (i = 0, 1, 2). \tag{2}$$

Such transformations as (1), (2), one-to-one and birational between two loci of the same dimension will be termed *birational transformations.* When this term is used it will be understood unless otherwise stated that the transformation is not defined outside the loci in question. The aggregate of curves which can be put into birational correspondence with a given curve and therefore with each other is called a *class* of algebraic curves. A given class defines the aggregate of birational transformations B which connect any pair of the class. This aggregate constitutes a groupoid. If a pair of the class is given B is uniquely determined unless one curve, and therefore every curve of the class, admits birational transformations into itself other than the identity which is not true of the general curve.

11. Linear series on an algebraic curve. Complete and special series. Under the transformation B in 10 line sections of the transformed curve correspond to the variable part of the intersections of $f(x)$ with curves of the net. We consider in general the sections of an irreducible curve $f(x)$ by the members of a linear system. First let those members of the system which contain f as a factor be dropped, leaving a linear system $|C_\nu|$ of dimension r. The curves of $|C_\nu|$ will have in general certain fixed intersections with f and in addition certain n intersections variable with the parameters of the system. We say then that $|C_\nu|$ cuts $f(x)$ in a *linear series* g_r^n whose ∞^r sets of n points on $f(x)$ are in one-to-one correspondence with the curves of the system. If those curves of $|C_\nu|$ which pass through a generic point of $f(x)$ pass necessarily through $\mu - 1$ other points of $f(x)$, the n intersections divide into n/μ variable sets of μ points. If $\mu > 1$ the g_r^n is *composite*; if $\mu = 1$ it is *simple.*

If the system $|C_\nu|$ is determined by the members $(\alpha_i x)^\nu = 0$, the mapping,

$$y_i = (\alpha_i x)^\nu \quad (i = 0, 1, 2, \cdots, r), \tag{1}$$

establishes a birational correspondence between $f(x)$ and a curve $C_r^n(y)$ of order n in S_r when g_r^n is simple; or a $(\mu, 1)$ correspondence between $f(x)$ and a $C_r^{n/\mu}$ when g_r^n is composite. We have then the theorem (cf. [64] Chap. III).

(2) *The aggregate of curves in hyperspace, space, the plane, and the n-fold projective line (n-sheeted Riemann surface), which are in birational correspondence with a given curve $f(x)$ is represented by the aggregate of simple linear series g_r^n $(r \geqq 1)$ cut out on $f(x)$ by linear systems.*

If some of the curves in (1) are dropped, a $g_{r'}^n$ $(r' < r)$ is cut out on f which is said to be *contained in* g_r^n. The curve $C_{r'}^n$ in $S_{r'}$ is then a projection of C_r^n in S_r from $r - r'$ of the reference points in S_r. If the $S_{r-r'-1}$ determined by these reference points cuts C_r^n in k points, the order n of the variable part of g_r^n reduces to $n - k$. A g_r^n which is not contained in a linear series of greater dimension but of the same order n is said to be *complete*; and the corresponding C_r^n is said to be a *normal* curve, i. e. it is not the projection of a curve of the same order in a higher space.

As a consequence of a theorem of Noether (cf. [64] Chap. V) it may be proved that every complete g_r^n can be cut out on f by linear systems of *adjoint curves* of f, i. e. curves which pass through each s-fold point of f with multiplicity $s - 1$ at least. Moreover all the adjoints of given order cut out a complete g_r^n. If G is a set of a complete g_r^n the other sets are constructed by passing through the n points of G an adjoint of sufficiently high order l to contain G. This adjoint cuts f in a set H in addition to the $s(s-1)$ intersections imposed on the adjoint at each s-fold point of f. Then all the adjoints of order l on H cut f in the sets of the complete g_r^n which contains G. That g_r^n as thus constructed is unique is a consequence of the residue theorem.

If g_r^n is complete and one of its sets A is divided into two sets B, C with respectively n_1, n_2 ($n_1 + n_2 = n$) points, then B, C are *residues* of each other with respect to g_r^n. The sets A of g_r^n which contain B contain in addition ∞^{r_2} sets C which lie in a complete $g_{r_2}^{n_2}$, the linear series residual to B in g_r^n. Similarly the ∞^{r_1} sets B, residual to C in g_r^n, lie in a complete $g_{r_1}^{n_1}$. The residue theorem states that, if B_1, C_1 are any two sets of $g_{r_1}^{n_1}$, $g_{r_2}^{n_2}$ respectively, the residue of either set with respect to g_r^n is the complete series defined by the other. We say then that the series $g_{r_1}^{n_1}$, $g_{r_2}^{n_2}$ are residual with respect to each other in g_r^n.

Two sets A, A_1 which determine, and lie in, the same complete g_r^n are called *equivalent*. This is indicated by $A \equiv A_1$. The fact that A is made up of B and C is naturally written $A \equiv B + C$. The residue theorem then states that if $A \equiv B + C$ and $A \equiv A_1$, $B \equiv B_1$, $C \equiv C_1$, then $A_1 \equiv B_1 + C_1$, i. e. B_1 and C_1 make up a set $A_2 \equiv A_1$. We customarily understand then by $A \equiv B + C$ that the complete g_r^n defined by A is the sum of its complete residual constituents $g_{r_1}^{n_1}$, $g_{r_2}^{n_2}$ defined by B and C respectively.

If f is of order m and genus p, the adjoints of order $m - 3$ of f—the so-called φ-curves—cut out a complete g_{p-1}^{2p-2} called the *canonical series*. It is the only complete series on f of order $2p - 2$ and dimension $p - 1$. For any complete g_r^n,

$$(3) \qquad n - r \leqq p, \quad n - r = p - i \qquad (i \geqq 0).$$

If $i > 0$ the g_r^n is called *special*. Every special complete g_r^n is contained in the canonical series which is itself special with *index of speciality* $i = 1$. For given g_r^n, i is the number of linearly independent φ-curves on a set G of g_r^n. For example a quintic curve f with nodes at P_1, P_2 ($p = 4$) has for φ-curves conics on P_1, P_2 which cut f in the canonical series g_3^6 with index $i = 1$. The line pencils on the two nodes cut f in complete g_1^3's residual with respect to each other in g_3^6, each of which has index $i = 2$. For, a set of one g_1^3 is on two conics each made up of a line that cuts out the set and one of the two independent lines of the other pencil.

If the canonical series is composite its sets are composed of pairs of points ($\mu = 2$) and these pairs are in a g_1^2, i.e. the curve f is cut by a pencil of curves in a variable pair of points. If conversely the curve has a g_1^2 the sets of the canonical series are composed of $p-1$ variable sets of g_1^2 and the curve is called *hyperelliptic*. A *rational curve* ($p = 0$) has ∞^2 g_1^2's; an *elliptic curve* ($p = 1$) has ∞^1 g_1^2's; a hyperelliptic curve ($p > 1$) has a unique g_1^2. For example on a conic the g_1^2's are cut out by line pencils with vertex at each of the ∞^2 points of the plane; on an elliptic cubic the g_1^2's are cut out by pencils on the ∞^1 points of the curve; and on a quartic with a node ($p = 2$) the unique g_1^2 is cut out by the pencil on the node.

Under birational transformation from $f(x)$ to $f'(x')$ linear series pass into linear series of the same order and dimension and complete series pass into complete series. In particular the canonical series on f passes into the canonical series on f'.

12. The canonical curve. Birational moduli. By birational transformation of a given curve, $f(x)$, the number of constants which appear in its equation may be reduced. If M is the smallest number which can be obtained and if in this reduced case the constants which remain will give rise, as they vary, to birationally distinct types then these M constants are termed the *moduli* of the given curve under birational transformation. We sketch a proof that

(1) *The general non-hyperelliptic curve of genus* $p \geqq 3$ *has* $3p-3$ *birational moduli.*

The given curve has a unique canonical series cut out by the φ-curves. When we map from the plane by means of this linear system, setting

$$y_i = (\varphi_i\, x)^{m-3} \quad (i = 0, 1, \cdots, p-1),$$

$f(x)$ is birationally related to the *canonical curve* C_{p-1}^{2p-2} of order $2p-2$ in S_{p-1}. When $f(x)$ is given, the φ-system is determined uniquely but the projectivity between the φ-curves and the linear sections $y_i = 0$ of C_{p-1}^{2p-2} is at our disposal, i. e. the canonical curve is determined in S_{p-1} only to within a projectivity. If, by a birational transformation B, $f(x)$ passes into $f'(x')$ the intersections of the canonical adjoints $(\varphi x)^{m-3}$ with $f(x)$ pass into the intersections with $f'(x')$ of its canonical adjoints $(\varphi x')^{m-3}$. Thus under B we have

$$y_i = (\varphi_i x)^{m-3} = (\psi_i x')^{m'-3} \quad (f(x) = 0, \ \ f'(x') = 0)$$

and image points x, x' on f, f' determine the same point y on the canonical curve. Hence

(2) *All birationally equivalent plane curves have projectively equivalent canonical curves.*

The M absolute projective constants of C_{p-1}^{2p-2} are then birational moduli of $f(x)$. If we select $p-3$ points in general position on the canonical curve and thus adjoin $p-3$ constants, the curve, projected from these points into an S_2, becomes a curve of order $p+1$ and genus p with $p(p-3)/2$ double points and therefore $4p-6 = M+p-3$ absolute projective constants. Hence $M = 3p-3$ as stated in (1).

The problem of determining moduli of $f(x)$, i. e., constants attached to $f(x)$ and invariant under birational transformations, is thus reduced to the problem of determining absolute projective invariants of the canonical curve. For example, a plane sextic with 6 nodes ($p = 4$) has for canonical curve a space sextic C_3^6, the complete intersection of a quadric and a cubic surface. When the quadric is not a cone and we name its points by the binary parameters $t_0 : t_1$; $\tau_0 : \tau_1$ of the two generators on a point, a section by a cubic surface is a double binary form

$$(3) \qquad (a\,\tau)^3\,(\alpha\,t)^3 = 0$$

in the digredient binary variables t, τ. This has 16 coefficients whose 15 ratios may be reduced to $3p-3 = 9$ by digredient transformation of t, τ. The absolute invariants of the form under such linear transformation are the moduli. It is true that the number of independent ones is nine but the number of those which are distinct from the point of view of rational integral expression is of course much larger. We have then a large complete system of invariants connected by a corresponding large system of syzygies.

In the case of hyperelliptic curves there is a pencil with parameter λ which cuts out g_1^2 and $2p+2$ members for which the two points of a set coincide. The $2p-1$ independent double ratios of the $2p+2$ corresponding parameters λ are the birational moduli. For elliptic curves this modulus is the same for the ∞^1 g_1^2's. Rational curves, birationally equivalent to a line, have no birational moduli.

13. Moduli of a curve under Cremona transformation. In this and the following section applications of linear series are given. We observe that when $f(x)$ is mapped (cf. 12) by its φ-curves upon the canonical curve in S_{p-1} the plane of $f(x)$ is mapped (cf. 2) upon a rational surface, M_2 in S_{p-1}. We assume that $p \geqq 4$, that the curve is not hyperelliptic, and that the φ-curves constitute a simple linear system in the plane. Then M_2 lies in space or hyperspace, is covered only once in the mapping, and the canonical curve is a simple curve on M_2. As before both the curve C_{p-1}^{2p-2} and surface M_2 which contains it simply are determined only to within projective transformation. Also as before a Cremona transformation applied to the mapping system converts $f(x)$ into $f'(x')$ and the canonical adjoints of $f(x)$ into those of $f'(x')$ (after deletion from the latter of possible fixed parts which are P-curves of the transformation), while the point y in S_{p-1} determined by the transformed point x' remains fixed. Thus the relation of C_{p-1}^{2p-2}, M_2 to $f(x)$ is invariant under Cremona transformation in the plane. If the rational M_2 is mapped upon the plane in any way in such wise that y passes into x' the linear sections of M_2 become the curves of the mapping system and C_{p-1}^{2p-2} becomes a curve $f'(x')$ for which the mapping system must be the canonical adjoints. But two distinct mappings of M_2 upon a plane give rise to a Cremona transformation in the plane. Hence

(1) *The invariants or moduli of a non-hyperelliptic curve of genus $p \geqq 4$ whose canonical adjoints are a simple linear*

system are the simultaneous projective invariants in S_{p-1} *of the canonical curve and a particular rational surface on which the curve is not multiple.*

We have here a companion theorem to that of **12** (2) which also brings the matter of invariants of a plane curve under the planar Cremona group into the domain of projective invariants. It is useful as a means of indicating the relative efficacy of the birational and Cremona transformations in the way of reducing a given curve to a canonical type. We express the situation as follows:

(2) *The class of curves birationally equivalent to a given curve* $f(x)$ *of the type described in* (1) *divides into a number of subclasses under Cremona transformation. If the canonical curve in* S_{p-1} *admits no collineations these subclasses are in one-to-one correspondence with the aggregate of rational surfaces* M_2 *which contain* C_{p-1}^{2p-2} *simply.*

We give a few simple examples. If $p = 4$ the canonical curve is a C_3^6. It is on a unique quadric Q whence the curve and quadric have 9 projective absolute constants. A typical form of the plane curve is the quintic with two nodes. This has 10 absolute projective constants but only 9 Cremona moduli since one such quintic can be transformed by quadratic transformation with F-points at the nodes and at one arbitrary simple point into ∞^1 types which are projectively distinct. Again the C_3^6 is on ∞^4 cubic surfaces. The curve and one such surface M_2^3 have the 13 projective moduli which belong to Q and M_2^3. A type of the plane curve is the sextic with 6 nodes which likewise has 13 projective moduli. But the plane curve mapped by its canonical adjoints gives rise to Q and M_2^3 together with an isolated "sixer" (Cayley), i. e. a set of 6 skew lines on M_2^3 which arise from directions at the 6 nodes. Since there are 72 sixers on M_2^3 there are in the plane 72 6-nodal sextics which are projectively distinct but equivalent under Cremona transformation (cf. Chap. III). Thus the projective invariants of Q and M_2^3 rather than the projective invariants of only one of the 72 types of 6-nodal sextics, are the proper Cremona invariants of the curve.

Marletta[13] finds necessary arithmetical conditions for the equivalence of two curves under Cremona transformation when they already are birationally equivalent. These are arithmetical criteria which distinguish the great variety of types of rational surfaces on C_{p-1}^{2p-2}.

The plane curves excepted under (1) are such as bring in planar involutions and multiple correspondences. These will not be discussed further.

14. Residual linear series. Curves in determinant forms. Applications to the planar quartic. Let the complete g_r^n on f^m with sets A be cut out by adjoints of order ν in a linear system $|C_\nu|$. If A be divided, say $A = B + C$, the sets B and C define complete linear series, $g_{r_1}^{n_1}$ and $g_{r_2}^{n_2} (n_1 + n_2 = n)$, each cut out by curves of $|C_\nu|$. If B is a general set in $g_{r_1}^{n_1}$ the curves of $|C_\nu|$ on B form a system,

$$\zeta = \zeta_0 A_0 + \zeta_1 A_1 + \cdots + \zeta_{r_2} A_{r_2}$$

whose members ζ are in one-to-one correspondence with the sets of $g_{r_2}^{n_2}$. This correspondence is determined when r_2 independent sets C are associated with the reference ζ's and when a further set, independent of any $r_2 - 1$ already chosen, is associated with the unit ζ's. The curves A_i are then defined to within a constant common to all. If B' is a second set in $g_{r_1}^{n_1}$ the curves of $|C_\nu|$ on B' likewise are in correspondence with sets C and therefore in projective correspondence with the curves ζ. This system can be written as

$$\zeta' = \zeta_0 A_0' + \zeta_1 A_1' + \cdots + \zeta_{r_2} A_{r_2}'.$$

Curves in the systems ζ, ζ' with the same parameters ζ_i cut out the same set C of $g_{r_2}^{n_2}$. Then the determinant

$$\begin{vmatrix} A_i & A_j \\ A_i' & A_j' \end{vmatrix}$$

of order 2ν in x must either contain f^m as a factor or vanish identically. For the corresponding members of the two projective pencils, $\zeta_i A_i + \zeta_j A_j = 0$ and $\zeta_i A'_i + \zeta_j A'_j = 0$, meet in ∞^1 sets C which run over f^m. The pencils either generate f^m in the usual sense or they reduce to the same pencil after deletion from each of a fixed part. Such fixed parts determine a pencil which cuts out the variable sets determined by B, B'. We assume that this second case does not occur. The sets B of $g_{r_1}^{n_1}$ can be put into similar correspondence with parameters $\eta_0, \cdots, \eta_{r_1}$. Then the curve of $|C_\nu|$ which cuts out a set $B(\eta)$ and a set $C(\zeta)$ will be

$$(1) \qquad \begin{gathered}(\alpha x)^\nu (b\eta)(c\zeta) = \sum A_{ij}\,\eta_i\,\zeta_j = 0 \\ (i = 0, \cdots, r_1;\ j = 0, \cdots, r_2).\end{gathered}$$

We interpret y, η as dual coördinates in S_{r_1}; z, ζ as dual coördinates in S_{r_2}, leaving x, ξ for dual coördinates in the plane of f^m. For given η and variable ζ, (1) defines a system which cuts out a fixed set of $g_{r_1}^{n_1}$ and a variable set of $g_{r_2}^{n_2}$, and vice versa.

In the excluded case the form (1) factors for every x into two factors linear respectively in η and ζ; in the general case it factors in this way when x is on f^m. Then, for given η, (1) is satisfied by all ζ's whose sets in $g_{r_2}^{n_2}$ contain x and these sets are independent of η. Thus, for each x on f^m, (1) becomes $(y\eta)\cdot(z\zeta) = 0$ and the form sets up a mapping of points x on f^m upon the birationally related curves $C_{r_1}^{n_1}$, $C_{r_2}^{n_2}$ in S_{r_1}, S_{r_2} respectively. The two row determinants of the matrix $\|A_{ij}\|$ contain f^m as a factor, and its k-row determinants contain f^m as a $(k-1)$-fold factor. We shall be interested primarily in the case $r_1 = r_2$ in particular applications, some of which we proceed to develop.

The canonical curve of genus $p = 3$ is a non-singular plane quartic f^4. Cubics in the plane, adjoints of order 3, cut it in a complete g_9^{12}. A set of 6 points selected on f^4 and *not on a conic* determine a complete g_3^6 whose residual series in g_9^{12} is $g_3^{6'}$. The form (1) for this case is

$$(2) \qquad (\alpha x)^3 (b\eta)(c\zeta) = 0,$$

where η, ζ are planes in $S_3(y)$, $S_3(x)$ respectively. For x on f^4, (2) factors into $(y\eta)\cdot(z\zeta)=0$ and f^4 is mapped on the birationally related space sextics $C^6(y)$, $C^6(z)$. The curve f^4 itself has 6 moduli and the choice of g_3^6 introduces 3 more since 6 points on f^4 determine g_3^6 and one of its ∞^3 sets. Hence $C^6(y)$, $C^6(z)$ have 9 moduli or 24 projective constants and are general space curves of order 6 and genus 3 (cf.[64] p. 369). For given η and variable x equation (2) is that of a point $(z\zeta)=0$ which runs over a cubic surface on $C^6(z)$ the map of the plane by cubic curves on the set η of g_3^6.

We may look upon (2) as a correlation from $S(y)$ to $S(z)$ which for given η determines a point $(z\zeta)=0$. The dual form of this correlation has coefficients which are three row minors of $|A_{ij}|$ containing $(f^4)^2$ as a factor; i. e.

$$(3)\quad \begin{aligned}(\lambda x)(\beta y)(\gamma z) &= (\alpha x)^3(\alpha' x)^3(\alpha'' x)^3(bb'b''y)(cc'c''z)/(f^4)^2\\ &= \sum a_{ij}\, y_i\, z_j \qquad (i,j=0,\cdots,3),\end{aligned}$$

where primes, seconds, etc. indicate equivalent symbols. The form (3) has $3\cdot 4\cdot 4$ coefficients or 47 ratios and therefore $47-8-15-15=9$ absolute constants under digredient linear transformation of x, y, z. We recover the form (2) from (3) by rewriting (3) in dual form

$$(4)\quad (\alpha x)^3(b\eta)(c\zeta) = (\lambda x)(\lambda' x)(\lambda'' x)(\beta\beta'\beta''\eta)(\gamma\gamma'\gamma''\zeta)\cdot k$$

(k a numerical factor). In (3) for variable x there is a net of correlations. The values of x for which the correlation is singular, i. e., for which $|a_{ij}|=0$, are those for which the dual form (2) factors into the singular points in either space. Hence

$$(5)\quad \begin{aligned}f^4 &= 24\,|a_{ij}|\\ &= (\lambda x)(\lambda' x)(\lambda'' x)(\lambda''' x)(\beta\beta'\beta''\beta''')(\gamma\gamma'\gamma''\gamma''')=0.\end{aligned}$$

We have thus attained an end which we frequently shall seek; namely an expression for a given geometric configuration, such as f^4 and the residual g_3^6's on it, by means of an algebraic form (3) whose coefficients are unrestricted.

For given y and variable x in (3) the planes in $S(z)$ turn about a point z whose equation is

$$(6) \qquad (\lambda\lambda'\lambda'')(\beta y)(\beta' y)(\beta'' y)(\gamma\gamma'\gamma''\zeta) = 0.$$

Similarly for given z the planes in $S(y)$ turn about the point y

$$(7) \qquad (\lambda\lambda'\lambda'')(\beta\beta'\beta''\eta)(\gamma z)(\gamma' z)(\gamma'' z) = 0.$$

These are equations in either direction of the cubic Cremona transformation T between $S(y)$ and $S(z)$ determined by the F-curves $C^6(y)$ and $C^6(z)$. For given y, (6) is the equation of the image point $(z\zeta) = 0$; for given ζ, the equation of the cubic surface in $S(y)$ on $C^6(y)$ which is the image of the plane $(\zeta z) = 0$; in (7) these relations are reversed. The same homaloidal webs appear in (2) in parametric form; for given ζ in (2) and variable x the point $(\eta y) = 0$ runs over the cubic surface (6).

An expression of f^4 as a four-row determinant whose elements are linear forms in x appears in (5) and evidently every such expression determines a form (3) and is associated with one of the ∞^3 pairs of residual g_3^6's on f^4. A similar expression for f^4 as a two-row determinant whose elements are quadratic in x arises from two complete g_1^4's residual in the g_5^8 cut out by conics. The residual series determine a form

$$(8) \qquad (\alpha x)^2(b\tau)(\beta t) = \sum A_{ij}\, t_i \tau_j \qquad (i, j = 0, 1)$$

where t, τ are digredient binary variables. For the binary variables there is no duality. The form (8) has $6\cdot 2\cdot 2 - 1 = 23$ projective constants and $23 - 8 - 3 - 3 = 9$ absolute constants which arise from the 6 moduli of f^4 and the 3 involved in the choice of g_1^4. Again f^4 is the locus of points x for which (8), as a form in t, τ, factors whence

$$(9) \qquad f^4 = (\alpha x)^2(\alpha' x)^2(bb')(\beta\beta') = 2\,|A_{ij}| = 0.$$

A further development of this case is given in connection with the curve of genus four (cf. 50).

We examine the case of symmetry in (8) which then is the polarized form of a quadratic in τ,

$$(10)\qquad (\alpha x)^2 (b\tau)^2 = A_0\tau_0^2 + 2A_1\tau_0\tau_1 + A_2\tau_1^2 = 0.$$

Then

$$(11)\qquad f^4 = (\alpha x)^2 (\alpha' x)^2 (bb')^2 = 2\begin{vmatrix} A_0 & A_1 \\ A_1 & A_2 \end{vmatrix} = 0$$

is expressed as a symmetric determinant. The two residual g_1^4's have coincided into a single g_1^4 which is made up of the sets of contacts of a system of contact conics, the quadratic system (10). Two conics of the system with parameters τ, τ' have their 8 contacts on the conic $(\alpha x)^2 (b\tau)(b\tau') = 0$. Let K denote a set of the canonical series i. e. a line section. The g_5^8 cut out by conics contains sets A such that $2K \equiv A$. If the 8 points A are made up of 4 points B doubled, then $2B \equiv 2K$. An obvious case is $B \equiv K$ i. e. the line sections doubled are improper contact conics. A pair of double tangents furnishes a case for which $2B \equiv 2K$ but $B \not\equiv K$. Moreover in any contact system (10) there are six pairs of double tangents with parameters determined from

$$(12)\qquad (\alpha\alpha'\alpha'')^2 (b\tau)^2 (b'\tau)^2 (b''\tau)^2 = 0.$$

Since there are $14 \cdot 27 = 6 \cdot 63$ pairs of double tangents we find that

(13) *The quartic f^4 has 63 systems of proper contact conics each containing six pairs of double tangents.*

Denote by D_i the two contacts of a double tangent. If D_1, D_2 and D_3, D_4 are two pairs in the system (10) the four double tangents have their 8 contacts on a conic and are called a *syzygetic tetrad.* Any three of the four have their 6 contacts on a conic and are called a *syzygetic triad*; three with contacts not on a conic are an *azygetic triad.* A simple reckoning with the sets shows that the conic on the contacts of a syzygetic triad cuts f^4 in the contacts of a fourth double tangent so that a syzygetic triad can be enlarged to a syzygetic tetrad in one and only one way.

Any two pairs from the 6 pairs of a contact system determine a syzygetic tetrad but the tetrad can be divided into pairs in three ways whence there are $63 \cdot 15/3 = 21 \cdot 15$ syzygetic tetrads and $21 \cdot 15 \cdot 4 = 28 \cdot 45$ syzygetic triads. Since $28 \cdot 9 \cdot 13$ triads can be formed from the 28 double tangents we conclude that

(14) *The* 28 *double tangents of* f^4 *contain* $28 \cdot 45$ *syzygetic triads and* $28 \cdot 72$ *azygetic triads.*

We consider now the systems of contact cubics of f^4. If C_i denotes a set of contacts then $2C_i \equiv 3K$. If C_i is on a conic, and the system therefore a *syzygetic* system, this conic cuts out a further pair $\varDelta$ such that $C_i + \varDelta \equiv 2K$. Hence $2\varDelta \equiv K$ and $\varDelta$ is the pair of contacts of a double tangent. Conics on $\varDelta$ cut out the g_3^6 determined by C_i. This maps f^4 into a curve on a quadric surface which cuts each generator three times and has a double point with generators as nodal tangents. The space sextic is then a special form of the space sextic of genus four (cf. **12** (3)) with an actual node. There are 28 syzygetic systems, one for each double tangent, and each system contains, according to (14), 45 syzygetic triads of double tangents as degenerate members.

When the set C_i of contacts are not on a conic and the system of contact cubics is *azygetic* we have the particular case of (2) for which $B \equiv C$ and $2B \equiv A$. Since the two g_3^6's now coincide, the spaces $S(y)$, $S(z)$ coincide and η, ζ are planes in the same space. Then (2) is merely the polarized form of the quadric,

$$(2^\circ) \qquad (\alpha x)^3 (b\eta)^2 = 0,$$

which furnishes for variable η the contact cubics of the system. Now (3) also is the polar form of

$$(3^\circ) \qquad (\lambda x)(\beta y)^2 = 0$$

which for variable x is a general net of quadrics in $S(y)$ whose planar equation is (2°). The cubic Cremona transformation T is now an involution whose pairs y, y' are apolar to the net (3°). The quartic f^4 is the locus of points x for which the quadric (3°) has a node at y, and the square of this node, $(\lambda y)^2 = 0$, is given by (2°). The equation of f^4 is

(5°) $f^4 = (\lambda x)(\lambda' x)(\lambda'' x)(\lambda''' x)(\beta \beta' \beta'' \beta''')^2 = 24\,|a_{ij}| = 0,$

where $|a_{ij}|$ is a symmetric determinant.

If the base points of the net are $p_1, \cdots, p_8$, or the set P_8^3, a pencil of the net, determined as x runs over a line ξ in the plane, lies on an elliptic space quartic curve $E(\xi)$ on P_8^3. The sextic locus, $C^6(y)$, the map of f^4 by g_3^6 and the curve of F points of the involution T, is the locus of nodes of quadrics of the net or the locus of nodes of curves $E(\xi)$. To the four intersections of ξ with f^4 there correspond the four nodes of nodal quadrics in the pencil on $E(\xi)$. If two of these nodal quadrics coincide, their nodes coincide at the double point of a nodal $E(\xi)$ which corresponds to a tangent ξ of f^4. To a double tangent ξ of f^4 there corresponds a binodal $E(\xi)$. Since this must be degenerate, and there cannot be two conics on P_8^3, it must consist of a line on, say $p_1 p_2$, or line $(12)^1$, and a cubic curve $(345678)^3$ which meet in the two points on $C^6(y)$ which correspond to the contacts of ξ. Thus the 28 pairs of contacts of double tangents of f^4 map into the 28 pairs of points where $C^6(y)$ is cut by the 28 lines $(p_i p_j)^1$ $(i, j = 1, \cdots, 8)$. Since plane sections of $C^6(y)$ map contacts of the system g_3^6, a plane $(p_i p_j p_k)^1$ cuts $C^6(y)$ in the map of the contacts of an azygetic triad of double tangents. There are 56 such planes and therefore 56 azygetic triads in an azygetic contact system. On comparison with (14) it appears that

(15) *The quartic f^4 has 28 syzygetic and 36 azygetic systems of contact cubics. For each system of the latter kind the quartic admits an expression as a symmetric four-row determinant.*

We recur in Chap. IV to the relations among these systems.

The study of loci expressible as determinants goes back to Hesse ([33] 1855) and has since frequently been renewed

(Cf.[50] II 2, Chap. 31). Evidently this is possible only if the locus contains incomplete intersections defined by the vanishing of a line of first minors. Noether (Cf.[50] II 2, p. 929) has proved that a general surface in S_3 of order $\mu > 3$ has only complete intersections. Hence a general quaternary form of order $\mu > 3$ cannot be expressed as a determinant nor can any general form in more variables. These results of the geometers have been overlooked in certain recent articles[28,29].

15. Congruence of sets of points under regular Cremona transformation in space and hyperspace. The groups $G_{m,k}$, $g_{m,k}$, and $e_{m,k}$. In space we define the *regular group* of Cremona transformation to be that group generated by collineations C and the involutorial cubic transformation A_{1234},

$$A_{1234}: \quad x_i' = 1/x_i \qquad (i = 0, \cdots, 3), \tag{1}$$

with F-points at the reference points p_1, p_2, p_3, p_4 and fixed point at the unit point p_5. This transformation, or its more general type $CA_{1234}C'$, has properties entirely analogous to those of the quadratic transformation in the plane. It has four F-points $p_1, \cdots, p_4$ and four inverse F-points $q_1, \cdots, q_4$ such that the ∞^2 directions at p_i correspond to the points on a P-surface, the plane $(q_j q_k q_l)^1$, and vice versa $(i, j, k, l = 1, \cdots, 4)$. It is determined by these two sets of F-points and one ordinary corresponding pair p_5, q_5. If then p_6, q_6 is any other corresponding pair the two sets of six points

$$\begin{matrix} p_1, p_2, p_3, p_4, p_5, p_6, \\ q_1, q_2, q_3, q_4, q_6, q_5 \end{matrix}$$

are projective in the order indicated. Hence A_{1234} is determined to within projectivities by its F-points alone and this is true of any regular Cremona transformation. The element A_{1234} carries planes into cubic surfaces with nodes at $q_1, \cdots, q_4$. The six lines $(q_i\, q_j)^1$ are F-curves since they are on all the surfaces of the homaloidal web. The existence of these F-curves however is a necessary consequence of the existence of the F-points. They all are of the *second kind* i. e. directions at a point of the F-curve correspond to points on a P-curve which is fixed as the point runs over the F-curve. This P-curve is itself an F-curve of the inverse transformation corresponding in a similar way to the original F-curve. S. Kantor[39] has studied these regular transformations in S_3 under the name of transformations without F-curves of the first kind. In hyperspace S_k the author has developed their properties ([17] II §§ 4, 5).

In a linear space S_k an algebraic locus of dimension r and order n will be called a *manifold* or *variety*; more specifically an M_r^n or V_r^n. A manifold of dimension one will ordinarily be called a curve C^n; and of dimension two a surface F^n. At the other extreme however, a manifold of dimension $k-1$, which of course is defined by a single equation, will be called a *spread*.

The definition of the transformation (1) and the regular group can be extended immediately to S_k and there they have like properties. Two sets of m points in S_k, P_m^k, Q_m^k are *congruent* under $A_{1,\ldots,k+1}$ if $p_1, \cdots, p_{k+1}$ and $q_1, \cdots, q_{k+1}$ are corresponding F-points of A and A^{-1} while $p_{k+2}, q_{k+2}; \cdots; p_m$, q_m are ordinary corresponding pairs. They are congruent under regular Cremona transformation in S_k if they are congruent under a sequence of elements A.

A spread in S_k with singularity complex γ at P_m^k is transformed into a spread with complex γ' at the congruent set Q_m^k. Taking account of the ordering of the points and of the fact that congruence under A suffices to define congruence in general it is clear that

(2) *The various types of regular transformations in S_k are in one-to-one correspondence with the double cosets of the group $g_{m,k}$, generated by the permutation group Π of $\gamma_1, \cdots, \gamma_m$ and $A_{1,\ldots,k+1}$, with respect to Π; where*

$$A_{1,\cdots,k+1}:\quad \begin{aligned} \gamma_0' &= k\gamma_0 - \textstyle\sum_j \gamma_j && (j=1,\cdots,k+1),\\ \gamma_i' &= (k-1)\gamma_0 - \textstyle\sum_j \gamma_j + \gamma_i && (i=1,\cdots,k+1),\\ \gamma_h' &= \gamma_h && (h=k+2,\cdots,m). \end{aligned}$$

If the set P_m^k is mapped upon a point P in a space $\Sigma_{k(m-k-2)}$ by first bringing it into a canonical form similar to that of P_m^2 in **7**(1) and P_m^1 in **8**(1), the various points P obtained in Σ by reordering P_m^k in S_k are conjugate under a Cremona $G_{m!}$ in $\Sigma_{k(m-k-2)}$ whose generators are given in ([17] I § 7). If P_m^k, Q_m^k are congruent under $A_{1,\cdots,k+1}$, the points P, Q of Σ are conjugate under the inversion of the coördinates. The $G_{m!}$ and this inversion generate the Cremona group $G_{m,k}$ in $\Sigma_{k(m-k-2)}$. As before

(3) *If the ordered sets P_m^k, Q_m^k are congruent under regular Cremona transformation in S_k, their representative points P, Q in $\Sigma_{k(m-k-2)}$ are conjugate under an operation τ of the Cremona group $G_{m,k}$ which in general is simply isomorphic with the linear group $g_{m,k}$.*

Further properties of types arise from the invariant quadratic and linear form of $g_{m,k}$ which are

$$(4)\qquad \begin{aligned} Q &= (k-1)\gamma_0^2-\gamma_0^2- \cdots -\gamma_m^2, \\ L &= (k+1)\gamma_0-\gamma_1- \cdots -\gamma_m. \end{aligned}$$

These will be developed as needed.

The elliptic norm-curve in S_k, an E^{k+1}, brings to light a useful form of $g_{m,k}$. An E^{k+1} on the $(k+1)$ F-points of $A_{1,\cdots,k+1}$ is transformed into a similar E'^{k+1} on the inverse F-points. This is birationally equivalent to E^{k+1} and therefore projective to it. If then we follow $A_{1,\cdots,k+1}$ by a collineation which sends E'^{k+1} back into E^{k+1} the set P_m^k on E^{k+1} goes into a congruent set Q_m^k on E^{k+1} and the elliptic parameters u_i' of the points of Q_m^k are expressed in terms of the parameters u_i of P_m^k by the linear congruences ([17] II § 6)

$$(5)\qquad A_{1,\cdots,k+1}: \quad \begin{aligned} u_i' &\equiv u_i-2(u_1+\cdots+u_{k+1})/(k+1) \\ &\qquad (i=1,\cdots k+1), \\ u_j' &\equiv u_j-(k-1)(u_1+\cdots+u_{k+1})/(k+1) \\ &\qquad (j=k+2,\cdots,m). \end{aligned}$$

The ambiguity which arises from the submultiple of a period merely leads to the projectively equivalent sets Q_m^k on E^{k+1}.

This element A and the permutations of the u's generate a group $e_{m,k}$ simply isomorphic with $g_{m,k}$. It is the form this latter group takes in the invariant linear space L. The quadratic invariant takes the form

$$(6) \quad [(k+1)^2 - m(k-1)](u_1^2 + \cdots + u_m^2) + (k-1)(u_1 + \cdots + u_m)^2.$$

For sets P_9^2, P_9^5 and P_8^3 this reduces to an invariant linear form $u_1 + \cdots + u_m$. For a discussion of these sets we shall find the group $e_{m,k}$ particularly effective.

The *discriminant conditions* on the set P_m^k are those which imply the coincidence of two points either in P_m^k itself or in some set Q_m^k congruent to P_m^k under regular transformation. The *P-spreads of the set* P_m^k are those loci of dimension $k-1$ which either are the ∞^{k-1} directions about a point of P_m^k or correspond on P_m^k to such directions in a congruent set.

16. Associated sets of points. Apolar matrices have been mentioned in I in connection with apolar linear systems of curves. The product of a row of the one with a row of the other was an apolarity condition. The emphasis here will be laid on the columns.

In S_k with linear spreads ξ let the set P_m^k of m points in S_k be given by their individual equations,

$$(p_1 \xi) = 0, \quad (p_2 \xi) = 0, \quad \cdots, \quad (p_m \xi) = 0.$$

Any $k+2$ are linearly dependent and the m points are therefore connected by $m-k-1$ independent relations,

$$(1) \quad q_{1i}(p_1 \xi) + q_{2i}(p_2 \xi) + \cdots + q_{mi}(p_m \xi) \equiv 0 \quad (i = 0, \cdots, m-k-2).$$

If η is a linear spread in S_{m-k-2} and if the relations (1) are multiplied in order by $\eta_0, \cdots, \eta_{m-k-2}$ and added, a single relation,

$$(2) \quad (q_1 \eta) \cdot (p_1 \xi) + (q_2 \eta) \cdot (p_2 \xi) + \cdots + (q_m \eta) \cdot (p_m \xi) \equiv 0,$$

is obtained which is an identity in both ξ and η. If the coördinates of $p_1, \cdots, p_m$ are the m columns of one matrix, the coördinates of $q_1, \cdots, q_m$ are the m columns of the apolar matrix.

Two sets of points, P_m^k in S_k and Q_m^{m-k-2} in S_{m-k-2}, are termed *associated sets* if their coördinates satisfy (2). If P_m^k is given, the relations (1) may be replaced by any $m-k-1$ independent combinations so that the set Q_m^{m-k-2} is determined only to within projective transformation. The symmetry of (2) in both sets shows that their relation is mutual and is unaltered by linear transformation of either set.

The properties of such associated sets have been developed by the author (cf. [17] I §§ 1, 2; [21]). We recapitulate some of these.

(a) The set P_m^1 on a line is associated with Q_m^{m-3} in S_{m-3}, and the linear set is projective to the set of parameters of the points q on the rational norm-curve R^{m-3} defined by Q_m^{m-3}.

If $m-1$ of the points q are put, in any order, at a given basis in S_{m-3}, the last point takes $m!$ positions, depending on the order, which are conjugate under the Cremona $G_{m,1}$.

(b) If r points (say $p_1, \cdots, p_r$) of P_m^k are selected, and a further group of s points (say $q_{r+1}, \cdots, q_{r+s}$) of the associated Q_m^{m-k-2} are also selected, and if the further $m-r-s$ points of each set are projected upon a lower space from either selected group then the projected sets P_{m-r-s}^{k-r}, $Q_{m-r-s}^{m-k-s-2}$ are also associated.

A combination of (a) and (b) leads to:

(c) The members of the pencil of S_{k-1}'s on $k-1$ of the points p each of which is on one of the $m-k+1$ other points p are projective to the parameters of the $m-k+1$ complementary points q on the rational norm-curve R^{m-k-2} which they define.

The proportionality of complementary determinants of the apolar matrices leads to irrational relations among the determinants which we will illustrate here only in special cases (cf. [17] I § 2 (19)) but to which we shall return in connection with theta relations. A product of determinants formed from P_m^k is prefaced by P, from Q_m^{m-k-1} by Q. For associated sets P_4^1, Q_4^1 the determinant products $P(ij)(kl)$ and $Q(ij)(kl)$ are proportional. From the determinant identity

$$P(12)(34)+P(13)(42)+P(14)(23) = 0$$

for P_4^1 we obtain the irrational relation,

$$\begin{aligned} (3) \quad & [P(12)(34)\cdot Q(12)(34)]^{1/2}+[P(13)(42)\cdot Q(13)(42)]^{1/2} \\ & \qquad +[P(14)(23)\cdot Q(14)(23)]^{1/2}=0, \end{aligned}$$

which is the necessary and sufficient condition for the association (or in this case the projectivity) of the two sets.

For associated P_5^2, Q_5^1 the line pencil from p_5 to the other four points is cut by a line in a P_4^1 associated to the Q_4^1 which omits q_5, whence the sets satisfy three-term relations of the following type:

$$(4) \qquad \sum_3 [P(ij5)(kl5)\cdot Q(ij)(kl)]^{1/2} = 0 \qquad (i, j, k, l = 1, 2, 3, 4).$$

For associated P_6^2, Q_6^2 the sections of the line pencils from p_5 to the first four points, and from q_6 to the first four points are associated P_4^1, Q_4^1 whence

$$(5) \qquad \sum_3 [P(ij5)(kl5)\cdot Q(ij6)(kl6)]^{1/2} = 0.$$

Here however there is a new type of relation which arises from the four-term planar determinant identity; namely

$$(6) \qquad \sum_4 [P(ijk)(l56)\cdot Q(ijk)(l56)]^{1/2} = 0.$$

For associated P_7^3, Q_7^2 there are then the two types,

$$\begin{aligned} (7) \quad & \sum_3 [P(ij67)(kl67)\cdot Q(ij5)(kl5)]^{1/2} = 0, \\ & \sum_4 [P(ijk7)(l567)\cdot Q(ijk)(l56)]^{1/2} = 0. \end{aligned}$$

If the set P_m^k is put in canonical form its representative point P in $\Sigma_{k(m-k-2)}$ is the same as the representative point Q in $\Sigma_{k(m-k-2)}$ of the associated set Q_m^{m-k-2} provided the latter set is taken in inverted order ([17] I § 6). Hence

(8) *The Cremona groups $G_{m,k}$ and $G_{m,m-k-2}$ in $\Sigma_{k(m-k-2)}$ coincide, and the linear groups $g_{m,k}$, $g_{m,m-k-2}$ are simply isomorphic. In this isomorphism the groups Π correspond identically and the generator $A_{1,\ldots,k+1}$ of $g_{m,k}$ corresponds to the generator $A_{k+2,\cdots,m}$ of $g_{m,m-k-2}$.*

From this identity of $G_{m,k}$ and $G_{m,m-k-2}$ there follows that

(9) *The vanishing of a discriminant condition on the set P_m^k implies the vanishing of a discriminant condition on the associated Q_m^{m-k-2}.*

For example the 63 discriminant conditions on Q_7^2 may be indicated by the types $(12)^0$, $(123)^1$, $(123456)^2$ which exemplify respectively 21, 35, and 7 conditions. The corresponding conditions on P_7^3 are respectively $(12)^0$, $(4567)^1$, $(1234567^2)^2$. These latter require respectively that two points coincide; that four are in a plane; and that there is a quadric cone with node at p_7 and on the other points.

The associated sets, P_{2k+2}^k and Q_{2k+2}^k, may lie in the same space and, in special cases, may coincide in the identical order. They then will be termed *self-associated.* Hence we call the P_8^3 which is the base of a net of quadrics a self-associated rather than, as customary, an associated set. The cases in which the two sets coincide in other than the identical order are discussed for $k = 2$ in ([17] I § 1) and for $k = 3$ in ([49]).

17. Special coördinate systems. In a study of geometrical configurations attached to rational curves we frequently meet with forms symmetrical in a number of binary variables, which properly interpreted give rise to interesting loci. We illustrate the procedure in space S_3 which is sufficiently typical of S_n $(n = 2, 3, \cdots)$.

The rational norm-curve in S_3, the twisted cubic C^3, of order 3 and class 3, has points and (osculating) planes whose coördinates, for proper reference basis in space and on the curve, in terms of a binary parameter $t_0 : t_1$, are

$$\begin{array}{llll} x_0 = t_0^3, & x_1 = 3\,t_0^2\,t_1, & x_2 = 3\,t_0\,t_1^2, & x_3 = t_1^3; \\ \xi_0 = t_1^3, & \xi_1 = -t_1^2\,t_0, & \xi_2 = t_1\,t_0^2, & \xi_3 = -t_0^3. \end{array} \tag{1}$$

The plane $\xi(t)$ osculates C^3 at the point $x(t)$.

The binary cubic

$$(\alpha\,t)^3 = a_0\,t_0^3 + 3\,a_1\,t_0^2\,t_1 + 3\,a_2\,t_0\,t_1^2 + a_3\,t_1^3 = 0 \tag{2}$$

then represents either a plane $a_0\,x_0 + a_1\,x_1 + a_2\,x_2 + a_3\,x_3 = 0$ or a point $a_3\,\xi_0 - 3\,a_2\,\xi_1 + 3\,a_1\,\xi_2 - a_0\,\xi_3 = 0$. Hence the equations,

$$\begin{array}{llll} x_0 = -a_3, & x_1 = 3\,a_2, & x_2 = -3\,a_1, & x_3 = a_0; \\ \xi_0 = a_0, & \xi_1 = a_1, & \xi_2 = a_2, & \xi_3 = a_3, \end{array} \tag{3}$$

furnish, for given cubic in (2), the point x such that the parameters of the three planes of C^3 on x are roots of $(\alpha\,t)^3$; and the plane ξ which cuts C^3 in three points whose parameters are the same roots. The coefficients $a_0, \cdots, a_3$ in (3) may of course be replaced by symmetric combinations of the roots or linear factors of $(\alpha\,t)^3$. When these all become equal equations (3) reduce to (1).

A given cubic determines therefore a point x or plane ξ which correspond in the null system set up by C^3 (a polarity in even spaces). The plane ξ of $(\alpha t)^3$ and the point x of $(\beta t)^3$ are incident if $(\alpha\beta)^3 = 0$, i. e. if the cubics are apolar. The four planes ξ of $(\alpha\,t)^3, \cdots, (\delta\,t)^3$ are on a point if $0 = \Delta = (\alpha\beta)\,(\alpha\gamma)\,(\alpha\delta)\,(\beta\gamma)\,(\beta\delta)\,(\gamma\delta)$; the four points x of the same cubics are coplanar if $9\Delta = 0$.

A surface, $(\delta x)^n = 0$, of order n in x will by using (3) be converted into a form of order n in the coefficients of the variable cubic $(\alpha t)^3$. If the coefficients be replaced by symmetric combinations of the roots t_1, t_2, t_3 the form becomes a symmetric form of order n in the binary variables,

$$(\delta\,x)^n = (a_1\,t_1)^n\,(a_2\,t_2)^n\,(a_3\,t_3)^n = 0. \tag{4}$$

This we call the *parametric equation* of the surface; if t_1, t_2, t_3 satisfy the symmetric form, then the point x of intersection of the planes t_1, t_2, t_3 of C^3 is on the surface. The surface is called the *parametric surface* or *spread* of the symmetric form. It occurs frequently in Meyer's Apolarität[44].

A second procedure is to use the completely polarized form, $(\delta x_1)(\delta x_2)\cdots(\delta x_n)$, of $(\delta x)^n$ and replace x_j by t_j from (1). Again a symmetric form,

$$(5) \qquad (a_1 t_1)^3 (a_2 t_2)^3 \cdots (a_n t_n)^3 = 0,$$

is obtained. If $t_1, \cdots, t_n$ satisfy (5) the n points of C^3 with these parameters are apolar to the surface $(\delta x)^n$. If a cubic

$$(\alpha' t)^3 = (\alpha'' t)^3 = \cdots = (\alpha^{(n)} t)^3$$

is such that

$$(a_1 \alpha')^3 (a_2 \alpha'')^3 \cdots (a_n \alpha^{(n)})^3 = 0$$

the point x is on $(\delta x)^n = 0$, and this surface is called the *apolarity surface* or *spread* of the symmetric form (5). Thus a given symmetric form represents either a parametric spread or an apolarity spread and these in general are in different spaces. For example the symmetric form $(a_1 t_1)^4 (a_2 t_2)^4 (a_3 t_3)^4 = 0$ has for parametric spread a quartic surface in S_3 referred to C^3; and for apolarity spread a cubic spread in S_4 referred to C^4. Some examples are studied minutely in ([14]).

Another special coördinate system in S_3 is based on the existence of a proper quadric Q. This with generators t, τ isolated in the simplest fashion has the following point, plane, and parametric equations,

$$(6) \qquad x_0 x_3 - x_1 x_2 = 0; \qquad \xi_0 \xi_3 - \xi_1 \xi_2 = 0;$$

$$Q: \quad \begin{aligned} x_0 &= -\tau_0 t_1, & x_1 &= \tau_0 t_0, & x_2 &= \tau_1 t_1, & x_3 &= -\tau_1 t_0; \\ \xi_0 &= \tau_1 t_0, & \xi_1 &= \tau_1 t_1, & \xi_2 &= \tau_0 t_0, & \xi_3 &= \tau_0 t_1. \end{aligned}$$

A bilinear form

$$(a\tau)(\alpha t) = a_{00} \tau_0 t_0 + a_{01} \tau_0 t_1 + a_{10} \tau_1 t_0 + a_{11} \tau_1 t_1 = 0,$$

represents either a point or a plane (pole and polar as to Q) by virtue of

$$(7) \quad \begin{aligned} x_0 &= a_{10}, & x_1 &= a_{11}, & x_2 &= a_{00}, & x_3 &= a_{01}; \\ \xi_0 &= -a_{01}, & \xi_1 &= a_{00}, & \xi_2 &= a_{11}, & \xi_3 &= -a_{10}. \end{aligned}$$

If $(a\tau)(\alpha t) = 0$ represents a plane, the values t, τ determine in (6) points on the plane section of Q; if a point, they determine planes on the point section of Q.

If for two bilinear forms, $(a\tau)(\alpha t)$ and $(b\tau)(\beta t)$, the invariant $(ab)(\alpha\beta)$ vanishes then the forms as points, or as planes, are an apolar pair of Q; but as point and plane, are incident. Four forms represent four planes on a point, or four points on a plane, if

$$(8)\qquad \begin{vmatrix} (bc)(ad) & (ca)(bd) \\ (\beta\gamma)(\alpha\delta) & (\gamma\alpha)(\beta\delta) \end{vmatrix} = \begin{vmatrix} (ca)(bd) & (ab)(cd) \\ (\gamma\alpha)(\beta\delta) & (\alpha\beta)(\gamma\delta) \end{vmatrix} = \begin{vmatrix} (ab)(cd) & (bc)(ad) \\ (\alpha\beta)(\gamma\delta) & (\beta\gamma)(\alpha\delta) \end{vmatrix} = 0.$$

A surface $(\delta x)^n = 0$ may be replaced from (7) by a form of degree n in the coefficients of the bilinear form; or it may be polarized n times and the variables be replaced from (6) by n pairs of variables t, τ; and the coefficients a_{ij} be introduced later by an apolarity process. In general the quaternary notation is replaced by a double binary notation (cf.[24] for examples). The next coördinate system of this character would occur in S_5 in connection with a binary-ternary notation.

CHAPTER II

TOPICS IN THETA FUNCTIONS

In the discussion of geometric configurations defined by sets of points in the plane and in space we shall have occasion to use relations which exist among theta functions in p variables as well as groups which are associated with them. The brief resumé of these matters which we proceed to give may be supplemented from the accounts of Stahl ([67] Chaps. 5–8), Krazer[41], and Krazer-Wirtinger[42].

18. Definition, periodic properties, and characteristics of the theta functions. We adopt for the constantly recurring exponential function the notation:

$$(1) \qquad e^z = E|z|.$$

The general theta series in p variables $u_1, \cdots, u_p$ has the form

$$(2) \qquad \vartheta(u) = \sum_M E|(am)^2 + 2(mu)|$$

in which $(am)^2$ and (mu) are quadratic and linear forms in the integers $m_1, \cdots, m_p$ of summation and $\sum_M$ indicates a summation over all positive and negative values of these integers. Explicitly

$$(3) \qquad \begin{aligned} (am)^2 &= a_{11} m_1^2 + 2 a_{12} m_1 m_2 + a_{22} m_2^2 + \cdots + a_{pp} m_p^2, \\ (mu) &= m_1 u_1 + m_2 u_2 + \cdots + m_p u_p. \end{aligned}$$

The $p(p+1)/2$ constants a_{ij} are the *moduli* of $\vartheta(u)$. Separated into real and imaginary parts,

$$(4) \qquad a_{ij} = r_{ij} + i s_{ij}.$$

The necessary and sufficient condition for the absolute convergence of (2) for all finite values of $u_1, \cdots, u_p$ is that the real quadratic form $(rm)^2$ be a definite and negative form in p independent variables. The distributive properties of the forms (3) expressed by

$$(5)\qquad \begin{aligned}(a,\ m+g)^2 &= (am)^2+2(am)(ag)+(ag)^2,\\ (m+g,\ u) &= (mu)+(gu),\end{aligned}$$

are necessary to verify relations given below.

The p quantities $v_1, \cdots, v_p$ form a *simultaneous period* of $\vartheta(u)$ if $\vartheta(u+v) = \vartheta(u)$; a *simultaneous quasiperiod if* $\vartheta(u+v) = k\,\vartheta(u)$. Either type will be referred to as a *period* of the theta function. It is then easily verified that $\vartheta(u)$ has $2p$ distinct periods, namely

$$(6)\qquad \begin{matrix}\pi i, & 0, & 0, & \cdots, & 0\\ 0, & \pi i, & 0, & \cdots, & 0\\ \cdot & \cdot & \cdot & \cdot & \cdot\\ 0, & 0, & 0, & \cdots, & \pi i\end{matrix}\ ; \qquad \begin{matrix}a_{11}, & a_{21}, & \cdots, & a_{p1}\\ a_{12}, & a_{22}, & \cdots, & a_{p2}\\ \cdot & \cdot & \cdot & \cdot\\ a_{1p}, & a_{2p}, & \cdots, & a_{pp}\end{matrix} \qquad (a_{ij} = a_{ji}).$$

From these by multiplication with *integers* $\lambda_1, \cdots, \lambda_p$; $\varkappa_1, \cdots, \varkappa_p$ respectively we construct the general period

$$(7)\qquad v = (a\varkappa)\,a+\lambda\pi i$$

for which

$$(8)\qquad \vartheta(u+(a\varkappa)a+\lambda\pi i) = \vartheta(u)\cdot E\,|-(a\varkappa)^2-2(\varkappa u)\,|.$$

This is proved by comparing exponents on the two sides when, for $\vartheta(u)$, $m+\varkappa$ replaces m. They are

$$\begin{aligned}&(am)^2+2(m,\ u+(a\varkappa)a)\\ &= (a,\ m+\varkappa)^2+2(m+\varkappa,\ u)-(a\varkappa)^2-2(\varkappa u).\end{aligned}$$

If we set

$$u_1 = v_1+iw_1, \cdots, u_p = v_p+iw_p$$

and interpret $v_1, \cdots, v_p, w_1, \cdots, w_p$ as rectangular coördinates in a linear real space R_{2p} then every system of finite complex values $u_1, \cdots, u_p$ defines a finite point u in R_{2p} and conversely. In particular the $2p$ periods of $\vartheta(u)$

define $2p$ points in R_{2p} which, with the origin $u = 0$, form a *proper* $(2p+1)$-point in R_{2p}, the analogue of a proper triangle in the plane. Hence the $2p$ strokes from the origin to the $2p$ period points serve as the axes of a coördinate system in R_{2p}. Thus for every point $u = c$ in R_{2p} there will exist *real* numbers $g_1, \cdots, g_p; h_1, \cdots, h_p$ such that

$$(9)\quad \begin{aligned} c &= \{g, h\} = \{g_1, \cdots, g_p;\ h_1, \cdots, h_p\} = (ag)\,a + h\pi i, \text{ i.e.} \\ c &= (ag)\,a_1 + h_1\pi i, \cdots, c_p = (ag)\,a_p + h_p\pi i. \end{aligned}$$

The $2p$ numbers g, h are called the *period characteristic* of $u = c$. All points u whose period characteristics satisfy

$$(10)\qquad 0 \leqq g < 1, \quad 0 \leqq h < 1$$

lie in the *initial period cell* C_{00} with *initial vertex* at $u = 0$. The 2^{2p} vertices of the initial cell are obtained from values $g, h = 0, 1$.

Any point c can be expressed uniquely as

$$(11)\qquad c = \{g, h\} = \{\varkappa, \lambda\} + \{g', h'\}$$

where $\varkappa, \lambda$ as always are integers and g', h' satisfy (10). For all values of g', h' subject to (10), the point c in (11) runs over the period cell $C_{\varkappa,\lambda}$; and the point $\{g, h\}$ of $C_{\varkappa,\lambda}$ is said to be *congruent* to the point $\{g', h'\}$ of the initial cell $C_{0,0}$. From (8) for $u = \{g', h'\}$ we find that

$$(12)\quad \vartheta(\{g, h\}) = \vartheta(\{g', h'\}) \cdot E\,|-(a\varkappa)^2 - 2(a\varkappa)(ag) - 2\pi i(\varkappa h')|.$$

Hence the behavior of $\vartheta(u)$ in any cell $C_{\varkappa,\lambda}$ can be determined when its behavior in $C_{0,0}$ is known.

In the original theta function we make a change of origin for the variables, which does not affect convergence, by replacing u by $u + c$. If c is expressed as in (9) and if *in addition* we multiply by $E\,|(ag)^2 + 2(g, u + h\pi i)|$, the general term becomes $E\,|(a, m+g)^2 + 2(m+g, u + h\pi i)|$. The resulting function is called the *theta function with characteristic* $[g, h] = [g_1, \cdots, g_p; h_1, \cdots, h_p]$. Such characteristics $[g, h]$

attached to the theta functions are to be distinguished from the *period characteristics* $\{g, h\}$ and will be called *theta characteristics*. Thus

(13) *The function*

$$\vartheta[g, h](u) = \vartheta[g_1, \cdots, g_p; h_1, \cdots, h_p](u_1, \cdots, u_p)$$

is defined by the series

$$\vartheta[g, h](u) = \sum_M E|(a, m+g)^2 + 2(m+g, u+h\pi i)|$$

and is connected with the original theta function by the relation

$$\text{(A)} \quad \vartheta[g, h](u) = \vartheta(u+\{g, h\}) \cdot E|(ag)^2 + 2(g, u+h\pi i)|.$$

Other indispensable formulae which follow from this definition are:

$$\text{(B)} \quad \begin{aligned} &\vartheta[g, h](u+\{\varkappa, \lambda\}) \\ &\quad = \vartheta[g, h](u) \cdot E|-(a\varkappa)^2 - 2(\varkappa u) + 2(g\lambda - h\varkappa)\pi i|; \end{aligned}$$

$$\text{(C)} \quad \begin{aligned} &\vartheta[g, h](u+\{g', h'\}) \\ &\quad = \vartheta[g+g', h+h'](u) \cdot E|-(ag')^2 - 2(g', u+h\pi i+h'\pi i)|; \end{aligned}$$

$$\text{(D)} \quad \vartheta[g+\varkappa, h+\lambda](u) = \vartheta[g, h](u) \cdot E|2(g\lambda)\pi i|;$$

$$\text{(E)} \quad \vartheta[g, h](-u) = \vartheta[-g, -h](u).$$

Of these (B) gives the periodic properties of $\vartheta[g, h](u)$; (C) the effect of a change of origin; and (D) indicates that integer changes in the characteristic at most produce multiplication by a constant. It would suffice then to consider *ab initio* only functions with *reduced* characteristics, $0 \leqq g < 1$, $0 \leqq h < 1$, though others would eventually occur by using (C).

19. Theta functions of higher order. Theta relations and theta zeros. Functions $2p$-tuply periodic. A *theta function of order* n (n a positive integer) *with characteristic* $[g, h]$ is defined to be a uniform function, $\vartheta_n[g, h](u)$, of $u_1, \cdots, u_p$, regular for all finite values of u, which satisfies for arbitrary integers $\varkappa, \lambda$ the equation

$$\text{(1)} \quad \begin{aligned} &\vartheta_n[g, h](u+\{\varkappa, \lambda\}) \\ &\quad = \vartheta_n[g, h](u) \cdot E|-n(a\varkappa)^2 - 2n(\varkappa u) + 2(g\lambda - h\varkappa)\pi i|. \end{aligned}$$

If the moduli a_{ij} of the function are to be indicated we write it $\vartheta_n[g, h](u)_a$. When $n = 1$, this is the function $\vartheta[g, h](u)$ (cf. 18 (B)). We verify, again by the use of 18 (B), that $\vartheta_n[g, h](u) = (\vartheta[g, h](u))^n$ is a function $\vartheta_n[ng, nh](u)$; and furthermore that the functions,

$$
(2) \quad \begin{gathered}
\vartheta[(g+\varrho)/n,\, h]\,(nu)_{na}, \qquad \vartheta[g,\, (h+\sigma)/n]\,(u)_{a/n}, \\
\vartheta^n\,[(g+\varrho)/n,\, (h+\sigma)/n]\,(u)_a,
\end{gathered}
$$

where ϱ, σ are integers, all are functions $\vartheta_n[g, h](u)_a$. The following theorem, with obvious extension to any number of factors, is an immediate consequence of the defining equation (1):

(3) *The product of two theta functions of orders n_1, n_2 with characteristics $[g, h]$, $[g', h']$ respectively, is a theta function of order n_1+n_2 and characteristic $[g+g', h+h']$.*

It may be proved (cf. Krazer [41] p. 40) that the most general function, $\vartheta_n[g, h](u)$, can be expressed linearly with constant coefficients in terms of a set of n^p particular functions, e. g. those of the first type in (2) for which $\varrho = 0, 1, \cdots, n-1$; and that these n^p functions are themselves linearly independent. Hence

(4) *Between any n^p+1 functions, $\vartheta_n[g, h](u)$, there must exist at least one linear homogeneous relation with coefficients which are constant with respect to u.*

By examining an expression linear in n^p such particular functions we derive the relation

$$
(5) \quad \begin{aligned}
\vartheta_n[g, h](u+\{g', h'\}) &= \vartheta_n[g+ng', h+nh'](u) \\
&\quad \times E\,|-n(ag')^2-2(g', nu+nh'\pi i+h\pi i)|.
\end{aligned}
$$

Thus g', h' may be so chosen that $[g+ng', h+nh'] = [0, 0]$ whence

(6) *The aggregate of theta functions of order n and characteristic $[g, h]$ may be converted into the aggregate of theta functions of order n and zero characteristic by a common change of origin and multiplication by a common exponential factor.*

This theorem permits the use of the simpler zero characteristic in certain problems such as the determination of the zeros in a period cell. Again from the definition there follows:

(7) *A function,* $\vartheta_n[g+\varkappa, h+\lambda](u)$, *where* $\varkappa$, λ *are integers, is also a function* $\vartheta_n[g, h](u)$.

We ask for the number of solutions or zeros of the p equations,

$$(8) \qquad \vartheta_{n_1}(u-e_1) = 0, \cdots, \vartheta_{n_p}(u-e_p) = 0,$$

where the thetas are general functions of the orders indicated and with the same characteristic which may, according to (6), be $[0, 0]$. It is understood that $e_i = e_{i1}, e_{i2}, \cdots, e_{ip}$. Then (cf. Krazer[41] p. 43).

(9) *The* p *theta functions* (8) *of orders* $n_1, \cdots, n_p$ *and the same characteristic vanish simultaneously at* $N = n_1 \cdot n_2 \cdot \cdots \cdot n_p \cdot p!$ *points in a period cell whose sum satisfies the congruences:*

$$u \equiv n_1 \cdot n_2 \cdot \cdots \cdot n_p \cdot (p-1)!\,(e1)\,e.$$

Here $(e1)e = (e_1 + \cdots + e_p)e$ and $e_i e_j = e_{ij} \mp e_{ji}$.

Any two distinct theta functions of the same order and characteristic such as two of the types given in (2) will have, according to (1), each period of the theta functions as a quasi-period with the same multiplicative factor and thus their ratio will be $2p$-tuply periodic in the strict sense. It may be proved conversely (cf. Krazer[41] Chap. 4) that any $2p$-tuply periodic function of p variables with no essential singularities at a finite distance, can, after proper linear transformation of the variables, be expressed as a quotient of two theta functions of like order and characteristic.

20. The half-periods. The odd and even theta functions. We ask for such points u in the initial period cell as will satisfy the congruence $2u \equiv \{0, 0\}$ or the equality $2u = \{\varkappa, \lambda\}$. This requires that $u = \{\varkappa/2, \lambda/2\}$ where $0 \leqq \varkappa/2 < 1$, $0 \leqq \lambda/2 < 1$ and gives rise to the 2^{2p} solutions $\varkappa = 0, 1$; $\lambda = 0, 1$. These 2^{2p} values of u in the initial cell are called the *half-periods.* They are in general *proper* but the particular one for which $\varkappa = \lambda = 0$ is *improper* since it is a trivial period. We give these half periods a distinctive notation as follows:

$$(1)\quad \{\varepsilon\}_2 = \{\varepsilon, \varepsilon'\}_2 = \{\varepsilon_1/2, \cdots, \varepsilon_p/2;\ \varepsilon_1'/2, \cdots, \varepsilon_p'/2\} \quad (\varepsilon, \varepsilon' = 0, 1).$$

In terms of a similar characteristic $[\eta]_2$ the 2^{2p} theta functions of first order are defined by

$$(2)\quad \vartheta[\eta]_2(u) = \sum_M E|(a, m+\eta/2)^2 + 2(m+\eta/2, u+\eta'\pi i/2)|.$$

By specializing the relations in 18 we find that

$$(3)\quad \vartheta[\eta]_2(u+\{2\varkappa\}_2) = \vartheta[\eta]_2(u)\cdot E|-(a\varkappa)^2 - 2(\varkappa u) + (\eta\varkappa' - \eta'\varkappa)\pi i|,$$

$$(4)\quad \vartheta(u+\{\varepsilon\}_2) = \vartheta[\varepsilon]_2(u)\cdot E|-(a\varepsilon)^2/4 - (\varepsilon, u+\varepsilon'\pi i/2)|,$$

$$(5)\quad \vartheta[\eta]_2(u+\{\varepsilon\}_2) = \vartheta[\eta+\varepsilon]_2(u) \cdot E|-(a\varepsilon)^2/4 - (\varepsilon u) - (\varepsilon, \eta'+\varepsilon')\pi i/2|,$$

$$(6)\quad \vartheta[\eta+2\zeta]_2(u) = \vartheta[\eta]_2(u)\cdot E|(\eta\zeta')\pi i|,$$

$$(7)\quad \vartheta[\eta]_2(-u) = \vartheta[-\eta]_2(u).$$

Formula (3) expresses the usual periodic property. Formula (6) shows that two functions $\vartheta[\varepsilon]_2(u)$ and $\vartheta[\eta]_2(u)$ whose characteristics $[\varepsilon]_2, [\eta]_2$ are such that $[\varepsilon] \equiv [\eta]$ mod. 2 will themselves differ only by a constant factor and are not *essentially distinct*. The 2^{2p} essentially distinct functions $\vartheta[\eta]_2(u)$ are obtained from values $\eta = 0, 1$; $\eta' = 0, 1$. Formulae (4), (5) show that the 2^{2p} essentially distinct functions are obtained by adding the 2^{2p} half-periods to the argument of any one. Thus the 2^{2p} functions form a homogeneous set.

According to (7), $\vartheta(u) = \vartheta[0]_2(u)$ is an even function. For any other function, $\vartheta[\eta]_2(u)$, we have from (7) and (6) that

$$\vartheta[\eta]_2(-u) = \vartheta[-\eta]_2(u) = \vartheta[-\eta+2\eta]_2(u)\cdot E|-(\eta\eta')\pi i| = (-1)^{(\eta\eta')}\,\vartheta[\eta]_2(u).$$

Thus the function is even or odd according as $(\eta\eta')$ is even or odd. From an easy enumeration there results:

(8) *Of the 2^{2p} essentially distinct theta functions $\vartheta[\eta]_2(u)$ of first order and rational characteristic with denominator 2, the $2^{p-1}(2^p+1)$ functions whose characteristic satisfies the congruence*

$$\eta_1 \eta_1' + \eta_2 \eta_2' + \cdots + \eta_p \eta_p' \equiv 0 \quad \text{mod. } 2$$

are even and the remaining $2^{p-1}(2^p-1)$ *functions are odd.*

With respect to odd and even theta functions of any order the following theorem (Krazer [41] pp. 357–362) holds.

(9) *If* $\vartheta_n[g,h](u)$ *is an odd or even function of* u *it is necessary that* $[g,h]$ *in reduced form be one of the* 2^{2p} *characteristics* $[\eta]_2$. *If* E *is the number of linearly independent even functions with given characteristic* $[\eta]_2$, O *the number of odd functions, then for*

n *even*; $[\eta]_2 = [0]$,
$$E = (n^p + 2^p)/2, \qquad O = (n^p - 2^p)/2;$$
$[\eta]_2 \neq 0$,
$$E = n^p/2, \qquad O = n^p/2;$$
n *odd*; $[\eta]_2$ *even* i. e. $(\eta\eta') \equiv 0 \,\textit{mod.}\, 2$,
$$E = (n^p+1)/2, \qquad O = (n^p-1)/2;$$
$[\eta]_2$ *odd* i. e. $(\eta\eta') \equiv 1 \,\textit{mod.}\, 2$,
$$E = (n^p-1)/2, \qquad O = (n^p+1)/2.$$

21. Integral linear transformation of the periods, and of the period and theta characteristics. The vertices of the period cells were built up from the $2p$ strokes from the origin to the values of u given by the period scheme, 18 (6). We seek a new set of $2p$ strokes which will furnish the same network of vertices. It is convenient to begin with the following more general expression for the $2p$ periods in which they are divided into two sets of p periods:

$$(1) \qquad \begin{matrix} \omega_{11}, \omega_{21}, \cdots, \omega_{p1} \\ \omega_{12}, \omega_{22}, \cdots, \omega_{p2} \\ \cdot \quad \cdot \quad \cdot \quad \cdot \quad \cdot \quad \cdot \\ \omega_{1p}, \omega_{2p}, \cdots, \omega_{pp} \end{matrix}\;; \qquad \begin{matrix} \omega_{11}', \omega_{21}', \cdots, \omega_{p1}' \\ \omega_{12}', \omega_{22}', \cdots, \omega_{p2}' \\ \cdot \quad \cdot \quad \cdot \quad \cdot \quad \cdot \quad \cdot \\ \omega_{1p}', \omega_{2p}', \cdots, \omega_{pp}' \end{matrix}\,.$$

The $2p$ new periods $\bar\omega$, $\bar\omega'$ must then be integral combinations of ω, ω' with determinant ± 1, since the old periods must be similarly expressible in terms of the new ones. Hence

$$(2) \qquad \begin{aligned} \bar\omega_{ik} &= \textstyle\sum_\mu (\alpha_{k\mu}\, \omega_{i\mu} + \beta_{k\mu}\, \omega_{i\mu}'), \\ \bar\omega_{ik}' &= \textstyle\sum_\mu (\gamma_{k\mu}\, \omega_{i\mu} + \delta_{k\mu}\, \omega_{i\mu}'). \end{aligned} \qquad (\mu, i, k = 1, \cdots, p)$$

A point u of a cell with period characteristic g, g' with respect to ω, ω' will have a characteristic $\bar{g}, \bar{g}'$ with respect to $\bar{\omega}, \bar{\omega}'$. From

$$\begin{aligned} u = \{g, g'\} &= \sum_\nu (g_\nu \omega_{i\nu} + g'_\nu \omega'_{i\nu}) \\ = \{\bar{g}, \bar{g}'\} &= \sum_\nu (\bar{g}_\nu \bar{\omega}_{i\nu} + \bar{g}'_\nu \bar{\omega}'_{i\nu}), \end{aligned}$$

the contragredience of g, g' with ω, ω' is evident and we find that

$$(3) \qquad \begin{aligned} g_\mu &= \sum_\nu (\alpha_{\nu\mu} \bar{g}_\nu + \gamma_{\nu\mu} \bar{g}'_\nu), \\ g'_\mu &= \sum_\nu (\beta_{\nu\mu} \bar{g}_\nu + \delta_{\nu\mu} \bar{g}'_\nu). \end{aligned}$$

If these new periods are such as will serve, after proper linear transformation of the variables, to define new theta functions, the $4p^2$ integer coefficients in (2) must satisfy certain relations (cf. Krazer [41] p. 131). These are

$$(4) \qquad \begin{gathered} \sum_\varrho (\alpha_{\varrho i} \gamma_{\varrho j} - \alpha_{\varrho j} \gamma_{\varrho i}) = 0, \quad \sum_\varrho (\beta_{\varrho i} \delta_{\varrho j} - \beta_{\varrho j} \delta_{\varrho i}) = 0, \\ \sum_\varrho (\alpha_{\varrho i} \delta_{\varrho j} - \beta_{\varrho j} \gamma_{\varrho i}) = \begin{cases} 1 \text{ if } i = j \\ 0 \text{ if } i \neq j \end{cases}; \end{gathered}$$

from which there follows (Stahl [67] Chap. 8)

$$(5) \qquad \begin{gathered} \sum_\varrho (\gamma_{i\varrho} \delta_{j\varrho} - \gamma_{j\varrho} \delta_{i\varrho}) = 0, \quad \sum_\varrho (\alpha_{i\varrho} \beta_{j\varrho} - \alpha_{j\varrho} \beta_{i\varrho}) = 0, \\ \sum_\varrho (\alpha_{i\varrho} \delta_{j\varrho} - \beta_{i\varrho} \gamma_{j\varrho}) = \begin{cases} 1 \text{ if } i = j \\ 0 \text{ if } i \neq j \end{cases}. \end{gathered}$$

These relations may be stated in a very convenient way with respect to the transformation (3) on the period characteristics (Stahl [67] p. 331).

(6) *Two period characteristics, $\{g, g'\}$, $\{h, h'\}$, transformed as in* (3) *into $\{\bar{g}, \bar{g}'\}$, $\{\bar{h}, \bar{h}'\}$ respectively, have the simultaneous absolute invariant,*

$$\sum_i \begin{vmatrix} g_i & g'_i \\ h_i & h'_i \end{vmatrix} = \sum_i \begin{vmatrix} \bar{g}_i & \bar{g}'_i \\ \bar{h}_i & \bar{h}'_i \end{vmatrix}, \text{ i.e. } |g, h| = |\bar{g}, \bar{h}|.$$

Under integral linear transformation of the periods a linear theta function $\vartheta[g, g'](u)_a$ is transformed, to within an exponential factor, into a linear theta function, $\vartheta[\bar{g}, \bar{g}'](\bar{u})_{\bar{a}}$ for which (cf. Krazer [41] pp. 166, 180) the transformed theta characteristic $[\bar{g}, \bar{g}']$ is expressed in terms of the original theta characteristic $[g, g']$ by

$$(7) \qquad \begin{aligned} \bar{g}_\mu &= \sum_\nu (\alpha_{\mu\nu} g_\nu - \beta_{\mu\nu} g'_\nu + \alpha_{\mu\nu} \beta_{\mu\nu}/2), \\ \bar{g}'_\mu &= \sum_\nu (-\gamma_{\mu\nu} g_\nu + \delta_{\mu\nu} g'_\nu + \gamma_{\mu\nu} \delta_{\mu\nu}/2). \end{aligned}$$

An obvious peculiarity is that the coefficients α, β, γ, δ of the original transformation (2) do not occur homogeneously. This complication is removed for congruence modulo 2 in **24**.

22. The finite geometry defined by the reduced half periods. Under integral linear transformation of the periods a half period with characteristic $\{\varepsilon\}_2$ is transformed into one with characteristic $\{\bar{\varepsilon}\}_2$ (cf. **21** (3)). If all such half periods both original and transformed, are reduced (modulo the periods) to half periods in the initial period cells then it is sufficient to take the transformation from $\varepsilon, \varepsilon'$ to $\bar{\varepsilon}, \bar{\varepsilon}'$ with coefficients reduced modulo 2. The effect of this from the standpoint of group theory may be expressed as follows:

(1) *The group I* (**21** (3)) *of integral linear transformations on the period characteristics contains an invariant subgroup I_2 which consists of those transformations which reduce modulo* 2 *to the identity. The factor group of I_2 under I is represented by the elements* (3) *which are distinct when reduced mod.* 2.

According to **21** (6) this group is defined by the invariant $|g, h|$. It also leaves the improper half period $\{g, g'\} = \{0, 0\}$ unaltered. If then we represent the proper reduced half periods with $\varepsilon, \varepsilon' = 0, 1$ as points in the finite geometry (mod. 2) of an S_{2p-1}, the group of the reduced half periods becomes a collineation group with an invariant of the type given. Thus (cf. [16] p. 247):

(2) *Under integral linear transformation of the periods of the theta function in p variables the reduced proper half period characteristics are transformed like the points of a finite space S_{2p-1}* (*mod.* 2) *under the group G_{NC}, isomorphic with the factor group in* (1), *of collineations which leaves unaltered the proper null system*

$$C = (x_1 y'_1 - x'_1 y_1) + (x_2 y'_2 - x'_2 y_2) + \cdots + (x_p y'_p - x'_p y_p).$$

This null system coördinates to a point x, x' in S_{2p-1} its incident null space, S_{2p-2}, whose equation in variables y, y' is given by $C=0$. We identify a period characteristic $\{\varepsilon, \varepsilon'\}_2$ with the point $x=\varepsilon'$, $x'=\varepsilon$ or with the null space, $\sum_\nu(\varepsilon_\nu x_\nu+\varepsilon'_\nu x'_\nu)\equiv 0 \pmod{2}$, of this point in C. According to (2), properties of sets of period characteristics, invariant under period transformation, are projective properties of the corresponding sets of points in S_{2p-1} with reference to the given null system C. We outline first these projective properties and translate them later to period characteristics.

The following facts result from easy enumerations (cf. [16] § 2):

(3) *The number of points in a finite space S_k (mod. 2) is*

$$H_k = 2^{k+1}-1.$$

Given n linearly independent S_{k-1}'s in S_k the number of points on m of the S_{k-1}'s but not on the remaining $n-m$ is 2^{k-n+1} if $m<n$; and is H_{k-n} if $m=n$.

In S_k $(k+1)$ linearly independent points constitute a *point reference basis* and the $(k+1)$ S_{k-1}'s on each set of k points constitute an *S_{k-1} reference basis.* Either basis determines the other and the two make up a *reference basis* in S_k.

(4) *The number, R_k, of reference bases in S_k is*

$$R_k = 2^k H_k R_{k-1}/(k+1) = 2^{k(k+1)/2} H_k H_{k-1}\cdots H_1/(k+1)!.$$

(5) *For $h<k$ the number $H_k^{(h)}$ of spaces S_h in S_k is*

$$H_k^{(h)} = H_k H_{k-1}\cdots H_{k-h}/H_h H_{h-1}\cdots H_1.$$

We shall be concerned mainly with a space S_{2p-1} of odd dimension with coördinates $x_1, \cdots, x_p, x'_1, \cdots, x'_p$. This reference basis determines a unique space, $\sum(x+x')=0$, not on the reference points. The $2p+1$ spaces are dependent but any $2p$ of them are independent. If one of the spaces is isolated, there is a unique point (cf. (3)) on the one space and not on any of the remaining $2p$ spaces. Hence the set of $2p+1$ spaces determines a set of $2p+1$ points each set ordered with respect to the other. Either set constitutes a *basis* in S_{2p-1} and the two make up a *self-dual basis.* The peculiarity of the odd dimension is that corresponding point and S_{2p-2} of a self-dual basis are incident.

(6) *The number of self-dual bases in* S_{2p-1} *is*

$$N_B = R_{2p-1}/(2p+1) = 2^{p(2p-1)} H_{2p-1} H_{2p-2} \cdots H_1/(2p+1)!.$$

The order of the collineation group G_N *in* S_{2p-1} *is* $N = (2p+1)!\, N_B$. *This group is augmented by* N *correlations to the correlation group of order* $2N$ *in* S_{2p-1}.

Those correlations for which corresponding point and S_{2p-2} are incident — the so-called null systems — are of particular interest. They all are of one type and each may be determined by the fact that it interchanges corresponding point and S_{2p-2} of a given self-dual basis. If C is given an enumeration of its invariant bases leads to the theorem:

(7) *The number of self-dual bases invariant under the null system* C *is*

$$N_{BC} = 2^{p^2} H_{2p-1} H_{2p-3} \cdots H_1/(2p+1)!$$

All proper null systems are conjugate under G_N *and each is unaltered by a group* G_{NC} (cf. (2)) *of order*

$$N_C = 2^{p^2} H_{2p-1} H_{2p-3} \cdots H_1.$$

Under G_N in S_{2p-1} two spaces of the same dimension are conjugate; under G_{NC} this is no longer true. Two linear spaces in S_{2p-1} are called *skew* if they have no points in common. If an S_{r-1} is determined by $x^{(1)}, \cdots, x^{(r)}$ the null S_{2p-2}'s of all points x in S_{r-1} meet in an S_{2p-r-1}, the *null space of* S_{r-1}, the common part of the null S_{2p-1}'s of the r points $x^{(i)}$. If the null space of S_{r-1} contains S_{r-1}, then S_{r-1} is called a *null* S_{r-1}. For example any line on x and in the null S_{2p-2} of x is a *null line*; any other line on x is an *ordinary line.* A null space of greatest dimension, an S_{p-1}, is called a *Göpel space.* The null space, S_{2p-r-1}, of a given S_{r-1} will in general have a space S_{m-1} in common with S_{r-1}, which will be called the *null subspace of* S_{r-1}. Then

(8) *Two spaces of the same dimension in S_{2p-1} are conjugate under G_{NC} if and only if their null subspaces have the same dimension. An S_{r-1} with a null subspace S_{m-1} ($r-m$ necessarily even) has a reference basis of the form*

$$y^{(1)}+z^{(1)};\quad y^{(2)}+z^{(2)},\ \cdots,\ y^{(m)}+z^{(m)},\ y^{(m+1)},\ \cdots,\ y^{(r)}$$

where the points y, z form part of a self-dual basis of C.

The number of spaces of various kinds and the order of the subgroup of G_{NC} which leaves a particular type unaltered have been determined by the author (cf. [16] pp. 250–251).

We indicate the translation from terms in the finite geometry to the corresponding terms in the exposition of the characteristic theory given by Krazer ([41] Chap. VII).

Point in S_{2p-1}.	Proper Per. Char.
Two points on a null (ordinary) line.	Two syzygetic (azygetic) Per. Char.
Points of a self dual basis of C.	Fundamental system (F.S.) of $2p+1$ Per. Char. (every pair azygetic).
(9) S_{r-1}.	Group E_r of Per. Char. of rank r.
Space in S_{2p-1} skew to S_{r-1}.	Group H conjugate to E_r.
Null subspace of S_{r-1}.	Syzygetic subgroup of E_r.
Null space of S_{r-1}.	Group adjoint to E_r.
Null S_{r-1}.	Syzygetic group E_r.
Göpel space.	Göpel group.

The following theorem (cf. [16] § 1) is of particular importance in the applications.

(10) *The group G_{NC} of the null system is generated by a complete conjugate set of H_{2p-1} involutions each attached to a point of S_{2p-1}. The involution I_x attached to x leaves x and every point on its null S_{2p-2} unaltered and interchanges the further pair of points on every ordinary line through x.*

23. The basis notation for the half periods. The $2p+1$ S_{2p-2}'s of a basis, B_{2p+2}, are linearly related. If their equations are respectively $y_i = 0$ $(i = 1, \cdots, 2p+1)$, this relation is $\sum y = 0$. In this superfluous coördinate system y, the $2p+1$ basis points are

$$(1) \qquad y_i = 0, \qquad y_j = 1 \; (j \neq i).$$

This i-th point $(i = 1, \cdots, 2p+1)$ is on $y_i = 0$ but on no other S_{2p-1} of the basis. If C has B_{2p+2} as an invariant basis its equation is

$$(2) \qquad C = y_1 y_1' + y_2 y_2' + \cdots + y_{2p+1} y_{2p+1}' = 0$$
$$(\sum y = 0, \sum y' = 0).$$

For $C_{y=y'} = \sum y_i^2 \equiv (\sum y_i)^2 \equiv 0 \pmod{2}$, and, due to $\sum y = 0$, the null S_{2p-2} of the point (1) is $y_i = 0$.

The point (1) will be named $P_{i,2p+2}$. The line joining two basis points contains a third point called a *residual* point of the basis. The residual point P_{ij} on the line through $P_{i,2p+2}$, $P_{j,2p+2}$ is $P_{ij} = P_{i,2p+2} + P_{j,2p+2}$ and has coördinates

$$(3) \qquad y_i = y_j = 1, \quad y_k = 0 \qquad (k \neq i, \; k \neq j).$$

The $2p+1$ points $P_{12}, P_{13}, \cdots, P_{1,2p+2}$ themselves evidently form an invariant basis B_1 of C; and from symmetry there exist similar bases $B_2, \cdots, B_{2p+1}$. A point P_{ij} is in the bases B_i, B_j and these two bases have no other common point. This set of $2p+2$ bases is symmetrical since any one of the bases and its residual points determine the set. The involution I_x (cf. **22** (10)) attached to the point $P_{i,2p+2}$ in (1) is $y_i' = y_i$, $y_k' = y_k + y_i \; (k \neq i)$. This interchanges B_i and B_{2p+2} and leaves each of the other $2p$ bases unaltered. The involution attached to P_{ij} in (3) is $y_i' = y_j$, $y_j' = y_i$, $y_k' = y_k$. This interchanges B_i and B_j and leaves the others unaltered. Hence (cf. [16] p. 271).

(4) *The $2p+1$ points of a basis and the $p(2p+1)$ residual points of the basis form a basis configuration BC, i.e., a set of $(p+1)(2p+1)$ points which can be divided in $2p+2$ ways into a basis and its residual points. Each point of BC is in two bases B_i, B_j and is denoted by P_{ij} $(i, j = 1, \cdots, 2p+2)$. The configuration is unaltered by*

a $G_{(2p+2)!}$, *symmetric on the* $2p+2$ *bases, a subgroup of* G_{NC}, *which is generated by the involutions attached to the configuration points. The involution attached to* P_{ij} *effects on the bases and the points an interchange of the subscripts.*

The points of S_{2p-1} not included among the points of BC will have $2k$ $(1 < k < p)$ coördinates $y = 1$ and the others zero. If the $2k$ unit coördinates are say the first ones the point is the sum $P_{12} + P_{34} + \cdots + P_{2k-1,2k}$ but it is equally well the sum $P_{2k+1,2k+2} + \cdots + P_{2p+1,2p+2}$. Evidently also the order in which the subscripts are paired in these sums is not material so that the point in question will be denoted by

$$P_{1234\cdots,2k} = P_{2k+1,\cdots,2p+2}.$$

Hence (cf. [18] § 1):

(5) *The points in the finite geometry are named by an even number of subscripts from the set* $1, 2, \cdots, 2p+2$ *in such wise that complementary sets of subscripts denote the same point. The linear condition is*

$$P_a + P_b = P_{ab}$$

where like subscripts in the sets a *and* b *cancel in* ab.

The following theorems are easily verified:

(6) *Two points* P_a, P_b *of* S_{2p-1} *are syzygetic or azygetic according as the sets of the subscripts,* a, b, *have an even or an odd number of subscripts in common.*

(7) *The number of basis configurations belonging to* C *is*

$$N_{BC} = 2^{p^2} H_{2p-1} H_{2p-3} \cdots H_1/(2p+2)!.$$

By virtue of (5) the basis notation reduces the construction of the linear space S_{2p-1} to a purely tactical process. According to (6) the same is true of the null-system C.

24. Theta characteristics as quadrics in the finite geometry. The 2^{2p} odd and even theta functions of the first order, $\vartheta[\eta]_2(u)$, are defined by characteristics $[g, g']$ with denominator 2 and numerators $\eta, \eta' = 0, 1$. If to these the period transformation of **21** is applied they become new theta functions of the same order and type whose characteristics $[\bar{\eta}]_2$ are connected with $[\eta]_2$ by the relations **21** (7)

$$
(1) \qquad \begin{aligned}
\bar{\eta}_\mu &= \sum_\nu (\alpha_{\mu\nu}\,\eta_\nu - \beta_{\mu\nu}\,\eta'_\nu + \alpha_{\mu\nu}\,\beta_{\mu\nu}),\\
\bar{\eta}'_\mu &= \sum_\nu (-\gamma_{\mu\nu}\,\eta_\nu + \delta_{\mu\nu}\,\eta'_\nu + \gamma_{\mu\nu}\,\delta_{\mu\nu}).
\end{aligned}
$$

In this transformation with integral coefficients the essentially distinct new characteristics $[\bar{\eta}]$ are obtained by reducing $\bar{\eta}, \bar{\eta}'$ modulo 2. Then

$$
(2) \qquad \alpha_{\mu\nu} \equiv \alpha^2_{\mu\nu}, \cdots; \qquad -\beta_{\mu\nu} \equiv \beta_{\mu\nu}, \cdots \quad (\text{mod. } 2),
$$

and (1) can be written

$$
(3) \qquad \begin{aligned}
\bar{\eta}_\mu &= \sum_\nu (\alpha^2_{\mu\nu}\,\eta_\nu + \beta^2_{\mu\nu}\,\eta'_\nu + \alpha_{\mu\nu}\,\beta_{\mu\nu}),\\
\bar{\eta}'_\mu &= \sum_\nu (\gamma^2_{\mu\nu}\,\eta_\nu + \delta^2_{\mu\nu}\,\eta'_\nu + \gamma_{\mu\nu}\,\delta_{\mu\nu}),
\end{aligned}
$$

in which the coefficients $\alpha, \beta, \gamma, \delta$ of the transformation occur *homogeneously*.

Then we observe that if the half periods (i. e. the points in the finite geometry) are transformed by **21** (3),

$$
(4) \qquad \begin{aligned}
x_\mu &= \sum_\nu (\alpha_{\nu\mu}\,\bar{x}_\nu + \gamma_{\nu\mu}\,\bar{x}'_\nu),\\
x'_\mu &= \sum_\nu (\beta_{\nu\mu}\,\bar{x}_\nu + \delta_{\nu\mu}\,\bar{x}'_\nu),
\end{aligned}
$$

the quadric

$$
(5) \qquad \sum_\nu (x_\nu\,x'_\nu + \eta_\nu\,x^2_\nu + \eta'_\nu\,x'^2_\nu)
$$

is transformed with the help of **21** (4), (5) into the quadric

$$
\sum_\nu (\bar{x}\,\bar{x}'_\nu + \bar{\eta}_\nu\,\bar{x}^2_\nu + \bar{\eta}'_\nu\,\bar{x}'^2_\nu)
$$

where $\bar{\eta}, \bar{\eta}'$ are expressed in terms of η, η' by precisely the equations (3). The 2^{2p} quadrics obtained from (5) by variation of $\eta, \eta' = 0, 1$ all have the same polar system

$$
\sum_\nu (x_\nu\,y'_\nu + x'_\nu\,y_\nu) \equiv 0,
$$

which coincides modulo 2 with the null system,

$$C = \sum_\nu (x_\nu\, y'_\nu - x'_\nu\, y_\nu) = 0,$$

of **22** (2). Any two of these quadrics with coefficients $[\eta]$, $[\zeta]$ respectively differ by the square of an S_{2p-2},

$$\sum_\nu [(\eta_\nu - \zeta_\nu) x_\nu^2 + (\eta'_\nu - \zeta'_\nu) x'^2_\nu] \equiv [(\eta_\nu + \zeta_\nu) x_\nu + (\eta'_\nu + \zeta'_\nu) x'_\nu]^2,$$

the null space of the point $x = \eta' + \zeta'$, $x' = \eta + \zeta$.

The particular quadric, $\sum_\nu x_\nu\, x'_\nu = 0$, is identified with the original even theta function, $\vartheta(u) = \vartheta[0](u)$. The $2^{2p} - 1$ remaining odd and even theta functions, $\vartheta[\eta]_2(u)$, are identified with the quadrics (5). If the half period $\{\varepsilon\}_2$ is identified as above with the point $x = \varepsilon'$, $x' = \varepsilon$, then the theta function $\vartheta[\eta + \varepsilon]_2(u) = k \cdot \vartheta[\eta]_2(u + \{\varepsilon\}_2)$ is to be identified with that quadric which arises from (5) when the square of the null space of the point x which corresponds to the half period $\{\varepsilon\}_2$ is added. Then (cf. [16] § 4)

(6) *Under integral linear transformation of the periods the* 2^{2p} *theta characteristics are permuted isomorphically with the* 2^{2p} *quadrics* (5) *in* S_{2p-1} (*mod.* 2), *whose polar systems coincide with the null system* C, *under the collination group* G_{NC} *of* C.

The 2^{2p} quadrics are of two types (cf. [27] Chap. 8; [16] § 3) according to the number of points which they contain. If

$$(7) \qquad E_p = 2^{p-1}(2^p + 1), \qquad O_p = 2^{p-1}(2^p - 1),$$

then there are E_p *E-quadrics* each containing $(E_p - 1)$ points, and O_p *O-quadrics* each containing $(O_p - 1)$ points. If to a quadric of either type there be added the squares of the null spaces of each of its points then all the quadrics of that type are obtained. It may be proved readily that the quadric (5) is an E-quadric if

$$(8) \qquad \eta_1\, \eta'_1 + \cdots + \eta_p\, \eta'_p \equiv 0 \ (\text{mod. } 2),$$

whence according to **20** (8) the E-quadrics correspond to the even and the O-quadrics to the odd theta functions. Many properties of these quadrics with respect to their incidence with linear spaces have been developed (cf. [16] §§ 3, 5).

From the relation $\vartheta[\eta+\varepsilon]_2(u) = k \cdot \vartheta[\eta]_2(u+\{\varepsilon\}_2)$ there follows that $\vartheta[\eta]_2(u)$ vanishes for $u = \{\varepsilon\}_2$ if $[\eta+\varepsilon]_2$ is the characteristic of an odd function. But $[\eta+\varepsilon]_2$ corresponds to an O-quadric either if $[\eta]_2$ corresponds to an O-quadric and $\{\varepsilon\}_2$ to a point on it, or if $[\eta]_2$ corresponds to an E-quadric and $\{\varepsilon\}_2$ to a point not on it. Hence

(9) *An even function vanishes for a half period if the corresponding E-quadric does not contain the corresponding point; an odd function vanishes for a half period if the corresponding O-quadric does contain the corresponding point.*

Thus it is clear from (7) that either type of function vanishes for the same number, $O_p = 2^{p-1}(2^p - 1)$, of half periods if for the odd functions the improper zero half period be included.

25. The basis notation for theta characteristics. To the coördinate system $y_1, \cdots, y_{2p+1}$ of **23** there is attached a definite quadric, $K = \sum y_i y_j = 0 \; (i, j = 1, \cdots, 2p+1; i < j)$, which is characterized by the fact that it contains none of residual points of the basis B_{2p+2}. The 2^{2p} quadrics then have equations of the form,

$$K + y_1^2 + \cdots + y_r^2 \equiv K + y_{r+1}^2 + \cdots + y_{2p+1}^2.$$

The following set of four theorems (cf. [18] § 1) may be verified without difficulty:

(1) *If B is the basis configuration of* **23** (4) *with bases $B_1, \cdots, B_{2p+2}$ every quadric Q belonging to C is uniquely determined by a separation of the bases of B into two complementary sets of $p+1-2k$ and $p+1+2k$ each. Q contains only those points P_{ij} of B which belong to bases B_i and B_j drawn from different sets. Q is an E- or an O-quadric according as k is even or odd. The particular quadric Q determined by the separation $B_1, \cdots, B_{p+1-2k}$; $B_{p+2-2k}, \cdots, B_{2p+2}$ will be denoted by*

$$Q_{1,\cdots,p+1-2k} = Q_{p+2-2k,\cdots,2p+2} = K + \sum y_i^2 = 0$$
$$(i = 1, \cdots, p+1-2k).$$

In accordance then with **24** (6) we shall adopt a similar notation for the 2^{2p} odd and even theta functions.

(2) *If the notation of* **23** (5) *for a point be adopted also for the square of the null space of the point then*

$$Q_a + Q_b = P_{ab}, \qquad Q_a + P_b = Q_{ab}$$

where like subscripts in a and b are cancelled in ab.

If P is interpreted as a half period, Q as a theta function, this theorem expresses the change in the function when the half period is added to the argument (cf. **24**).

(3) *If in* (2) Q_a, Q_b *are both E-, or both O-, quadrics the point* P_{ab} *lies on both; if however* Q_a, Q_b *are of different type,* P_{ab} *lies on neither.*

An easy test for type is given in (1), i. e. $k \equiv 0, 1 \pmod{2}$. In connection with **24** (9), (3) furnishes a criterion for the vanishing of a theta function for a given half period.

(4) *The involution* (cf. **22** (10)) *attached to the point* P_{ab} *in* (2) *interchanges the quadrics* Q_a, Q_b *if they are of the same type and leaves each unaltered if they are of opposite types.*

This theorem in conjunction with **22** (10) enables one to pass from a given basis configuration to any other and thereby to obtain all the conjugates of a given figure in the finite geometry or of a given arrangement of theta and period characteristics.

We illustrate the notation for the smaller values of p. When $p = 1$, $S_{2p-1} = S_1$ contains but three points which make up only one basis. Nevertheless the basis notation with $2p+2 = 4$ subscripts persists. We name the three points or proper half periods by $P_{12} = P_{34}$, $P_{13} = P_{24}$, $P_{14} = P_{23}$; the three even theta functions by $\vartheta_{12} = \vartheta_{34}$, $\vartheta_{13} = \vartheta_{24}$, $\vartheta_{14} = \vartheta_{23}$; and the odd theta function by $\vartheta = \vartheta_{1234}$. According to (3) and **24** (9) ϑ_{12} vanishes for $u = P_{12}$; while ϑ vanishes only for the zero half period. The involution attached to P_{12} leaves P_{12} unaltered and interchanges P_{13} and P_{14}. The three involutions generate a G_6 which is symmetric on the three half periods and three even functions, and which leaves the odd function invariant.

When $p=2$, the 15 points in S_3 make up a single basis configuration with bases $B_1, \cdots, B_6$ whence the 15 half periods are of the type P_{12}. There are 10 even functions (cf. (1) for $k=0$) of the type $\vartheta_{123}=\vartheta_{456}$; and 6 odd functions of type $\vartheta_1=\vartheta_{23456}$. The function $\vartheta_{123}=\vartheta_{456}$ vanishes for the 6 half periods of types P_{12}, P_{45}; the function ϑ_1 for the 5 proper half periods of type P_{12} as well as the zero half period. The modular group of period transformations of order 720 is simply isomorphic with the permutation group of the six bases.

When $p=3$ with $B=B_1, \cdots, B_8$ there are 28 half periods of type P_{12} and 35 of type $P_{1234}=P_{5678}$. There are 35 even functions of type $\vartheta_{1234}=\vartheta_{5678}$; 28 odd functions of type $\vartheta_{12}=\vartheta_{345678}$; and one even function of type $\vartheta=\vartheta_{12345678}$. The even theta function $\vartheta_{1234}=\vartheta_{5678}$ vanishes for 12 half periods of type P_{12} and 16 of type $P_{1235}=P_{4678}$; the even function ϑ for the 28 of type P_{12}; while the odd function ϑ_{12} vanishes for 12 of type P_{13}, 15 of type P_{1234}, and the zero half period. The basis configuration isolates one of the 36 even thetas and the modular group is of order 36.8!.

When $p=4$ and $B=B_1, \cdots, B_9, B_0$ the 255 half periods comprise 45 of type P_{12} and 210 of type P_{1234}. The even functions comprise 126 of type $\vartheta_{12345}=\vartheta_{67890}$ and 10 of type ϑ_1; the odd functions comprise 120 of type ϑ_{123}. The even function ϑ_{12345} vanishes for 20 half periods of types P_{12}, P_{67} and 100 of type P_{1678}; the even function ϑ_1 vanishes for 36 of type P_{23} and 84 of type P_{1234}; the odd function ϑ_{123} vanishes for 21 of type P_{14}, 63 of type P_{1245}, 35 of type P_{4567}, and for the zero half period. The number of basis configurations is $2^8.9.51$ (cf. 23 (7)) and the order of the modular group is $2^8.3^3.17.10!$.

The basis notation has been frequently used and, in early cases at least, could hardly have been overlooked for tactical reasons. However, it has usually been developed from the standpoint of theta characteristics. It is clear from the geometric aspect that its real origin is in the basis configuration which is composed of period characteristics alone. An important advantage of the geometric approach is the marked difference between a theta characteristic (a proper quadric) and a period characteristic (a point or the square of its null space).

26. Systems of theta characteristics. Projection and section of the null system. When the 2^{2p} theta characteristics are represented by the proper quadrics

$$Q[\eta, \eta'] \equiv \sum_{\nu} (x_{\nu} x'_{\nu} + \eta_{\nu} x_{\nu}^2 + \eta'_{\nu} x'^2_{\nu})$$

it is clear that the theta characteristics alone do not form a linear system. For

$$Q[\eta, \eta'] + Q[\zeta, \zeta'] \equiv \sum_{\nu} \{(\eta_{\nu} + \zeta_{\nu}) x_{\nu}^2 + (\eta'_{\nu} + \zeta'_{\nu}) x'^2_{\nu}\}.$$

This sum of squares is the square of an S_{2p-2} and such spaces are permuted under G_{NC} like their null points, i. e., like period characteristics. But the 2^{2p} quadrics and the $2^{2p} - 1$ squared S_{2p-2}'s together constitute a linear system R_{2p} of dimension $2p$. The sum of j members of R_{2p} is a proper quadric or a squared S_{2p-2} according as the number of summands which are proper quadrics is odd or even. A linear system F_k of dimension k contained in R_{2p} may consist only of $2^{k+1} - 1$ squared S_{2p-2}'s and then behaves like a system of period characteristics. If however F_k contains one quadric it must contain 2^k quadrics and $2^k - 1$ squared S_{2p-2}'s. If these be replaced, as is convenient, by their null points, the null points fill up a linear space S_{k-1} called the allied space of F_k ([16] §§ 5, 6). Two systems F_k and F'_k with the same allied space and a common quadric coincide. Hence the 2^{2p} quadrics can be distributed in a single way into 2^{2p-k} systems F_k with a given common allied space. A *Göpel system* F_k has a null space as its allied S_{k-1}; a *Göpel system* has a Göpel space as its allied S_{p-1}.

We give a few examples of systems which contain quadrics. Of systems F_0 there are two types; an E-quadric, and an O-quadric. Of systems F_1 with two quadrics Q_a, Q_b and allied space $S_0 = P_{ab}$ there are three types according as the two quadrics are both E-quadrics, both O-quadrics, or an E-

and an O-quadric. In the first two types P_{ab} is on both quadrics; in the third, on neither. Systems F_2 determined by three quadrics Q_a, Q_b, Q_c contain a fourth quadric Q_{abc} and the allied space S_1 contains three points P_{ab}, P_{ac}, P_{bc}. They divide first into two types according as the three collinear points P are on a null or an ordinary line. The four quadrics are then called a *syzygetic* or an *azygetic tetrad* respectively. Since any three determine the fourth, the three are also termed a *syzygetic* or an *azygetic triad* as the case may be. At least two of the four quadrics, say Q_a Q_b are of the same type. Then in the syzygetic case the involution I_{ab} attached to P_{ab} interchanges Q_a, Q_b and leaves P_{ac}, P_{bc} each unaltered (cf. 22 (10), 25 (4)). Hence $Q_a + P_{ac} = Q_c$ and $Q_b + P_{ac} = Q_{abc}$ are also interchanged and are of the same type. Thus the system contains 4, 2, or 0 E-quadrics. In the azygetic case I_{ab} interchanges Q_a, Q_b and also P_{ac}, P_{bc} whence $Q_c = Q_a + P_{ac} = Q_b + P_{bc}$ is invariant and similarly Q_{abc} is invariant. Thus Q_c and Q_{abc} are of different type and the system contains 1 or 3 E-quadrics. For $p = 3$ and a syzygetic tetrad with allied $S_1 = P_{12}, P_{34}, P_{1234}$ the three types of tetrads are exemplified by $\vartheta_{1357}, \vartheta_{2357}, \vartheta_{1457}, \vartheta_{2457}$; $\vartheta, \vartheta_{1234}, \vartheta_{12}, \vartheta_{34}$; $\vartheta_{13}, \vartheta_{23}, \vartheta_{14}, \vartheta_{24}$. In the azygetic case with $S_1 = P_{12}, P_{13}, P_{23}$ the two types of azygetic tetrads are exemplified by $\vartheta_{1456}, \vartheta_{2456}, \vartheta_{3456}, \vartheta_{78}$; $\vartheta, \vartheta_{12}, \vartheta_{13}, \vartheta_{23}$. In general two systems F_k and $F'_{k'}$ each containing proper quadrics are conjugate under the G_{NC} of period transformations if first their dimensions k, k' coincide; if second their allied spaces S_{k-1}, S'_{k-1} are conjugate (cf. 22 (8)); and if finally they have the same number of quadrics of each type E, O.

Doubly periodic functions of the second order can be constructed from the 2^{2p} functions by forming ratios in two way:

$$(1) \qquad \vartheta_a^2(u)/\vartheta_b^2(u), \qquad \vartheta_a(u)\,\vartheta_b(u)/\vartheta_c(u)\,\vartheta_{abc}(u).$$

The first ratio is an even function with characteristic $aa = bb = 0$ (cf. 19 (3)); the second has the characteristic $ab = cabc$ and is odd or even according as P_{ab}, P_{ac}, P_{bc} is a null line or an ordinary line.

A *fundamental set* (F. S.) of $2p+2$ quadrics is any set of $2p+2$ quadrics subject to no other relation than the vanishing of their sum and such that any three are azygetic. A discussion of these will be found in ([16] § 6) with references to the corresponding arithmetical development of Krazer ([41] Chap. 7). We note here merely that every F. S. is obtained by adding to one of its quadrics the squared S_{2p-2}'s of a self-dual basis of C and contains s O-quadrics where $s \equiv p$ mod. 4. A *normal* F. S. is one in which all or all but one of the quadrics are of the same type. According as $p \equiv 0, 1, 2, 3$ (mod. 4) the normal F. S. contains $2p+2$ E-quadrics; one O-quadric and $2p+1$ E-quadrics; $2p+2$ O-quadrics; one E-quadric and $2p+1$ O-quadrics. For example a normal F. S. for $p=1$ is $\vartheta, \vartheta_{12}, \vartheta_{13}, \vartheta_{14}$; for $p=2$ is $\vartheta_1, \cdots, \vartheta_6$; for $p=3$ is $\vartheta, \vartheta_{12}, \vartheta_{13}, \cdots, \vartheta_{18}$; and for $p=4$ is $\vartheta_1, \vartheta_2, \cdots, \vartheta_9, \vartheta_0$. The normal F. S. has usually been employed to introduce the basis notation.

An important means of setting up configurations in the finite geometry, or sets of characteristics, is the operation of *projection and section* of the null system C, or more explicitly C_p. A null system in ordinary projective space is non-singular only when the dimension of the space is odd. If in such a space S_{2p-1} a point P is isolated, the null lines on the point P, and therefore in the null S_{2p-2} of P, are cut by any S_{2p-2}, ζ, not on P in the points of an $S_{2\pi-1}$ $(\pi = p-1)$. If P', P'' are two points in $S_{2\pi-1}$ they will be defined to be syzygetic or azygetic in the null system C_π in $S_{2\pi-1}$ according as P', P'' are syzygetic or azygetic in C_p, i. e. according as the plane $PP'P''$ is or is not a null plane of C_p. The null system C_π as thus defined in $S_{2\pi-1}$ is the projection and section of C_p from P and by ζ respectively.

The subgroup g_P of G_{NC_p} which leaves P unaltered has the order (**22** (7)) $2^{p^2} H_{2p-3} \cdots H_1$. It has an invariant subgroup I of order 2^{2p-1} which leaves every null line on P unaltered. This subgroup I is abelian with involutorial elements (cf.[19] p. 326). The factor group of I with respect to g_P, i. e., the group of permutations on the null lines through P induced by g_P, is the group G_{NC_π} of the null system C_π in $S_{2\pi-1}$.

The 2^{2p-1} quadrics containing P are permuted in pairs by the involution attached to P. These comprise O_π pairs of O-quadrics and E_π pairs of E-quadrics. The quadrics of a pair have the same section by the null S_{2p-2} of P and the further section by ζ furnishes the O_π and E_π quadrics whose polar systems coincide with the null system C_π. These pairs of quadrics are called the *Steiner set* attached to the half period P.

This operation of projection and section may be repeated in the space $S_{2\pi-1}$ or a like result may be obtained in the original space by projection from a null line and section by an S_{2p-3} skew to the line (cf. [16] p. 261). Thus for $p = 3$ and projection from P_{78} there are six pairs of odd theta characteristics ϑ_{i7}, ϑ_{i8}, and ten pairs of even theta characteristics

$$\vartheta_{ijk7},\ \vartheta_{ijk8} = \vartheta_{lmn8},\ \vartheta_{lmn7} \qquad (i, j, \cdots, n = 1, \cdots, 6)$$

corresponding to the six odd and ten even theta characteristics for $p = 2$. On further projection from P_{56} there is one tetrad of odd theta characteristics ϑ_{57}, ϑ_{58}, ϑ_{67}, ϑ_{68}; and three tetrads of even theta characteristics

$$\vartheta_{ij57},\ \vartheta_{ij58},\ \vartheta_{ij67},\ \vartheta_{ij68} = \vartheta_{kl68},\ \vartheta_{kl67},\ \vartheta_{kl58},\ \vartheta_{kl57}$$
$$(i, j, k, l = 1, \cdots, 4)$$

as in the case $p = 1$.

27. Theta modular groups determined by sets of points. The linear group $g_{m,2}$ of 6, and $g_{m,k}$ of 15, associated with regular Cremona transformations whose F-points are found in a set P_m^k of m points in a projective space S_k, has elements with integer coefficients and determinants ± 1. Such a group, in general infinite, is the source of a variety of finite groups when the coefficients are reduced modulo r. The situation is similar to that of the reduction of the period transformations modulo 2 and may be expressed as follows (cf. [19] (1)).

(1) *The elements of $g_{m,k}$ whose coefficients reduce mod. r to those of the identity form an invariant subgroup (in general infinite) of $g_{m,k}$ whose factor group $g_{m,k}^{(r)}$ is finite. A re-*

presentation of this factor group consists of the elements of $g_{m,k}$ *which are distinct when reduced mod.* r.

These factor groups for $r = 2$ are intimately connected with the properties of sets of points under Cremona transformation. On the other hand the groups $g^{(2)}_{m,k}$ are isomorphic with the group G_{NC_p} for suitable choice of p, or with subgroups of G_{NC_p}. A complete discussion is found in ([19]) with 16 cases according as $m, k \equiv 0, 1, 2, 3$ (mod. 4). One case, $g_{10,2}$, will be discussed here in detail and for other cases references will be made as needed to the article cited.

The group $g_{10,2}$ is generated by the transpositions of $\gamma_1, \cdots, \gamma_{10}$ and the element A_{123} **(6** (1)**)**. With reduction mod. 2 the variables γ will be replaced by x and A_{123} has then the form

$$(2)\quad \begin{aligned} A_{123} : x'_i &= x_i + (x_0 + x_1 + x_2 + x_3) \quad (i = 0, \cdots, 3), \\ x'_j &= x_j \qquad\qquad\qquad (j = 4, \cdots, 10). \end{aligned}$$

Linear forms with coefficients reduced modulo 2 are of two types according as x_0 does or does not appear. Set

$$(3)\quad B_{12\cdots l} = x_1 + x_2 + \cdots + x_l; \quad C_{12\cdots l} = x_0 + x_1 + x_2 + \cdots + x_l.$$

Let $B(l)$ represent the aggregate of forms B with l terms, which of course are all conjugate under the transpositions alone; $C(l)$ the aggregate of forms C with l terms in addition to x_0. On applying A_{123} to $B(l)$, the form $B(l)$ is reproduced or is transformed into a form $C(l')$ according as $B(l)$ contains an even or odd number of the terms x_1, x_2, x_3. More precisely:

(4) *A form* $B(l)$ *is transformed by* A_{123} *into* $B(l)$, $C(l+1)$, $B(l)$, $C(l-3)$ *according as* $B(l)$ *has* 0, 1, 2, 3 *subcripts in common with* A_{123}. *A form* $C(l)$ *is transformed by* A_{123} *into* $B(l+3)$, $C(l)$, $B(l-1)$, $C(l)$ *according as* $C(l)$ *has* 0, 1, 2, 3 *subscripts in common with* A_{123}. *When either type is altered the subscripts of the transform are those of the original form and of* A_{123} *with like subscripts cancelled.*

Let b_1, b_2, b_3, b_4 denote those aggregates of forms $B(l)$ for which $l \equiv 1, 2, 3, 4$ (mod. 4) respectively; c_0, c_1, c_2, c_3 those aggregates of forms $C(l)$ for which $l \equiv 0, 1, 2, 3$ (mod. 4) respectively. According to (3) a form $B(l)$ has conjugates $C(l+1)$, $C(l-3)$ and these in turn have conjugates $B(l+4)$, $B(l)$, $B(l-4)$. Then as $l \equiv 1, 2, 3, 4 \pmod{4}$ there are four conjugate sets of linear forms namely:

$$(5) \qquad b_1,\, c_2;\; b_2,\, c_3;\; b_3,\, c_0;\; b_4,\, c_1.$$

Naturally the invariant linear form $x_0 + x_1 + \cdots + x_{10}$ (cf. **4** (3)) is excluded from the set c_2. The transformation will be determined if its effect upon the forms b_1 and c_0 is known since these include the individual variables.

In the basis notation of the finite geometry for $p = 5$ with $2p+2 = 12$ bases $B_1, \cdots, B_{10}, B_\alpha, B_\beta$ the linear forms of the first and third conjugate sets (5) will be identified with the points of S_{2p-1} as follows:

$$(6) \qquad b_1,\, c_2;\; b_3,\, c_0 = P_{1\beta},\, P_{12\alpha\beta};\; P_{123\beta},\, P_{\alpha\beta}.$$

The identification of b_1 with $P_{1\beta}$ means that B_i corresponds to $P_{i\beta}$, B_{ijklm} to $P_{ijklm\beta}, \cdots$; similarly for the other sets $(i, j, \cdots, m = 1, 2, 3, \cdots, 10)$. Furthermore the transposition (x_i, x_j) is identified with the involution attached to the point P_{ij}; and the element A_{123} with the involution attached to $P_{123\alpha}$. Then

(7) *The involutions attached to* P_{ij} *and* $P_{123\alpha}$ *permute the points* P *in* (6) *precisely as the corresponding transpositions* (x_i, x_j) *and* A_{123} *permute the corresponding linear forms.*

This is evident for the transpositions and is easily verified for A_{123} by comparing the effect of A_{123} as stated in (4) with that of the involution $P_{123\alpha}$ as given in **22** (10), **23** (6).

Since the forms (6) in b_1 and c_0 contain the individual variables the element of $g^{(2)}_{10,2}$ is completely determined when the effect on these forms is given. Since the points (6) contain all the points of the S_{2p-1} $(p = 5)$ the transformation of G_{NC_5} is completely determined when the effect on the points is given. Hence $g^{(2)}_{10,2}$ is simply isomorphic with the group in the finite geometry generated by the involutions P_{ij} and $P_{123\alpha}$. These generators transform P_{ij} into all the points of type P_{12}, $P_{1,\cdots,6}$, $P_{1,\cdots,10}$, $P_{123\alpha}$, $P_{1,\cdots,7,\alpha}$, i. e. into all the points not on the E-quadric $Q_{\alpha\beta}$ (cf. 25(3)). Such a complete conjugate set of generating involutions generates the subgroup of G_{NC_5} which leaves $Q_{\alpha\beta}$ invariant ([19] p. 325). The order of G_{NC_5} is $2^{25} H_9 H_7 H_5 H_3 H_1$ and the number of E-quadrics is $2^4(2^5+1)$. Hence

(8) *The group $g_{10,2}^{(2)}$ of transformations of $g_{10,2}$ reduced mod. 2 has the order $10!\, 2^{13} \cdot 31 \cdot 51$ and is simply isomorphic with that subgroup of the modular group G_{NC_5} which has an invariant even theta characteristic.*

This result is used later to prove that a ten-nodal rational plane sextic can be transformed by Cremona transformation into only $2^{13} \cdot 31 \cdot 51$ projectively distinct rational sextics.

28. A remarkable system of equations. Göpel invariants. F. Schottky ([63] § 5) has devised an interesting set of equations which has enabled him to simplify very materially the numerous relations which exist among the odd and even theta functions. In many cases (to which we shall add others) he finds that these relations are equivalent to well-known projective relations which connect the coördinates of the sets of points, P_m^k, discussed in Chapter I.

Schottky first attaches to each half-period, or point in the finite geometry, a constant, say the constant e_μ to the point P_μ, where μ is a certain set of subscripts in the basis notation. Secondly he attaches likewise to each of the 2^{2p} odd and even theta functions, or quadrics in the finite geometry, a constant, say the constant f_m to the quadric Q_m, where again m is an aggregate of subscripts in the basis notation.

The system of equations referred to then reads as follows:

$$(1) \qquad f_m = r \prod (e_\mu) \quad (P_\mu \text{ not on } Q_m).$$

The number of equations is the number, 2^{2p}, of quadrics Q_m. The quantity r is a factor of proportionality. The product $\prod$ is extended over the $2^{p-1}(2^p-1)$ points P_μ not on Q_m when Q_m is even; or over the $2^{p-1}(2^p+1)$ points P_μ not on Q_m when Q_m is odd.

For $p = 1$ the system reads

$$(2) \qquad f_{12} = re_{12}, \quad f_{13} = re_{13}, \quad f_{23} = re_{23}, \quad f = re_{12}\, e_{13}\, e_{23}.$$

For $p = 2$ typical equations are

$$(3) \qquad \begin{aligned} f_{123} &= re_{12}\, e_{13}\, e_{23} \cdot e_{45}\, e_{46}\, e_{56}, \\ f_1 &= re_{23}\, e_{24}\, e_{25}\, e_{26}\, e_{34}\, e_{35}\, e_{36}\, e_{45}\, e_{46}\, e_{56}. \end{aligned}$$

The solution of the system of equations (1) for the $2^{2p} - 1$ constants e_μ in terms of the 2^{2p} constants f_m is

$$(4) \qquad e_\mu = \left\{\prod(f_m) \Big/ \prod(f_n)\right\}^{1/2^{2p-1}} \quad (Q_n \text{ on } P_\mu,\ Q_m \text{ not on } P_\mu).$$

For, of the 2^{2p} quadrics Q, 2^{2p-1} are not, and 2^{2p-1} are, on P_μ. If then we insert on the right of (4) the values f given in (1), the r^{2p-1} in numerator and denominator cancel; and the factor e_μ occurs in each factor f of the numerator but in no factor of the denominator. Furthermore of the 2^{2p-1} quadrics on P_μ, 2^{2p-2} only are on $P_{\mu'}$ whence $e_{\mu'}$ occurs 2^{2p-2} times in the numerator and in the denominator.

For example the values of the constants e_μ in the cases $p = 1$ and $p = 2$ are:

$$(5) \qquad \begin{aligned} (p = 1) \quad e_{12} &= (f_{12}\, f / f_{13}\, f_{23})^{1/2}, \\ (p = 2) \quad e_{12} &= \left\{ \frac{f_3\, f_4\, f_5\, f_6\, f_{123}\, f_{124}\, f_{125}\, f_{126}}{f_1\, f_2\, f_{134}\, f_{135}\, f_{136}\, f_{145}\, f_{146}\, f_{156}} \right\}^{1/8}. \end{aligned}$$

Schottky uses for the value of f_m,

$$(6) \qquad f_m = c_m^4$$

where $c_m = \vartheta_m(0)$ if $\vartheta_m(u)$ is an *even* function. If however $\vartheta_m(u)$ is an *odd* function the interpretation of the constant c_m attached to it depends upon the number p of variables. We shall find that the constants e_μ can be interpreted as the *discriminant conditions* attached to a set of r points, P_r^k, in S_k (cf. 9, 15).

Of special importance are those products of the e_μ's whose conjugates under the group of period transformations contain the constants c_m attached to the odd functions (and thus as yet undefined) merely as a factor of proportionality. Such products arise from the aggregate of points in Göpel spaces whose types we proceed to tabulate.

$p = 2$: The Göpel lines (cf. **22**) or null lines, comprise
15 of type P_{12}, P_{34}, P_{56}.

$p = 3$: The 315 null lines comprise
210 of type P_{12}, P_{34}, P_{1234}; and
105 of type $P_{1234}, P_{1256}, P_{1278}$.
These combine into 135 Göpel planes which comprise
105 of type $P_{12}, P_{34}, P_{56}, P_{78}, P_{1234}, P_{1256}, P_{1278}$; and
30 of type $P_{1234}, P_{1256}, P_{3456}, P_{1357}, P_{2457}, P_{2367}, P_{1467}$.

$p = 4$: The 255.21 null lines comprise
18.35 of type P_{12}, P_{34}, P_{1234};
45.35 of type $P_{12}, P_{3456}, P_{7890}$;
90.35 of type $P_{1234}, P_{1256}, P_{3456}$.
The 255.45 null planes comprise
70.45 of type $P_{12}, P_{34}, P_{56}, P_{7890}, P_{1256}, P_{3456}, P_{1234}$;
105.45 of type $P_{1234}, P_{1256}, P_{1278}, P_{3456}, P_{3478}, P_{5678}, P_{90}$;
80.45 of type $P_{1234}, P_{1256}, P_{3456}, P_{1357}, P_{2457}, P_{2367}, P_{1467}$.
These combine into 255.9 Göpel S_3's which comprise
105.9 of type $P_{12}, P_{34}, P_{56}, P_{78}, P_{90}, P_{1234}, P_{1256}, P_{1278}, P_{1290}, P_{3456}, P_{3478}, P_{3490}, P_{5678}, P_{5690}, P_{7890}$;
150.9 of type $P_{1234}, P_{1256}, P_{3456}, P_{1357}, P_{2457}, P_{2367}, P_{1467}, P_{90}, P_{5678}, P_{3478}, P_{1278}, P_{2468}, P_{1368}, P_{1458}, P_{2358}$.

For $p = 1$ a Göpel space is merely a point P_{12} and the corresponding e_{12} has the value (cf. (5))

$$e_{12} = \left\{ \prod (f) \right\}^{1/2} / f_{13} f_{23}. \tag{7}$$

In the numerator there occurs the product of all the constants f_m; in the denominator those constants f_m for which Q_m is on P_{12}.

For $p = 2$ a typical Göpel space G is the line P_{12}, P_{34}, P_{56} and the corresponding product of e_μ's is

$$e_{12}\, e_{34}\, e_{56} = \left\{ \prod (f) \right\}^{1/8} / (f_{135} f_{136} f_{145} f_{146})^{1/2}. \tag{8}$$

Again there occur in the denominator those constants f_m for which Q_m contains the Göpel line. Indeed we observe that if Q_m contains G then, according to (4), f_m occurs once in the denominator for each factor e_μ on the left and once to account for it in $\prod(f)$. If Q_m does not contain G it meets G in one point only, and from (4) f_m appears once in the denominator but twice in the numerator.

In general a quadric ϑ_m either contains a Göpel space G and is an E-quadric or it contains $2^{p-1}-1$ points of G and omits the remaining 2^{p-1} points of G. Hence by the same argument

$$\text{(9)}\quad \prod G(e_\mu) = \left\{\prod(f)\right\}^{1/2^{2p-1}} \Big/ \left\{\prod(f_m)\right\}^{1/2^{p-1}} \quad (Q_m \text{ on } G),$$

where $\prod G$ is extended over the points of the Göpel space, $\prod(f_m)$ over the 2^p *even* quadrics Q_m which contain G, and $\prod(f)$ is the product of the 2^{2p} constants f.

If $G_1, G_2, \cdots$ is the set of all Göpel spaces in the finite geometry we define the products,

$$\text{(10)}\qquad \prod G_1(e_\mu), \quad \prod G_2(e_\mu), \quad \cdots,$$

to be the *Göpel invariants* of the functions. They constitute a conjugate set under the group of period transformations. Their ratios are expressed (cf. (9) and (6)) in terms of the zero values of the even thetas.

With respect to these Göpel invariants we prove the theorem:

(11) *If, for* $p = 1, 2, 3, 4$ *the three Göpel spaces,* G_1, G_2, G_3 *of dimension* $p-1$ *have a common null* S_{p-2}*, and if the functions are abelian theta functions, then there exists a three term relation,*

$$\prod G_1(e_\mu) \pm \prod G_2(e_\mu) \pm \prod G_3(e_\mu) = 0.$$

For $p = 1$ the three Göpel spaces S_0 have no common S_{-1} and the relation is merely

$$e_{12} \pm e_{13} \pm e_{23} = 0.$$

This (cf. (7), (4)) is a consequence of the relation (cf. 30 (4))

$$\vartheta_{12}^4 - \vartheta_{13}^4 + \vartheta_{23}^4 = 0.$$

For $p = 2$ the three Göpel lines with common $S_0 = P_{56}$ yield the relation

$$e_{12}\, e_{34}\, e_{56} \pm e_{13}\, e_{24}\, e_{56} \pm e_{14}\, e_{23}\, e_{56} = 0.$$

This, according to (8) and (6), is satisfied if

$$(\vartheta_{135}^2\, \vartheta_{136}^2 \cdot \vartheta_{145}^2\, \vartheta_{146}^2)^{-1} \pm (\vartheta_{145}^2\, \vartheta_{146}^2 \cdot \vartheta_{125}^2\, \vartheta_{126}^2)^{-1}$$
$$\pm (\vartheta_{125}^2\, \vartheta_{126}^2 \cdot \vartheta_{135}^2\, \vartheta_{136}^2)^{-1} = 0,$$

or

$$\vartheta_{125}^2\, \vartheta_{126}^2 \pm \vartheta_{135}^2\, \vartheta_{136}^2 \pm \vartheta_{145}^2\, \vartheta_{146}^2 = 0 \text{ (cf. 30 VI).}$$

For $p = 3$ the three Göpel planes with common null line P_{56}, P_{78}, P_{5678} are obtained by the process of projection and section from this line as explained in 26. The three points P_{14}, P_{24}, P_{34} of the resulting S_1 yield three sets of four points:

$$G_1:\ P_{14}, P_{23}, P_{1456}, P_{1478};$$
$$G_2:\ P_{24}, P_{13}, P_{2456}, P_{2478};$$
$$G_3:\ P_{34}, P_{12}, P_{3456}, P_{3478};$$

which with the common null line make up respectively the three Göpel planes. The three E-quadrics Q_{14}, Q_{24}, Q_{34} of S_1 yield three sets of four E-quadrics,

$$Q_{i457}, Q_{i458}, Q_{i467}, Q_{i468} \qquad (i = 1, 2, 3), \tag{12}$$

all of which are on the common null line. Since in the $S_1\,(p = 1)$ Q_{i4} is on P_{j4}, P_{k4} $(i, j, k = 1, 2, 3)$ but not on P_{i4}, the four quadrics (12) contain G_j and G_k but not G_i. If then we set

$$\sigma_i = \vartheta_{i457}\, \vartheta_{i458}\, \vartheta_{i467}\, \vartheta_{i468}$$

there follows from (9) and (6) that

$$\prod G_1(e_\mu) : \prod G_2(e_\mu) : \prod G_3(e_\mu) = 1/\sigma_2\sigma_3 : 1/\sigma_3\sigma_1 : 1/\sigma_1\sigma_2$$
$$= \sigma_1 : \sigma_2 : \sigma_3.$$

The relation (11) then follows from $\sigma_1 \pm \sigma_2 \pm \sigma_3 = 0$ (45 (A)).

For $p = 4$ and three Göpel S_3's with a common null plane with points

$$P_{56},\ P_{78},\ P_{90},\ P_{5678},\ P_{5690},\ P_{7890},\ P_{1234}$$

there are three sets of eight quadrics

$$Q_{i4579},\ Q_{i4570},\ Q_{i4589},\ Q_{i4580},\ Q_{i4679},\ Q_{i4670},\ Q_{i4689},\ Q_{i4680}$$
$$(i = 1, 2, 3)$$

and three corresponding theta products

$$\sigma_i = \vartheta_{i4579} \cdots \vartheta_{i4680}$$

such that according to (9) and (6)

$$\prod G_1(e_\mu) : \prod G_2(e_\mu) : \prod G_3(e_\mu) = (\sigma_2\,\sigma_3)^{-1/2} : (\sigma_3\,\sigma_1)^{-1/2} : (\sigma_1\,\sigma_2)^{-1/2}$$
$$= (\sigma_1)^{1/2} : (\sigma_2)^{1/2} : (\sigma_3)^{1/2}.$$

Schottky ([59] § 4) shows that for the abelian theta functions

$$(\sigma_1)^{1/2} \pm (\sigma_2)^{1/2} \pm (\sigma_3)^{1/2} = 0$$

which verifies (11) (cf. also 57 (15)).

29. A determination of sign. The coefficients of the theta relations of the second order which are derived in the next section are determined by substituting certain half periods in the relation. When only the squares of the odd and even thetas occur the formula 20 (5) yields

$$(1) \qquad \vartheta^2[\eta]_2\,(u + \{\varepsilon\}_2) = f(\varepsilon) \cdot (-1)^{(\varepsilon\eta')} \cdot \vartheta^2[\eta + \varepsilon]_2\,(u).$$

The factor $f(\varepsilon)$ divides out of the relation leaving the sign $(-1)^{(\varepsilon\eta')}$ to be determined. Since however only the ratio of two such signs, say for $\vartheta^2[\eta]_2$ and $\vartheta^2[\zeta]_2$, is necessary it suffices to find $(-1)^{(\varepsilon,\,\eta'+\zeta')} = (-1)^{(\varepsilon\varkappa')}$ where $\{\varkappa\}_2$ is the half period characteristic which is the sum of the two theta characteristics $[\eta]_2$, $[\zeta]_2$.

The sign $(-1)^{(\varepsilon\varkappa')}$ associated with the half periods $\{\varepsilon\}_2 = \{\varepsilon, \varepsilon'\}_2$ and $\{\varkappa\}_2 = \{\varkappa, \varkappa'\}_2$ depends only on the first part of the characteristic $\{\varepsilon, \varepsilon'\}_2$ and the second part of $\{\varkappa, \varkappa'\}_2$. It is then the same as that derived from $\{\varepsilon, 0\}_2$ and $\{0, \varkappa'\}_2$. In the finite geometry the 2^p-1 points $\{\varepsilon, 0\}_2$ $(\varepsilon = 0, 1)$ lie in a Göpel space G_1, and the 2^p-1 points $\{0, \varkappa'\}_2$ $(\varkappa' = 0, 1)$ lie in a Göpel space G_2 which is skew to G_1. Any half period $\{\lambda, \lambda'\}_2$ can be expressed uniquely as $\{\lambda, 0\}_2 + \{0, \lambda'\}_2$. Geometrically any point P is on a unique line $P_1 P_2$ where P_1 on G_1 is the projection of P upon G_1 from G_2 and P_2 on G_2 is the projection of P upon G_2 from G_1. If in the basis notation P_m corresponds to $\{\varepsilon\}_2$ and P_n to $\{\varkappa\}_2$ we shall set

$$(2) \qquad (-1)^{(\varepsilon\varkappa')} = (P_m, P_n).$$

Then (P_m, P_n) is $+1$ or -1 according as the projection of P_m upon G_1 from G_2 is syzygetic or azygetic with the projection of P_n upon G_2 from G_1.

The sign (P_m, P_n) will depend not merely on the sets of subscripts but also on the way the Göpel spaces G_1, G_2 are selected. A simple selection (cf. [18] § 3) is that in which G_1, G_2 are defined by the two sets of $p+1$ points, each linearly dependent,

$$(3) \qquad \begin{aligned} G_1:&\quad P_{12}, P_{34}, P_{56}, \cdots, P_{2p+1, 2p+2}; \\ G_2:&\quad P_{23}, P_{45}, P_{67}, \cdots, P_{2p+2, 1}. \end{aligned}$$

Then a point such as P_{1347} has projections on G_1, G_2 which are obtained as follows:

$$\begin{aligned} P_{1347} &= P_{1223455667} = (P_{12} + P_{56}) + (P_{23} + P_{45} + P_{67}) \\ &= P_{1256} + P_{234567}. \end{aligned}$$

We prove first that

$$(4) \qquad (P_{ij}, P_{ij}) = (-1)^{i+j+1}.$$

For if i, j are integers in the natural order separated by $a_1, \cdots, a_r$ then

$$P_{ij} = P_{ia_1} + P_{a_1 a_2} + P_{a_2 a_3} + \cdots + P_{a_r j}$$

where alternate points are in the same space G_1 or G_2. A point of the sum is azygetic to its adjacent points and syzygetic to all the others. If i and j have not the same parity, the first and last point of the sum are in the same G and each point in the other G' is azygetic to two of the points of the sum in G. If i and j are of the same parity the last point of the sum, say in G, is azygetic only to the preceding one in G' and the sign is negative.

Let

$$\varepsilon_{i_1 i_2 \cdots i_r} = +1, -1 \tag{5}$$

according as $i_1, i_2, \cdots, i_r$ is an even or odd permutation from the natural order of these integers; and let

$$\varepsilon_{i_1 i_2 \cdots i_r | j_1 j_2 \cdots j_s | k_1 k_2 \cdots k_t |} = +1, -1 \tag{6}$$

according as $i_1 \cdots i_r\ j_1 \cdots j_s\ k_1 \cdots k_t$ is an even or odd permutation from the natural order *after* the sets $i_1 \cdots i_r$; $j_1 \cdots j_s$; $k_1 \cdots k_t$ have themselves been arranged in natural order. Then

$$(P_{ij}, P_{jk}) = (-1)^{j+1} \varepsilon_{ijk}. \tag{7}$$

To prove this let provisionally $i<j<k$ which gives rise to six cases,

$$1^\circ \qquad (P_{ij}, P_{jk}) = (-1)^{j+1}.$$

For if j is even the point P_1, the projection of P_{ij} on G_1 from G_2 ends with $P_{j-1,j}$; and the point P_2, the projection of P_{jk} on G_2 from G_1 begins with $P_{j,j+1}$ whence $(P_1, P_2) = -1$. But if j is odd P_1 ends with $P_{j-2,j-1}$ and P_2 begins with $P_{j+1,j+2}$, and $(P_1, P_2) = +1$. This situation is reversed in

$$2^\circ \qquad (P_{jk}, P_{ij}) = (-1)^{j}.$$

The remaining four cases depend on (4), 1°, 2°.

$$\begin{aligned} 3^\circ \quad (P_{ij}, P_{ik}) &= (P_{ij}, P_{ij}+P_{jk}) \\ &= (-1)^{i+j+1} \cdot (-1)^{j+1} \quad = (-1)^{i}. \\ 4^\circ \quad (P_{ik}, P_{ij}) &= (P_{ij}+P_{jk}, P_{ij}) \\ &= (-1)^{i+j+1} \cdot (-1)^{j} \quad = (-1)^{i+1}. \end{aligned}$$

$$5^\circ \quad (P_{ik}, P_{jk}) = (P_{ij}+P_{jk}, P_{jk}) = (-1)^{j+1} \cdot (-1)^{j+k+1} = (-1)^k.$$

$$6^\circ \quad (P_{jk}, P_{ik}) = (P_{jk}, P_{ij}+P_{jk}) = (-1)^{j} \cdot (-1)^{j+k+1} = (-1)^{k+1}.$$

Inspection shows that the cases $1^\circ, \cdots, 6^\circ$ all are comprised under (7).

Finally

$$(8) \qquad (P_{ij}, P_{kl}) = \varepsilon_{ij\,|\,kl}.$$

For $(P_{ij}, P_{kl}) = (P_{ij}, P_{ki}+P_{il}) = (-1)^{i+1}\,\varepsilon_{jik}\cdot(-1)^{i+1}\,\varepsilon_{jil}$. But $\varepsilon_{jik}\,\varepsilon_{jil} = \varepsilon_{ij\,|\,kl}$.

The formulae (4), (7), (8) complete the determination of the required sign for half periods with two indices in the basis notation. This is sufficient for the relations ($p=1, 2$) of the next section. Since for any value of p the generic point can be expressed as a sum of points with two indices these formulae suffice to determine the required sign for any two points. We give without proof the general result:

$$(9) \qquad \begin{aligned} &(P_{i_1 i_2 \cdots i_{2r-\alpha} j_1 j_2 \cdots j_\alpha},\ P_{j_1 j_2 \cdots j_\alpha k_1 k_2 \cdots k_{2s-\alpha}}) \\ &= (-1)^{j_1+j_2+\cdots+j_\alpha+e}\ \varepsilon_{i_1\cdots i_{2r-\alpha}\,|\,j_1\cdots j_\alpha\,|\,k_1\cdots k_{2s-\alpha}} \\ &(e=1 \text{ if } \alpha\equiv 1, 2 \bmod. 4;\ e=0 \text{ if } \alpha \equiv 0, 3 \bmod. 4). \end{aligned}$$

30. Theta relations of the second order ($p=2$). In the case of the elliptic thetas ($p=1$) with one odd and three even functions of the first order the squares give rise to four even functions of the second order and zero characteristic of which according to **20** (9) only two are linearly independent. The six products pair off into three sets of two. The products in each pair have the same characteristic $[\eta]_2 \neq 0$ but opposite parity. Hence no relations exist among the products.

As in the preceding section we identify the three points in the finite geometry with the half periods as follows:

$$(1) \quad P_{12}=P_{34}=\{1,0\}_2;\ P_{23}=P_{14}=\{0,1\}_2;\ P_{13}=P_{24}=\{1,1\}_2.$$

For the functions the basis and the characteristic notations are related as follows:

$$(2)\quad \begin{aligned} &\vartheta_{24}(u) = \vartheta_{13}(u) \equiv \vartheta_{0,0}(u); \quad \vartheta_{14}(u) = \vartheta_{23}\ (u) \equiv \vartheta_{1,0}(u); \\ &\vartheta_{34}(u) = \vartheta_{12}(u) \equiv \vartheta_{0,1}(u); \quad \vartheta\,(u) \ \ = \vartheta_{1234}(u) \equiv \vartheta_{1,1}(u). \end{aligned}$$

The four relations among any three of the four squares are

$$(3)\quad \begin{aligned} &(-1)^i\,\vartheta_{i4}^2\,\vartheta_{i4}^2(u) + (-1)^j\,\vartheta_{j4}^2\,\vartheta_{j4}^2(u) + (-1)^k\,\vartheta_{k4}^2\,\vartheta_{k4}^2(u) = 0, \\ &\varepsilon_{ij}\,\vartheta_{k4}^2\,\vartheta^2\,(u) = \vartheta_{i4}^2\,\vartheta_{j4}^2\,(u) - \vartheta_{j4}^2\,\vartheta_{i4}^2\,(u) \\ &\qquad\qquad (i, j, k = 1, 2, 3). \end{aligned}$$

The coefficients are easily checked. In the first relation (3) let $u = P_{i4}$. The first term vanishes and from **29** (1) the other two are proportional to

$$(-1)^j\,\vartheta_{j4}^2\,\vartheta_{k4}^2 + (P_{i4},\,P_{i4})\,(-1)^k\,\vartheta_{k4}^2\,\vartheta_{j4}^2$$

which vanishes since $(P_{i4},\,P_{i4}) = (P_{jk},\,P_{jk}) = (-1)^{j+k+1}$ (cf. **29** (4)). In the second relation (3) let $u = P_{i4}$. The last term vanishes and the others contribute $\varepsilon_{ij}\,\vartheta_{k4}^2\,\vartheta_{i4}^2 = (P_{i4},\,P_{j4})\,\vartheta_{i4}^2\,\vartheta_{k4}^2$. But $(P_{i4},\,P_{j4}) = (-1)^{4+1}\,\varepsilon_{i4j} = \varepsilon_{ij}$. The coefficients in the right member are checked by setting $u = 0$. The value $u = 0$ in the first relation gives rise to the modular relation

$$(4)\quad (-1)^i\,\vartheta_{i4}^4 + (-1)^j\,\vartheta_{j4}^4 + (-1)^k\,\vartheta_{k4}^4 = 0.$$

In the case $p = 2$ the number of relations is much greater. We give here for the first time in the basis notation a complete set of three and four term relations. The 16 theta squares are connected by 240 four term relations which read as follows:

$$\text{I.}\ (-1)^i\,\vartheta_{imn}^2\,\vartheta_{imn}^2\,(u) + (-1)^j\,\vartheta_{jmn}^2\,\vartheta_{jmn}^2\,(u) + (-1)^k\,\vartheta_{kmn}^2\,\vartheta_{kmn}^2\,(u) + (-1)^l\,\vartheta_{lmn}^2\,\vartheta_{lmn}^2\,(u) = 0,$$

$$\text{II.}\ (-1)^i\,\varepsilon_{imn}\,\vartheta_{imn}^2\,\vartheta_i^2\,(u) + (-1)^j\,\varepsilon_{jmn}\,\vartheta_{jmn}^2\,\vartheta_j^2\,(u) + (-1)^k\,\varepsilon_{kmn}\,\vartheta_{kmn}^2\,\vartheta_k^2(u) + (-1)^l\,\varepsilon_{lmn}\,\vartheta_{lmn}^2\,\vartheta_l^2\,(u) = 0,$$

$$\text{III.}\ \vartheta_{imn}^2\,\vartheta_n^2\,(u) + (-1)^j\,\varepsilon_{imj}\,\vartheta_{jmn}^2\,\vartheta_{ijn}^2\,(u) + (-1)^k\varepsilon_{imk}\vartheta_{kmn}^2\vartheta_{ikn}^2(u) + (-1)^l\,\varepsilon_{iml}\,\vartheta_{lmn}^2\,\vartheta_{iln}^2(u) = 0,$$

$$\text{VI.}\quad \vartheta^2_{imn}\,\vartheta^2_{jmn}(u) - \vartheta^2_{jmn}\,\vartheta^2_{imn}(u) + (-1)^k\,\varepsilon_{ijk}\,\vartheta^2_{kmn}\,\vartheta^2_l(u) + (-1)^l\,\varepsilon_{ijl}\,\vartheta^2_{lmn}\,\vartheta^2_k(u) = 0.$$

To prove these we first observe (**20** (9)) that any five theta squares are linearly dependent. If to the four given in I we add $\vartheta^2_i(u)$ then in the linear relation connecting the five the coefficient of $\vartheta^2_i(u)$ must vanish, since, for $u = P_{mn}$, $\vartheta^2_i(u)$ does not vanish whereas the four vanish. To check the coefficients in I set $u = P_{ij}$. The last two terms vanish and the first two are proportional to

$$(-1)^i\,\vartheta^2_{imn}\,\vartheta^2_{jmn} + (-1)^j\,(P_{ij},\,P_{ij})\,\vartheta^2_{imn}\,\vartheta^2_{jmn}$$

which also vanishes (cf. **29** (4)).

The other types are derived from I. Replace u in I by $u + P_{mn}$. The identity becomes

$$(-1)^i\,\vartheta^2_{imn}\,\vartheta^2_i(u) + (-1)^j\,(P_{mn},\,P_{ij})\,\vartheta^2_{jmn}\,\vartheta^2_j(u) + \cdots = 0.$$

Since $(-1)^j\,\varepsilon_{mn\,|\,ij}/(-1)^i = (-1)^j\,\varepsilon_{jmn}/(-1)^i\,\varepsilon_{imn}$, the relation II has the form given. Again in I replace u by $u + P_{im}$. Then

$$(-1)^i\,\vartheta^2_{imn}\,\vartheta^2_n(u) + (-1)^j(P_{im},\,P_{ij})\,\vartheta^2_{jmn}\,\vartheta^2_{ijn}(u) + \cdots = 0.$$

Since $(P_{im},\,P_{ij}) = (-1)^{i+1}\,\varepsilon_{mij} = (-)^i\,\varepsilon_{imj}$, the relation III is verified. Finally in I replace u by $u + P_{ij}$ and divide by $(-1)^i$. Then

$$\begin{aligned}\vartheta^2_{imn}\,\vartheta^2_{jmn}(u) &+ (-1)^{i+j}\,(P_{ij},\,P_{ij})\,\vartheta^2_{jmn}\,\vartheta^2_{imn}(u)\\ &+ (-1)^{i+k}\,(P_{ij},\,P_{ik})\,\vartheta^2_{kmn}\,\vartheta^2_l(u)\\ &+ (-1)^{i+l}\,(P_{ij},\,P_{il})\,\vartheta^2_{lmn}\,\vartheta^2_k(u) = 0\end{aligned}$$

which reduces to IV.

With reference to the Kummer surface (cf. **32**) I, $\cdots$, IV represent the 240 relations which exist among the 16 sets of 6 tropes on the 16 respective nodes. On the node $u = 0$ there are the 6 odd tropes any four of which are related as in II. On each of the 15 nodes $u = P_{mn}$ the 6 tropes are composed of 2 odd and 4 even ones. The four even ones are connected by the relation I. Any three of the four even ones and any one of the two odd ones are related as in III. The two odd ones and any two of the four even ones are related as in IV. Thus I, $\cdots$, IV comprise the $15+15+120+90=240$ relations mentioned.

If we set $u=0$ in these relations the types II and IV vanish identically but the types I and III yield the modular relations (cf. [17] I p. 193)

$$\text{V.}\ (-1)^i \vartheta_{imn}^4+(-1)^j \vartheta_{jmn}^4+(-1)^k \vartheta_{kmn}^4+(-1)^l \vartheta_{lmn}^4=0,$$

$$\text{VI.}\ (-1)^j \varepsilon_{imj} \vartheta_{jmn}^2 \vartheta_{ijn}^2+(-1)^k \varepsilon_{imk} \vartheta_{kmn}^2 \vartheta_{ikn}^2 +(-1)^l \varepsilon_{iml} \vartheta_{lmn}^2 \vartheta_{iln}^2=0.$$

The 120 products of pairs of the 16 thetas are also connected by a set of 120 three term relations. With respect to an isolated proper half period P_{mn} the 16 functions divide into 8 pairs such that the characteristic of the product of members of a pair is that of P_{mn}. Of these products four are odd and four are even, namely

$$(5)\quad \begin{aligned} &\vartheta_i(u)\,\vartheta_{imn}(u),\ \vartheta_j(u)\,\vartheta_{jmn}(u),\ \vartheta_k(u)\,\vartheta_{kmn}(u),\ \vartheta_l(u)\,\vartheta_{lmn}(u);\\ &\vartheta_{ijn}(u)\,\vartheta_{kln}(u),\ \vartheta_{ikn}(u)\,\vartheta_{jln}(u),\ \vartheta_{iln}(u)\,\vartheta_{jkn}(u),\ \vartheta_m(u)\,\vartheta_n(u).\end{aligned}$$

Any three of each set of four are linearly related (cf. **20** (9)). Thus for each P_{mn} there is a set of eight linear relations.

In the determination of these relations there are two points of difficulty. The first is that the sign $(-1)^{(\varepsilon\varkappa')}$ discussed in **29** is, for a function of the first order, a factor $(e^{\pi i/2})^{(\varepsilon\varkappa')}$. The second arises from the formula (**20** (6)) which states that

$$(6)\qquad \vartheta[\eta+2\zeta]_2(u)=(-1)^{(\eta\zeta')}\,\vartheta[\eta]_2(u).$$

Thus the situation in question is to a certain extent arithmetical whereas the basis notation is purely geometric or tactical. The results given below in VII, VIII, IX were, so far as the signs are concerned, deduced from lists of particular formulae which are not reproduced here. The basis notation used to obtain the lists was the specific one

$$(7) \qquad \{11;11\}_2 \equiv \{P_{56}, P_{12}; P_{45}, P_{23}\}$$

by which it is to be understood that

$$P_{56} = \{10; 00\}, \text{ etc.}$$

The 120 desired relations then appear as follows:

(8) *With a basis notation defined as in* (7) *and* i, j, k *three indices in the natural order,*

$$\text{VII.} \quad \vartheta_{iln}\vartheta_{jkn}\cdot\vartheta_i(u)\vartheta_{imn}(u) - \vartheta_{jln}\vartheta_{ikn}\cdot\vartheta_j(u)\vartheta_{jmn}(u) + \vartheta_{kln}\vartheta_{ijn}\cdot\vartheta_k(u)\vartheta_{kmn}(u) = 0.$$

If i, j, k, l *are four indices in the natural order,*

$$\text{VIII.} \quad \vartheta_{ijn}\vartheta_{kln}\cdot\vartheta_{ijn}(u)\vartheta_{kln}(u) - \vartheta_{ikn}\vartheta_{jln}\cdot\vartheta_{ikn}(u)\vartheta_{jln}(u) + \vartheta_{iln}\vartheta_{jkn}\cdot\vartheta_{iln}(u)\vartheta_{jkn}(u) = 0.$$

The determinant of two columns in the order given from the matrix

$$\text{IX.} \quad \left\| \begin{matrix} \vartheta_m(u)\,\vartheta_n(u) & \vartheta_{ijn}(u)\,\vartheta_{kln}(u) & \vartheta_{ikn}(u)\,\vartheta_{jln}(u) & \vartheta_{iln}(u)\,\vartheta_{jkn}(u) \\ 0 & \vartheta_{ijn}\,\vartheta_{kln} & \vartheta_{ikn}\,\vartheta_{jln} & \vartheta_{iln}\,\vartheta_{jkn} \end{matrix} \right\|$$

is equal to the determinant of the remaining two columns in the order given.

Thus of the eight relations among the products (5) four are given by VII, one by VIII, and three by IX. By setting $u=0$ in VIII the modular relation VI is again obtained with a determination of sign which is different from, but nevertheless consistent with, that of VI.

These nine relations are much simplified by using the equations of Schottky (**28** (3), (6)). Set

$$(9) \qquad (ijk\cdots) = e_{ij}\,e_{ik}\cdots e_{jk}\cdots.$$

Then the equations read

$$(10) \qquad \begin{aligned} f_{123} &= c_{123}^4 = \vartheta_{123}^4(0) = r\,(123)(456), \\ f_1 &= c_1^4 \phantom{= \vartheta_{123}^4(0)} = r\,(23456). \end{aligned}$$

Schottky takes, for the value of c_1, the initial term in the development of the odd function ϑ_1 for fixed values of $u = u_1, u_2$, say $u^{(0)}$. He also modifies the theta functions by constant factors as follows:

$$\vartheta_1(u) = c_1\,\sigma_1(u), \qquad \vartheta_{123}(u) = c_{123}\,\sigma_{123}(u). \tag{11}$$

Then in terms of the e_{ij}, $\sigma_i(u)$, $\sigma_{ijk}(u)$ the nine relations are

$$\begin{aligned}
\text{I}^\circ \quad & (156)(234)\,\sigma_{156}^2 \pm (256)(134)\,\sigma_{256}^2 \pm (356)(124)\,\sigma_{356}^2 \\
& \qquad \pm (456)(123)\,\sigma_{456}^2 = 0, \\
\text{II}^\circ \quad & (234)\,\sigma_1^2 \pm (134)\,\sigma_2^2 \pm (124)\,\sigma_3^2 \pm (123)\,\sigma_4^2 = 0, \\
\text{III}^\circ \quad & (15)(234)\,\sigma_1^2 \pm (26)(34)\,\sigma_{126}^2 \pm (36)(24)\,\sigma_{136}^2 \\
& \qquad \pm (46)(23)\,\sigma_{146}^2 = 0, \\
\text{IV}^\circ \quad & (34)[\sigma_{156}^2 - \sigma_{256}^2] = \pm (12)\,[(35)(36)\,\sigma_4^2 \pm (45)(46)\,\sigma_3^2], \\
\text{V}^\circ \quad & (156)(234) \pm (256)(134) \pm (356)(124) \pm (456)(123) = 0, \\
\text{VI}^\circ \quad & (12)(34) \pm (13)(42) \pm (14)(23) = 0, \\
\text{VII}^\circ \quad & (23)\,\sigma_1\,\sigma_{156} \pm (31)\,\sigma_2\,\sigma_{256} \pm (12)\,\sigma_3\,\sigma_{356} = 0, \\
\text{VIII}^\circ \quad & (12)(34)\,\sigma_{126}\,\sigma_{346} \pm (13)(24)\,\sigma_{136}\,\sigma_{246} \\
& \qquad \pm (14)(23)\,\sigma_{146}\,\sigma_{236} = 0, \\
\text{IX}^\circ \quad & \sigma_{245}\,\sigma_{246} - \sigma_{145}\,\sigma_{146} = \pm (12)(34)\,\sigma_5\,\sigma_6 .
\end{aligned}$$

Here we have given only a typical relation in each set and have made no determination of sign.

The relations II°, VII°, IX° are given by Schottky ([63] pp. 270 –271). He uses the relation VII° to determine the $(ij) = e_{ij}$. Since (cf. (10), (11)) $\sigma_{ijk}(0) = 1$, the initial terms in this relation are those of

$$e_{23}\,\sigma_1 \pm e_{31}\,\sigma_2 \pm e_{12}\,\sigma_3 = 0. \tag{12}$$

Similarly for the fixed values $u^{(0)}$ the initial terms of $\sigma_1, \sigma_2, \sigma_3$ are 1 whence

$$e_{23} \pm e_{31} \pm e_{12} = 0. \tag{13}$$

The 20 relations of this kind imply that

$$e_{ij} = (e_i - e_j). \tag{14}$$

The equations (12) then imply that the initial terms of the odd $\sigma_1, \cdots, \sigma_6$ are linear forms in u_1, u_2 whose roots are $e_1, \cdots, e_6$. But the initial terms of the odd $\vartheta_1, \cdots, \vartheta_6$ are six linear forms projective to the six factors of the fundamental sextic which defines the functions of genus two (cf. [6.1] p. 481). Thus $e_1, \cdots, e_6$ are projective to the roots of the fundamental sextic.

31. Theta relations in the characteristic notation. If the moduli a_{ij} for which $i \neq j$ are all zero the function $\vartheta[g, h](u)_a$ breaks up into a product of elliptic thetas, namely

$$\vartheta[g, h](u)_a$$
$$(1) \qquad = \vartheta[g_1, h_1](u_1)_{a_{11}} \cdot \vartheta[g_2, h_2](u_2)_{a_{22}} \cdot \cdots \cdot \vartheta[g_p, h_p](u_p)_{a_{pp}}.$$

We may then regard the general theta as a symbolic non-commutative product of the elliptic thetas. In many cases formulae derived for the elliptic thetas may be extended by this symbolic multiplication to higher values of p. The process must however be used with caution. For example $\vartheta[1, 1](u_1)_{a_{11}}$ and $\vartheta[1, 1](u_2)_{a_{22}}$ vanish for $u_1 = u_2 = 0$ while $\vartheta[1\,1, 1\,1](u_1, u_2)_a$ does not.

The operations,

$$I_{\{\varepsilon\}_2} \qquad u' = u + \{\varepsilon\}_2,$$

where $\{\varepsilon\}_2$ is any one of the $2^{2p}-1$ proper half periods, together with the identity, $u' = u$, form an abelian group, $G_{2^{2p}}$, of order 2^{2p} and type $(1, 1, \cdots, 1)$. A simple set of generators consists of $I_{\varepsilon_i}, I_{\varepsilon'_i}$ $(i = 1, \cdots, p)$ where I_{ε_i} is that operation $I_{\{\varepsilon\}_2}$ for which all the $\varepsilon, \varepsilon'$ except ε_i are zero. The p generators I_ε generate a G_{2^p}; the p generators $I_{\varepsilon'}$ a G'_{2^p}; and $G_{2^{2p}}$ is the direct product of G_{2^p} and G'_{2^p}.

According to the formula **20** (5),

$$\vartheta^2[\eta]_2(u + \{\varepsilon\}_2) = f(a, u, \varepsilon) \cdot (-1)^{(\varepsilon\eta')} \cdot \vartheta^2[\eta + \varepsilon]_2(u),$$

an operation of $G_{2^{2p}}$ permutes the theta squares, to within a factor common to all, and in certain cases changes the sign. To eliminate the permutation so far as possible we construct in the elliptic case four functions $X_{j_1}^{i_1}(i_1, j_1 \equiv 0, 1 \bmod. 2)$ linear in the theta squares; namely,

$$
(2)\quad
\begin{array}{lllllll}
X_0^0(u) = & \vartheta^2[0,0]_2(u) + \vartheta^2[0,1]_2(u) + \vartheta^2[1,0]_2(u) + \vartheta^2[1,1]_2(u), \\
X_1^0(u) = & \vartheta^2[0,0]_2(u) - \quad " \quad + \quad " \quad - \quad " \quad , \\
X_0^1(u) = & \vartheta^2[0,0]_2(u) + \quad " \quad - \quad " \quad - \quad " \quad , \\
X_1^1(u) = & -\vartheta^2[0,0]_2(u) + \quad " \quad + \quad " \quad - \quad " \quad .
\end{array}
$$

These have the property that $I_{\varepsilon_1'}$ converts $X_{j_1}^{i_1}$ into $(-1)^{j_1} X_{j_1}^{i_1}$ and I_{ε_1} converts $X_{j_1}^{i_1}$ into $X_{j_1+1}^{i_1}$. The theta squares are in turn expressed in terms of the $X_{j_1}^{i_1}$ as follows:

$$
(2')\quad
\begin{array}{ll}
2^2\,\vartheta^2[0,0]_2(u) = & X_0^0(u) + X_1^0(u) + X_0^1(u) - X_1^1(u), \\
2^2\,\vartheta^2[0,1]_2(u) = & X_0^0(u) - \quad " \quad + \quad " \quad + \quad " \quad , \\
2^2\,\vartheta\;[1,0]_2(u) = & X_0^0(u) + \quad " \quad - \quad " \quad + \quad " \quad , \\
2^2\,\vartheta^2[1,1]_2(u) = & X_0^0(u) - \quad " \quad - \quad " \quad - \quad " \quad .
\end{array}
$$

An immediate consequence is:

(3) *For general values of* p, *the* 2^{2p} *functions* $X_j^i(u)$ $(i, j \equiv 0,1$ mod. 2) *defined by the non-commutative symbolic product*

$$X_{j_1 j_2 \cdots j_p}^{i_1 i_2 \cdots i_p} = X_{j_1}^{i_1}(u_1) \cdot X_{j_2}^{i_2}(u_2) \cdot X_{j_3}^{i_3}(u_3) \cdots X_{j_p}^{i_p}(u_p)$$

and the 2^{2p} *functions* $\vartheta^2[\eta, \eta']_2(u)$ *with the symbolic expression* (1), *are expressed in terms of each other by symbolic multiplication of factors whose values are given in* (2) *and* (2′). *Any set of* 2^p *functions* X_j^i *with fixed superscripts* i *is transformed into itself by* $G_{2^{2p}}$ *according to the law*

$$u' \equiv u + \{\varepsilon, \varepsilon'\}_2; \qquad \varrho\, X_j^{i'} = (-1)^{(\varepsilon' j)}\, X_{j+\varepsilon}^{i}.$$

Thus G'_{2^p} merely changes the signs of the X_j^i whereas G_{2^p} merely permutes the X_j^i.

We now prove that

(4) *If any one of the* 2^p *sets of* 2^p *functions* X *with fixed superscripts contains* 2^p *functions* X_j *which are linearly*

related with constant coefficients then each X_j *vanishes identically. At least one set of* 2^p *functions* X_j *does not vanish and is linearly independent. Any other set is proportional to this one with constant factor of proportionality.*

Suppose that a linear relation of the form

$$\sum_j a_{j_1 \cdots j_p} X_{j_1 \cdots j_p}(u) \equiv 0$$

exists. On applying $I_{\varepsilon'_1}$ a new relation is obtained in which the terms with $j_1 = 1$ are changed in sign. By adding and subtracting the two relations a pair of relations is obtained in each of which the subscript j_1 is fixed. Proceeding similarly with $I_{\varepsilon'_2}$, etc., we find eventually 2^p relations of the form $a_{j_1 \cdots j_p} X_{j_1 \cdots j_p} \equiv 0$. Since not all of the coefficients are zero at least one of the functions X_j vanishes identically. On applying to this function the operations I_ε it is transformed into each one of the set X_j whence the entire set vanishes identically. Not every set could vanish identically since not every theta square vanishes identically. If then $X^{i_1 \cdots i_p}_{j_1 \cdots j_p} \not\equiv 0$ no member of the set X^i is identically zero and the set is linearly independent. Since all the other X's are of the second order and zero characteristic any other X must be linearly expressible in terms of the 2^p in the set X^i, i. e.

$$a X^k_{l_1 \cdots l_p}(u) = \sum_j a_{j_1 \cdots j_p} X^i_{j_1 \cdots j_p}(u) \quad (k \neq i).$$

Apply $I_{\varepsilon'_1}$ to this and add to or subtract from the old relation the new one thus obtained according as l_1 is even or odd. The result is an identical relation

$$a X^k_{l_1 \cdots l_p} = \sum_{j_2 \cdots j_p} a_{l_1 j_2 \cdots j_p} X^i_{l_1 j_2 \cdots j_p}.$$

Proceeding similarly with $I_{\varepsilon'_2}$, etc., we obtain finally $a X^k_l(u) = a_l X^i_l(u)$ and by applying the I_ε the proportionality of the sets X^k and X^i with factor $a : a_l$ is proved.

In order to exhibit the common part of the 2^p sets X^i and to obtain the factors of proportionality certain formulae (cf. Krazer[41] pp. 364–366) are necessary which here are proved ab initio. The exponent of a general term of the product

$$\vartheta[\eta,\eta']_2(u)\cdot\vartheta[\eta,\eta']_2(v)$$

is

$$(a, m+\eta/2)^2+(a, n+\eta/2)^2+2(m+\eta/2,\ u+\eta'\pi i/2) \\ +2(n+\eta/2,\ u+\eta'\pi i/2).$$

This may be rewritten as

$$(a, m+n+\eta)^2/2+(a, m-n)^2/2+(m+n+\eta,\ u+v+\eta'\pi i) \\ +(m-n,\ u-v).$$

If now we introduce new letters, μ,ν, of summation from

$$\begin{aligned} m+n &= 2\mu+\alpha, \\ m-n &= 2\nu+\alpha, \end{aligned} \qquad (\alpha=0,1)$$

then all values of m, n are obtained from all values of μ,ν for $\alpha=0,1$. Again the exponent can be written as

$$(a, 2\mu+\alpha+\eta)^2/2+(a, 2\nu+\alpha)^2/2+(2\mu+\alpha+\eta,\ u+v+\eta'\pi i) \\ +(2\nu+\alpha,\ u-v)$$

$$=\sum_{\alpha=0,1}[2(a, \mu+\alpha/2+\eta/2)^2+2(\mu+\alpha/2+\eta/2,\ u+v) \\ +2(a,\nu+\alpha/2)^2+2(\nu+\alpha/2,\ u-v)+(2\mu+\alpha+\eta,\ \eta'\pi i)].$$

The exponential $e^{2(\mu\eta')\pi i}=1$ and $e^{(\alpha+\eta,\eta'\pi i)}=(-1)^{(\alpha+\eta,\eta')}$. If now α is held (η,η' also fixed) the summation for μ,ν yields a product of two thetas. Summing finally for α we have

$$\vartheta[\eta,\eta']_2(u)\cdot\vartheta[\eta,\eta']_2(v) \\ =\sum_{\alpha=0,1}(-1)^{(\alpha+\eta,\eta')}\,\vartheta[\alpha+\eta,0]_2(u+v)_{2a}\cdot\vartheta[\alpha,0]_2(u-v)_{2a}.$$

This becomes, on replacing $\alpha+\eta$ by a new α,

$$\vartheta[\eta,\eta']_2(u)\cdot\vartheta[\eta,\eta']_2(v) \\ =\sum_{\alpha=0,1}(-1)^{(\alpha\eta')}\cdot\vartheta[\alpha+\eta,0]_2(u-v)_{2a}\cdot\vartheta[\alpha,0]_2(u+v)_{2a}. \tag{5}$$

For $u=v$, (5) becomes

$$(5') \quad \begin{aligned} &\vartheta^2[\eta, \eta']_2(u) \\ &= \sum_{\alpha=0,1} (-1)^{(\alpha\eta')} \cdot \vartheta[\alpha+\eta, 0]_2(0)_{2a} \cdot \vartheta[\alpha, 0]_2(2u)_{2a}. \end{aligned}$$

The equations (5) may be solved as follows. Multiply by $(-1)^{(\beta\eta')}$ and sum for $\eta'=0, 1$, arranging the terms on the right with reference to α. Then

$$\begin{aligned} &\sum_{\eta'} (-1)^{(\beta\eta')} \vartheta[\eta, \eta']_2(u) \cdot \vartheta[\eta, \eta']_2(v) \\ &= \sum_{\alpha} \vartheta[\alpha+\eta, 0]_2(u-v)_{2a} \cdot \vartheta[\alpha, 0]_2(u+v)_{2a} \cdot \sum_{\eta'} (-1)^{(\alpha-\beta, \eta')}. \end{aligned}$$

Now $\sum_{\eta'} (-1)^{(\alpha-\beta,\eta')}$ is 2^p if $\alpha=\beta$, but vanishes if $\alpha \neq \beta$. Setting then $\alpha=\beta$ and replacing β later by α we have

$$(6) \quad \begin{aligned} &2^p \vartheta[\alpha+\eta, 0]_2(u-v)_{2a} \cdot \vartheta[\alpha, 0]_2(u+v)_{2a} \\ &= \sum_{\eta'} (-1)^{(\alpha\eta')} \vartheta[\eta, \eta']_2(u) \vartheta[\eta, \eta']_2(v). \end{aligned}$$

For $u=v$, (6) becomes

$$(6') \quad \begin{aligned} &2^p \vartheta[\alpha+\eta, 0]_2(0)_{2a} \cdot \vartheta[\alpha, 0]_2(2u)_{2a} \\ &= \sum_{\eta'} (-1)^{(\alpha\eta')} \vartheta^2[\eta, \eta']_2(u). \end{aligned}$$

An important formula is derived from (5) by replacing u, v by $u+v$, $u-v$:

$$(7) \quad \begin{aligned} &\vartheta[\eta, \eta']_2(u+v) \cdot \vartheta[\eta, \eta']_2(u-v) \\ &= \sum_{\alpha} (-1)^{(\alpha\eta')} \cdot \vartheta[\alpha, 0]_2(2u)_{2a} \cdot \vartheta[\alpha+\eta, 0]_2(2v)_{2a}. \end{aligned}$$

We specialize some of these formulae for $p=1$. Let

$$(8) \quad \begin{aligned} Z_0 &= \vartheta[0, 0]_2(2u)_{2a}, & Z_1 &= \vartheta[1, 0]_2(2u)_{2a}; \\ z_0 &= \vartheta[0, 0]_2(0)_{2a}, & z_1 &= \vartheta[1, 0]_2(0)_{2a}. \end{aligned}$$

Then (5′) yields

$$(9) \quad \begin{aligned} \vartheta^2[0,0]_2(u) &= z_0 Z_0 + z_1 Z_1, & \vartheta^2[1,0]_2(u) &= z_1 Z_0 + z_0 Z_1, \\ \vartheta^2[0,1]_2(u) &= z_0 Z_0 - z_1 Z_1, & \vartheta^2[1,1]_2(u) &= z_1 Z_0 - z_0 Z_1. \end{aligned}$$

From these in turn (5′) can be obtained by symbolic multiplication.

The values of the $X_{j_1}^{i_1}$ (cf. (2)) are therefore

$$(10)\qquad \begin{aligned} X_0^0 &= 2(z_0+z_1)\,Z_0, & X_0^1 &= 2(z_0-z_1)\,Z_0,\\ X_1^0 &= 2(z_0+z_1)\,Z_1, & X_1^1 &= 2(z_0-z_1)\,Z_1. \end{aligned}$$

We have then the theorem:

(11) *If*

$$Z_{\eta_1\cdots\eta_p} = \vartheta[\eta_1\cdots\eta_p,\ 0\ 0\cdots 0]_2\,(2u)_{2a}$$

and

$$z_{\eta_1\cdots\eta_p} = \vartheta[\eta_1\cdots\eta_p,\ 0\cdots 0]\,(0)_{2a}$$

and if, in non-commutative symbols,

$$Z_{\eta_1\cdots\eta_p} = Z_{\eta_1}\cdot Z_{\eta_2}\cdot\cdots\cdot Z_{\eta_p},\quad z_{\eta_1\cdots\eta_p} = z_{\eta_1}\cdot z_{\eta_2}\cdot\cdots\cdot z_{\eta_p}$$

and $\alpha_{i_1\cdots i_p} = \alpha_{i_1}\cdot\alpha_{i_2}\cdot\cdots\cdot\alpha_{i_p}$ *where*

$$\alpha_0 = z_0+z_1,\qquad \alpha_1 = z_0-z_1;$$

then the values of the 2^{2p} *functions* X *in terms of the functions* Z *are*

$$X_{j_1\cdots j_p}^{i_1\cdots i_p} = 2^p\cdot\alpha_{i_1\cdots i_p}\cdot Z_{j_1\cdots j_p}.$$

The formula (7) specialized for $p=1$ yields

$$(12)\qquad \begin{aligned} \vartheta[0,0]_2(u+v)\cdot\vartheta[0,0]_2(u-v) &= Z_0(u)\cdot Z_0(v)+Z_1(u)\cdot Z_1(v),\\ \vartheta[0,1]_2(u+v)\cdot\vartheta[0,1]_2(u-v) &= Z_0(u)\cdot Z_0(v)-Z_1(u)\cdot Z_1(v),\\ \vartheta[1,0]_2(u+v)\cdot\vartheta[1,0]_2(u-v) &= Z_0(u)\cdot Z_1(v)+Z_1(u)\cdot Z_0(v),\\ \vartheta[1,1]_2(u+v)\cdot\vartheta[1,1]_2(u-v) &= Z_0(u)\cdot Z_1(v)-Z_1(u)\cdot Z_0(v). \end{aligned}$$

These again by symbolic multiplication give rise to (7).

32. Theta manifolds; the Kummer surface and generalizations; modular families. The aggregate of theta functions of order n with given characteristic, say the zero characteristic (**19** (6)), can be expressed linearly with constant coefficients in terms of n^p such functions (**19** (4)). If n^p linearly independent functions are chosen as the homogeneous point coördinates of a point P in a space S_{n^p-1} then, for variation of $u_1, \cdots, u_p$, the point P describes a normal theta manifold $M_{n,p}$ of dimension p. Any manifold of this dimension with the property that the coördinates of its variable point can be expressed uniformly as $2p$-tuply periodic functions of p variables is either $M_{n,p}$ or a projection of it (**19**). The n^p functions may be so chosen that E are even and O are odd ($E+O=n^p$) (cf. **20** (9)). In this section only the case $n=2$ for which the functions are even is discussed. The manifold in question is then a projection of $M_{2,p}$ from the linear space defined by the odd functions (cf. [52] Chap. 17, §1).

The even functions of the second order and zero characteristic, which include the theta squares, can all be expressed in terms of the 2^p functions

$$(1) \qquad Z_{\eta_1 \cdots \eta_p}(u_1, \cdots, u_p) \quad (\eta_1, \cdots, \eta_p = 0, 1)$$

of **31** (11). We set

$$(2) \qquad \nu(p) = 2^p - 1, \qquad m(p) = 2^{p-1} \cdot p!$$

As the u's vary the point Z runs over a manifold K_p of dimension p in $S_{\nu(p)}$. Since p functions linear in Z have $2^p \cdot p!$ simultaneous zeros (**19** (9)) which divide into pairs $\pm u$ yielding the same point Z, the order of K_p is $m(p)$. For $p=2$ this is the long known Kummer quartic surface K_2^4. For general p the properties of the manifold $K_p^{m(p)}$ whose existence was pointed out by Klein have been developed by Wirtinger [74].

Associated with K_p there is a dual manifold $W_p(v)$ whose elements are $S_{\nu(p)-1}$'s with coördinates

$$(3) \qquad W_{\eta_1 \cdots \eta_p}(v_1, \cdots, v_p) = Z_{\eta_1 \cdots \eta_p}(v_1, \cdots, v_p).$$

Evidently W^p is the polar reciprocal of K_p in the quadric

$$(4) \qquad \sum\nolimits_\eta Z^2_{\eta_1 \cdots \eta_p} = 0.$$

The group $G_{2^{2p}}$ of transformations, $u' = u + \{\varepsilon\}_2$, will, according to **31** (3), (11), give rise to a collineation group whose expression in Z coördinates and dual expression in W coördinates is

$$(5) \qquad I_\varepsilon: \quad \begin{aligned} u' &= u + \{\varepsilon, \varepsilon'\}_2; \quad Z'_j = (-1)^{(\varepsilon' j)} Z_{j+\varepsilon}; \\ W'_j &= (-1)^{(\varepsilon' j)} W_{j+\varepsilon} \qquad (\varepsilon, \varepsilon' = 0, 1). \end{aligned}$$

Since u' as well as u gives rise to a point of K_p, or an $S_{\nu(p)-1}$ of W_p, these manifolds are each individually unaltered by the collineation $G_{2^{2p}}$. The polar system of (4), a correlation of period two, is invariant under $G_{2^{2p}}$ whence this polarity and $G_{2^{2p}}$ generate an abelian correlation group, $\Gamma_{2 \cdot 2^{2p}}$, of type $(1, 1, \cdots, 1)$ whose 2^{2p} correlations interchange K_p and W_p. An equation of the polarity (4) (cf. **31** (7) for $[\eta, \eta'] = [0, 0]$) is

$$(6) \qquad \vartheta(u - v)\, \vartheta(u + v) = 0.$$

By combining this with the parametric form of (5) the equations of the 2^{2p} correlations appear as

$$(7) \qquad \vartheta[\varepsilon, \varepsilon']_2 (u - v) \cdot \vartheta[\varepsilon, \varepsilon']_2 (u + v) = 0.$$

Their explicit equations in terms of the coördinates Z, W are furnished by **31** (7) or arise by symbolic multiplication from the equations **31** (12).

The correlations of $\Gamma_{2 \cdot 2^{2p}}$, of period two, are either polarities or null systems. The null systems arise from symbolic products with an odd number of factors of the fourth type in **31** (12) i. e. from the odd functions in (7). There are then $E_p = 2^{p-1}(2^p + 1)$ polarities and $O_p = 2^{p-1}(2^p - 1)$ null systems in $\Gamma_{2 \cdot 2^{2p}}$. The 2^{2p} points of K_p and the 2^{2p} $S_{\nu(p)-1}$'s of W_p which form a conjugate set under $\Gamma_{2 \cdot 2^{2p}}$ make up a configuration $(2^{2p})_{O_p}$ such that each of the points ($S_{\nu(p)-1}$'s) is on O_p of the $S_{\nu(p)-1}$'s (points).

An especially important configuration of this type arises from the proper and zero half periods. Wirtinger [74] describes its properties as follows. The 2^{2p} half period points are multiple points of $K_p^{m(p)}$ of order 2^{p-1} and the 2^{2p} half period $S_{\nu(p)-1}$'s touch $K_p^{m(p)}$ along a manifold $M_{p-1}^{m(p)/2}$. With respect to W_p the configuration presents a dual behavior. Finally Wirtinger shows that with respect to $K_p^{m(p)}$ the $S_{\nu(p)-1}$'s of W_p are characterized by the fact that each touches K_p along a manifold $M_{p-2}^{m(p)/4}$ which lies in an $S_{\nu(p)-p-1}$, i. e. each cuts K_p in an $M_{p-1}^{m(p)}$ on which $M_{p-2}^{m(p)/4}$ is a locus of double points. Of the spaces of W_p the ∞^{p-2} whose $M_{p-2}^{m(p)/4}$'s pass through P on K_p contain the tangent space S_p of K_p at P and precisely $m(p)/4$ of these pass through an arbitrary R_{p-3} whence the K_p and W_p may be regarded as dual forms of the same locus. For $p=2$, K_p is the Kummer surface with 16 nodes determined by the half periods including $u=0$ and W_p is a quartic envelope of planes with 16 double planes similarly determined. The double planes of W_p are *tropes* of K_p, the plane sections of K_p determined by the theta squares, which touch K_p along conics. The plane v of W_p touches K_p at an M_0^1 or point whence W_p and K_p are dual forms of the same surface.

The involutorial elements of $G_{2^{2p}}$ are all of the same projective type. For $p=1$ the three involutions and their fixed elements with multipliers ± 1 are:

$$
\begin{aligned}
u' &= u+\{0,1\}_2: \quad \begin{matrix} Z_0' = Z_0 \\ Z_1' = -Z_1 \end{matrix}; \quad \begin{matrix} Z_0' = +Z_0 \\ Z_1' = -Z_0 \end{matrix};\\
(8)\quad u' &= u+\{1,0\}_2: \quad \begin{matrix} Z_0' = Z_1 \\ Z_1' = Z_0 \end{matrix}; \quad \begin{matrix} Z_0'+Z_1' = +(Z_0+Z_1) \\ Z_0'-Z_1' = -(Z_0-Z_1) \end{matrix};\\
u' &= u+\{1,1\}_2: \quad \begin{matrix} Z_0' = Z_1 \\ Z_1' = -Z_0 \end{matrix}; \quad \begin{matrix} Z_0'-iZ_1' = +i(Z_0-iZ_1) \\ Z_0'+iZ_1' = -i(Z_0+iZ_1). \end{matrix}
\end{aligned}
$$

The corresponding canonical forms of the collineations and of their spaces of fixed points for general p are obtained from these by symbolic multiplication. We observe that each collineation has linear spaces, $S_{\nu(p-1)}(\pm)$, of fixed points and is the harmonic perspectivity determined by these two skew spaces. Two involutions defined by $\{\delta\}_2$ and $\{\varepsilon\}_2$ are interchangeable and either transforms the other into itself. The two fixed spaces of either are invariant under the other and they either are interchanged or each is invariant. The following theorem states the situation:

(9) (a) *If* $\{\delta\}_2$, $\{\varepsilon\}_2$ *are azygetic, the six spaces,* $S_{\nu(p-1)}(\pm)$, *of fixed points of the collineations defined by* $\{\delta\}_2$, $\{\varepsilon\}_2$, $\{\delta+\varepsilon\}_2$, *are all skew to each other. The line from a point of one space belonging to* $\{\delta\}_2$ *across the pair belonging to* $\{\varepsilon\}_2$ *is incident with all six. The locus of the* $\infty^{2^{p-1}-1}$ *such lines is an* $M^{2^{p-1}}_{2^{p-1}}$ *of the type defined by a matrix of two rows and* 2^{p-1} *columns. Such a manifold is ruled in two ways, the one ruling consisting of* $\infty^{2^{p-1}-1}$ *lines, the cross ruling of* ∞^1 $S_{2^{p-1}-1}$*'s. The six fixed spaces belong to the cross ruling and meet each line in three harmonic pairs.*

(b) *If* $\{\delta\}_2$, $\{\varepsilon\}_2$, $\{\delta+\varepsilon\}_2$ *are syzygetic there exist four skew spaces* $S_{\nu(p-2)}$ *such that the six* $S_{\nu(p-1)}$*'s which contain the respective pairs of the four are the six fixed spaces. Two complementary pairs belong to the same half period.*

The theorem requires proof only for sample azygetic and syzygetic pairs since all azygetic and all syzygetic pairs are conjugate under the modular group to be introduced presently. For an azygetic pair take that for which $\delta_1' = 1$ and $\varepsilon_1 = 1$ while all the other δ's and ε's are zero. The six fixed spaces are precisely those of (8) with arbitrary $\eta_2, \cdots, \eta_p$. The matrix in question is

$$(10) \qquad \left\| \begin{matrix} Z_{0\eta_2\cdots\eta_p} \\ Z_{1\eta_2\cdots\eta_p} \end{matrix} \right\| = 0 \qquad (\eta_2, \cdots, \eta_p = 0, 1).$$

The two rows define the fixed $S_{\nu(p-1)}$'s of $\{\delta\}_2$; and those of $\{\varepsilon\}_2$, $\{\delta+\varepsilon\}_2$ are defined by linear combinations of the rows with parameters $1:1$; $1:-1$ and $1:i$; $1:-i$ which proves the harmonic property. The elements of the two rows, or any two independent linear combinations, can not vanish simultaneously whence the spaces are skew.

As a sample of a syzygetic pair take $\delta_1' = 1$, $\varepsilon_2' = 1$ with the others zero. Then the six fixed spaces are

$$(11) \quad \begin{matrix} Z_{00} = Z_{01} = 0 \\ Z_{10} = Z_{11} = 0 \end{matrix}; \quad \begin{matrix} Z_{00} = Z_{10} = 0 \\ Z_{01} = Z_{11} = 0 \end{matrix}; \quad \begin{matrix} Z_{00} = Z_{11} = 0 \\ Z_{10} = Z_{01} = 0 \end{matrix};$$

with variable indices $\eta_3, \cdots, \eta_p$ to be supplied. The four spaces of (9(b)) are then those which contain respectively the four vertices of the tetrahedon $Z_{00}, Z_{10}, Z_{01}, Z_{11}$, with variable indices $\eta_3, \cdots, \eta_p$ to be supplied.

Of particular interest always are those points which take up, under the operations of a group, a number of positions smaller than the order of the group. These are necessarily fixed points of some of the elements. A point of K_p is fixed for the element, $\pm u' \equiv u + \{\varepsilon\}_2$, when $2u \equiv \{\varepsilon\}_2$ or $u \equiv \{\varepsilon\}_2/2$, i. e. u is a proper quartic period. If $\{\varepsilon\}_4$ is a proper quarter period for which $2\{\varepsilon\}_4 \equiv \{\varepsilon\}_2$ then there are 2^{2p} proper quartic periods, namely

$$\{\varepsilon\}_4 + \{\delta\}_2 \qquad (\delta, \delta' = 0, 1), \tag{12}$$

all of which when doubled are congruent to $\{\varepsilon\}_2$. Since $3\{\varepsilon\}_4 \equiv -\{\varepsilon\}_4$ the quarter periods $3\{\varepsilon\}_4$ and $\{\varepsilon\}_4$ determine the same point on K_p or more generally $\{\varepsilon\}_4 + \{\delta\}_2$ and $\{\varepsilon\}_4 + \{\delta + \varepsilon\}_2$ determine the same point on K_p. Hence

(13) *The two fixed $S_{\nu(p-1)}$'s of the collineation, $u' = u + \{\varepsilon\}_2$, each meet K_p in $2^{2(p-1)}$ points, each point being defined by the pair of proper quarter periods $\{\varepsilon\}_4 + \{\delta\}_2$, $\{\varepsilon\}_4 + \{\delta + \varepsilon\}_2$. The points P_δ and P_ζ on K_p, with parameters $\{\delta\}_2$ and $\{\zeta\}_2$ in (12), are in the same or different $S_{\nu(p-1)}$'s according as $\{\delta\}_2$, $\{\zeta\}_2$ are each syzygetic or each azygetic with $\{\varepsilon\}_2$. In a particular $S_{\nu(p-1)}$ the quarter period configuration of $2^{2(p-1)}$ points admits a group $G_{2^{2(p-1)}}$ which is the factor group with respect to I_ε of the $G_{2^{2p-1}}$ consisting of the elements $I_\varkappa$ for which $\{\varkappa\}_2$ is syzygetic to $\{\varepsilon\}_2$. This configuration and its group correspond in the finite geometry to projection and section from the point corresponding to $\{\varepsilon\}_2$. The configuration dual to the $2^{2(p-1)}$ points on $S_{\nu(p-1)}(+)$ is cut out on $S_{\nu(p-1)}(+)$ by the quarter period spaces of W_p which contain $S_{\nu(p-1)}(-)$.*

This distribution of the quarter period points between $S_{\nu(p-1)}(\pm)$ is a consequence of (9).

The collineation group, $G_{2^{2p}}$, has integer coefficients and therefore is independent of the moduli, a_{ij}, of the theta functions (1). As these moduli change the manifold K_p changes and describes a family, $F_p(a)$, of manifolds K_p each of which admits the group $G_{2^{2p}}$. We seek that group, $G(p)$, of collineations which transforms this family into itself. $G(p)$ must contain $G_{2^{2p}}$ as an invariant subgroup and must transform the 2^{2p-1} involutions of $G_{2^{2p}}$ in such wise that syzygetic and azygetic pairs are invariant. The order of $G(p)$ then is not greater than $2^{2p} \cdot N_C = 2^{2p} \cdot 2^{p^2} H_{2p-1} H_{2p-3} \cdots H_1$ (**22** (7)). On the other hand under integer linear transformation of the periods the theta squares, and therefore the functions (1) are permuted according to the law (Krazer[41] p. 181),

$$\vartheta^2[\eta, \eta']_2 (u)_a = C^2 \cdot e^{-2U} \cdot \vartheta^2 [\bar{\eta}, \bar{\eta}']^2 (u')_{a'}, \tag{14}$$

where C is constant with respect to u, and U is constant with respect to $[\eta, \eta']_2$ whence e^{-2U} figures as a proportionality factor. The $\bar{\eta}, \bar{\eta}'$ are given in terms of η, η' in **24** (1). Hence the elements of $G(p)$ not in $G_{2^{2p}}$ are precisely those which arise from transformation mod. 2 of the periods, and the order of $G(p)$ is precisely $2^{2p} \cdot N_C$. Since $G(p)$ can be defined to be the maximal collineation group which transforms $G_{2^{2p}}$ with integer coefficients into itself, $G(p)$ must likewise have numerical coefficients which are determined presently. G_{N_C} is generated by the involutions attached to the points in the notation of the finite geometry (**22** (10)). For $p = 1$, there are three such involutions each of which leaves one of the three elements (8) unaltered and interchanges the other two. Corresponding elements of $G(p)$ $(p = 1)$, to which we add the identity, are

		$J_{0,0}$	$J_{0,1}$	$J_{1,0}$	$J_{1,1}$
$J_{\{\varepsilon\}_2}$	$Z_0' =$	Z_0	Z_0	$Z_0 + iZ_1$	$Z_0 + Z_1$
	$Z_1' =$	$Z_1;$	$iZ_1;$	$iZ_0 + Z_1;$	$-Z_0 + Z_1.$

These modular involutions are of period 4 in the space $Z_0 : Z_1$ but, in each case

$$J_{i,k}^2 = I_{i,k} \tag{15}$$

so that in the factor group $J_{i,k}$ gives rise to an involution. We summarize the above:

(16) *The collineation $G_{2^{2p}}$ of K_p is an invariant subgroup of a collineation group $G(p)$ of order $2^{2p} \cdot N_C$ which leaves unaltered the family $F_p(a)$ of theta manifolds K_p which have the same $G_{2^{2p}}$ but variable moduli a_{ij}. The factor group of $G_{2^{2p}}$ with respect to $G(p)$, the modular group, is the permutation group of order N_C of the members of the family and it arises from the group of integer linear transformations (mod. 2) of the periods.*

We recall (29 (3)) that the half periods in the basis notation can be identified with those in the characteristic notation by the parallel schemes:

$$G_1\colon\ P_{12},\ P_{34},\ P_{56},\ \cdots,\ P_{2p-3,\,2p-2},\ P_{2p-1,\,2p},\ P_{2p+1,\,2p+2};$$
$$G_2\colon\ P_{23},\ P_{45},\ P_{67},\ \cdots,\ P_{2p-2,\,2p-1},\ P_{2p,\,2p+1},\ P_{2p+2,\,1};$$

$$(17)\quad \begin{aligned} G_1\colon\ & \varepsilon_1 = 1,\ \varepsilon_2 = 1,\ \varepsilon_3 = 1,\ \cdots,\ \varepsilon_{p-1} = 1,\ \varepsilon_p = 1, \\ & \qquad \varepsilon_1 = \varepsilon_2 = \cdots = \varepsilon_p = 1; \\ G_2\colon\ & \varepsilon_1' = \varepsilon_2' = 1,\ \varepsilon_2' = \varepsilon_3' = 1,\ \varepsilon_3' = \varepsilon_4' = 1,\ \cdots \\ & \cdots,\ \varepsilon_{p-1}' = \varepsilon_p' = 1,\ \varepsilon_p' = 1,\ \varepsilon_1' = 1. \end{aligned}$$

It is necessary only to observe that the syzygetic and azygetic relations are the same in the two arrangements. The permutation group of the bases is generated by the involutions attached to all the points of G_1 and G_2 except the last point of G_2. If to these generators we add one with four subscripts, say

$$(18)\qquad P_{2345},\quad \text{or}\quad \varepsilon_1' = \varepsilon_3' = 1,$$

the entire group is generated.

We consider now the element of period 4,

$$(19)\quad J(\varepsilon_1' = 1)\colon\quad Z'_{0\eta_2\cdots\eta_p} = Z_{0\eta_2\cdots\eta_p},\quad Z'_{1\eta_2\cdots\eta_p} = iZ_{1\eta_2\cdots\eta_p}.$$

This has the same spaces of fixed points as $J^2(\varepsilon_1' = 1) = I(\varepsilon_1' = 1)$. Since its square is in $G_{2^{2p}}$ the element J figures in the factor group as an involution. We write the involution I_δ (cf. (5)) in the form

$$Z'_{0\eta_2\cdots\eta_p} = (-1)^{(\eta\delta')} Z_{0\eta_2\cdots\eta_p+\delta_1\delta_2\cdots\delta_p},$$
$$Z'_{1\eta_2\cdots\eta_p} = (-1)^{(\eta\delta')} Z_{1\eta_2\cdots\eta_p+\delta_1\delta_2\cdots\delta_p}.$$

If $\delta_1 = 0$, δ is syzygetic to $\varepsilon'_1 = 1$, otherwise azygetic. But the transform of I_δ by J is I_δ if $\delta_1 = 0$ and is $I_{\delta+\varepsilon'_1}$ if $\delta_1 = 1$. Hence J transforms the involutions I_δ among themselves precisely as the involution attached to a point in the finite geometry transforms the points of the finite space. The required generating involutions of the modular group are then obtained by attaching multipliers $1, i$ respectively to the $\pm$ fixed spaces of the corresponding involutions of $G_{2^{2p}}$.

We apply the method set forth above to the specific case $p=2$ and the Kummer surface K_2^4. The group $G_{2^{2p}} = G_{16}$ is generated by

(20)

	$I_{00,10}$	$I_{00,01}$	$I_{10,00}$	$I_{01,00}$
$Z'_{00} =$	Z_{00}	Z_{00}	Z_{10}	Z_{01}
$Z'_{10} =$	$-Z_{10}$	Z_{10}	Z_{00}	Z_{11}
$Z'_{01} =$	Z_{01};	$-Z_{01}$;	Z_{11};	Z_{00}.
$Z'_{11} =$	$-Z_{11}$	$-Z_{11}$	Z_{01}	Z_{10}

The modular group $G_{16\cdot 6!}$ is generated by

(21)

	$J_{00,10}$	$J_{00,01}$	$J_{00,11}$	$J_{11,00}$	$J_{01,00}$
$Z'_{00} =$	Z_{00}	Z_{00}	Z_{00}	$iZ_{00} + Z_{11}$	$Z_{00} + iZ_{01}$
$Z'_{10} =$	iZ_{10}	Z_{10}	iZ_{10}	$iZ_{10} + Z_{01}$	$Z_{10} + iZ_{11}$
$Z'_{01} =$	Z_{01};	iZ_{01};	iZ_{01};	$Z_{10} + iZ_{01}$;	$iZ_{00} + Z_{01}$.
$Z'_{11} =$	iZ_{11}	iZ_{11}	Z_{11}	$Z_{00} + iZ_{11}$	$iZ_{10} + Z_{11}$

The generators are those attached to $P_{16}, P_{45}, P_{23}, P_{56}, P_{34}$ and that attached to $P_{2345} = P_{61}$ is in this case duplicated.

The four even functions Z_{ij} are of the second order. Their quadratic, cubic, and quartic combinations are of orders 4, 6, 8 with respectively (cf. **20** (9)) $(4^2+2^2)/2$, $(6^2+2^2)/2$, $(8^2+2^2)/2$ that are linearly independent. The number of such combinations is respectively $3\cdot4\cdot5/6$, $4\cdot5\cdot6/6$, $5\cdot6\cdot7/6$. Thus there must exist one quartic relation on the Z_{ij}, say

$$\sum a_{ijkl} Z_{00}^i Z_{10}^j Z_{01}^k Z_{11}^l = 0 \qquad (i+j+k+l=4).$$

This must be unaltered or at most changed in sign when the involution $I_{00,10}$ (20) is applied. Hence only terms of the same parity in the first subscript can occur; similarly for the second subscript. If these parities are $1, 0$, $j+l$ is odd and either $j>0$ or $l>0$. But then the line $Z_{10}=Z_{11}=0$ would lie on K^4 whereas, being a fixed line of $I_{00,10}$, it meets K^4 in only four points determined by quarter periods. Parities other than $0, 0$ would lead to a similar situation whence the terms must be made up of Z_{ij}^2 and the product $Z_{00}\,Z_{10}\,Z_{01}\,Z_{11}$. On applying also the permutations (20) the quartic relation must have the form

$$\begin{aligned} &\alpha_0\,(Z_{00}^4+Z_{10}^4+Z_{01}^4+Z_{11}^4)+2\,\alpha_{10}\,(Z_{00}^2\,Z_{10}^2+Z_{01}^2\,Z_{11}^2)\\ (22)\quad &+2\,\alpha_{01}\,(Z_{00}^2\,Z_{01}^2+Z_{10}^2\,Z_{11}^2)\\ &+2\,\alpha_{11}\,(Z_{00}^2\,Z_{11}^2+Z_{10}^2\,Z_{01}^2)+4\,\beta_0\,Z_{00}\,Z_{10}\,Z_{01}\,Z_{11}=0. \end{aligned}$$

If now the modular substitutions J in (21) are applied to this form it must be transformed into another of the same type in Z'_{ij} with coefficients α' which are linear in the α's. The modular group of order $6!$, the factor group of G_{16} with respect to $G_{16.6!}$, appears as a collineation group on the modular forms $\alpha_0, \cdots, \beta_0$. The explicit expressions of the generating involutions are

	$J_{00,10}$	$J_{00,01}$	$J_{00,11}$	$J_{11,00}$	$J_{01,00}$
$\alpha'_0=$	α_0	α_0	α_0	$\alpha_0-\alpha_{11}$	$\alpha_0-\alpha_{01}$
$\alpha'_{10}=$	$-\alpha_{10}$	α_{10}	$-\alpha_{10}$	$\alpha_{10}-\alpha_{01}-\beta_0$	$\alpha_{10}-\alpha_{11}-\beta_0$
(23) $\alpha'_{01}=$	α_{01};	$-\alpha_{01}$;	$-\alpha_{01}$;	$-\alpha_{10}+\alpha_{01}-\beta_0$;	$-3\,\alpha_0-\alpha_{01}$.
$\alpha'_{11}=$	$-\alpha_{11}$	$-\alpha_{11}$	α_{11}	$-3\,\alpha_0-\alpha_{11}$	$-\alpha_{10}+\alpha_{11}-\beta_0$
$\beta'_0=$	$-\beta_0$	$-\beta_0$	$-\beta_0$	$-2\,\alpha_{10}-2\,\alpha_{01}$	$-2\,\alpha_{10}-2\,\alpha_{11}$

The ratios of the five coefficients of K_2^4 in (22) are functions of the three moduli a_{ij} and they must be connected by one relation. The four linear conditions on the coefficients which require that K_2^4 have a node at $z_{ij}=Z_{ij}(0)$ determine the ratios of the coefficients in terms of the three ratios of z_{ij}. If the latter be eliminated the resulting modular relation turns out to be (cf. Hudson[36] p. 81)

(24) $\alpha_0^3 - \alpha_0(\alpha_{10}^2 + \alpha_{01}^2 + \alpha_{11}^2 - \beta_0^2) + 2\alpha_{10}\alpha_{01}\alpha_{11} = 0.$

The modular group, $G_{6!}$, in (23) is the collineation group which leaves this cubic spread (24) in S_4 unaltered. The lack of symmetry with respect to a group isomorphic with the permutation $g_{6!}$ is rectified in the next chapter.

33. **The theta manifold K_p ($p = 3$).** When $p = 3$ the generalized Kummer manifold

$$(1) \qquad Z_{ijk}(u_1, u_2, u_3) \qquad (i, j, k = 0, 1),$$

is a K_3^{24} in a linear space S_7. The quadratic, cubic and quartic combinations of the 8 Z_{ijk} of orders 4, 6, 8 are expressible respectively in terms of 36, 112, 260 independent functions. The number of such combinations is 36, 120, 330 respectively. Hence there are 8 cubic relations on the Z_{ijk} and 70 quartic relations which are identically satisfied for all values of u, i. e. K_3^{24} is on 8 cubic and 70 quartic spreads in S_7 which are linearly independent. If each of the 8 cubic relations be multiplied by each of the 8 variables, 64 of the quartic relations are obtained, leaving 8 quartic relations to be accounted for. Wirtinger ([76] §§ 17–23) shows that all identical relations among the functions Z_{ijk} are consequences of these quartic relations.

We shall be more concerned here with the cubic relations. We note, as for $p = 2$, that, due to the existence of the G_8' of changes of sign in the G_{64}, $u' = u + \{\varepsilon\}_2$, there may be derived from any one relation another in which the terms have the same parity in each of the three indices. Furthermore from a cubic relation in which the terms have one type of parity there is obtained, by the operations of the permutation G_8 in G_{64}, eight relations in which the eight possible types of parity occur. The possible terms of parity 0, 0, 0 are of three kinds, Z_{000}^3, $Z_{000}Z_{100}^2$, $Z_{100}Z_{010}Z_{110}$. Instead of using triple subscripts it is more convenient to set

$$(2) \quad \begin{array}{llll} Z_{000} = Z, & Z_{100} = Z_1, & Z_{010} = Z_2, & Z_{001} = Z_3, \\ Z_{111} = Z_7, & Z_{011} = Z_4, & Z_{101} = Z_5, & Z_{110} = Z_6, \end{array}$$

It may be observed that we have chosen the single subscript notation by numbering from $1, \cdots, 7$ the points of the plane in a finite geometry mod. 2. The seven linear triads then give rise to products like $Z_{100}\,Z_{010}\,Z_{110}$ of parity $0, 0, 0$. The most general cubic relation of parity $0, 0, 0$ has then the form

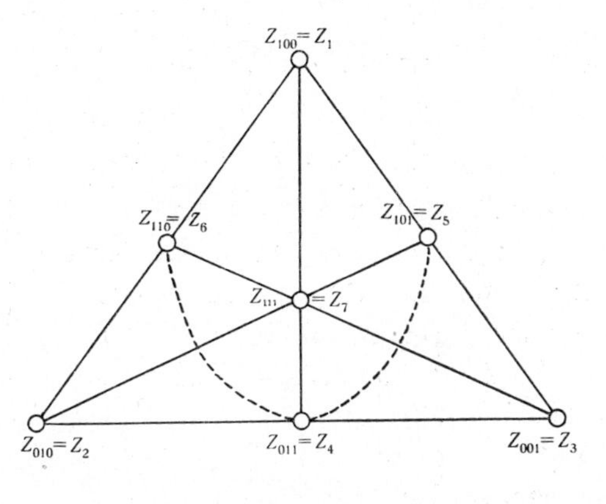

$$
\begin{aligned}
C \equiv\ & \alpha Z^3 + Z(\alpha_1 Z_1^2 + \cdots + \alpha_7 Z_7^2) \\
& + \alpha_{423}\, Z_4\, Z_2\, Z_3 + \alpha_{153}\, Z_1\, Z_5\, Z_3 + \alpha_{126}\, Z_1\, Z_2\, Z_6 \\
& + \alpha_{147}\, Z_1\, Z_4\, Z_7 + \alpha_{257}\, Z_2\, Z_5\, Z_7 + \alpha_{367}\, Z_3\, Z_6\, Z_7 \\
& + \alpha_{456}\, Z_4\, Z_5\, Z_6 = 0.
\end{aligned} \tag{3}
$$

The involutions of G_8 arise by addition of unity to one, two, or all of the subscripts. According to (2) they are in the new notation:

$$
\begin{aligned}
I_1 &= (ZZ_1)\,(Z_2 Z_6)\,(Z_3 Z_5)\,(Z_4 Z_7), \\
I_2 &= (ZZ_2)\,(Z_3 Z_4)\,(Z_1 Z_6)\,(Z_5 Z_7), \\
I_3 &= (ZZ_3)\,(Z_1 Z_5)\,(Z_2 Z_4)\,(Z_6 Z_7), \\
I_4 &= (ZZ_4)\,(Z_2 Z_3)\,(Z_5 Z_6)\,(Z_1 Z_7), \\
I_5 &= (ZZ_5)\,(Z_3 Z_1)\,(Z_6 Z_4)\,(Z_2 Z_7), \\
I_6 &= (ZZ_6)\,(Z_1 Z_2)\,(Z_4 Z_5)\,(Z_3 Z_7), \\
I_7 &= (ZZ_7)\,(Z_1 Z_4)\,(Z_2 Z_5)\,(Z_3 Z_6).
\end{aligned} \tag{4}
$$

We observe that I_j contains the pair (ZZ_j) and the other pairs are, in the finite plane, collinear with Z_j. On applying these to (3) seven new cubic relations, $C_1 = 0, \cdots, C_7 = 0$, are obtained with the same 15 coefficients α as $C = 0$. If the cubic relation of parity i, j, k is multiplied by Z_{ijk}, a quartic relation of parity $0, 0, 0$ results. The addition of the eight such quartic relations yields the following important quartic relation of parity $0, 0, 0$:

$$
\begin{aligned}
L^4 \equiv\ & \alpha(Z^4 + Z_1^4 + \cdots + Z_7^4) \\
& + 2\alpha_1 (Z^2 Z_1^2 + Z_2^2 Z_6^2 + Z_3^2 Z_5^2 + Z_4^2 Z_7^2) + \cdots \\
(5) \qquad & + 2\alpha_7 (Z^2 Z_7^2 + Z_1^2 Z_4^2 + Z_2^2 Z_5^2 + Z_3^2 Z_6^2) \\
& + 4\alpha_{423} (Z Z_4 Z_2 Z_3 + Z_7 Z_1 Z_5 Z_6) + \cdots \\
& + 4\alpha_{456} (Z Z_4 Z_5 Z_6 + Z_7 Z_1 Z_2 Z_3) = 0 .
\end{aligned}
$$

In L^4 the pairing of terms with coefficient α_j is given by I_j in (4); the products with coefficient α_{ijk} arise from a linear triad and the complementary tetrad in the finite plane. Since the arrangements all depend on collinearity in the finite plane, each of the 15 parts of L with a given coefficient is invariant under G_{64}.

We observe that $\partial / \partial Z [L^4] = 4C$. Since L^4 is invariant under G_8 also $\partial / \partial Z_j [L^4] = 4 C_j$. But $C = C_j = 0$ for points on K_3^{24}. Hence

(6) *There exists a unique quartic spread L^4 in S_7 which contains K_3^{24} as a double manifold. When the group G_{64} of K_3^{24} has the canonical form, the* 15 *coefficients α of L^4 (themselves modular forms whose ratios are modular functions) are subject to a set of* 63 *cubic relations which are in correspondence with the* 63 *half periods.*

The latter statement follows from the fact that the two S_3's of fixed points which belong to an involution I_ε of G_{64} meet K_3^{24} in quarter period points (**32** (13)) which form a Kummer configuration. Thus such an S_3 meets L^4 in a 16-nodal quartic surface with a $G_{2^{2p}}$ $(p = 2)$ and therefore a Kummer surface. But the coefficients of a Kummer surface satisfy the cubic relation **32** (24). For each of the 63 involutions in G_{64}, i. e. for each of the half periods, there occurs such a cubic modular relation.

It may well be that the cubic relations alone define K_3^{24} and that the eight additional quartic relations mentioned above are necessarily satisfied when $C = C_j = 0$. In any case the modular forms α are well defined and will be identified later (Chap. IV) with irrational invariants of the ternary quartic. For this generators of the modular group are necessary. The group G_{NC} of period transformations has the order $8\,!\,36$ and the group $G(p)\ (p=3)$ of order $8\,!\,36\cdot 64$ is generated by

	$J_{000,100}$	$J_{000,010}$	$J_{000,001}$	$J_{000,111}$	$J_{110,000}$	$J_{011,000}$	$J_{001,000}$	$J_{101,000}$
$Z'=$	Z	Z	Z	Z	$iZ+Z_6$	$iZ+Z_4$	$iZ+Z_3$	$iZ+Z_5$
$Z_1'=$	iZ_1	Z_1	Z_1	iZ_1	iZ_1+Z_2	iZ_1+Z_7	iZ_1+Z_5	iZ_1+Z_3
$Z_2'=$	Z_2	iZ_2	Z_2	iZ_2	Z_1+iZ_2	iZ_2+Z_3	iZ_2+Z_4	iZ_2+Z_7
$Z_3'=$	Z_3	Z_3	iZ_3	iZ_3	iZ_3+Z_7	Z_2+iZ_3	$Z+iZ_3$	Z_1+iZ_3
(7) $Z_4'=$	Z_4	iZ_4	iZ_4	Z_4	Z_5+iZ_4	$Z+iZ_4$	Z_2+iZ_4	iZ_4+Z_6
$Z_5'=$	iZ_5	Z_5	iZ_5	Z_5	iZ_5+Z_4	iZ_5+Z_6	Z_1+iZ_5	$Z+iZ_5$
$Z_6'=$	iZ_6	iZ_6	Z_6	Z_6	$Z+iZ_6$	Z_5+iZ_6	iZ_6+Z_7	Z_4+iZ_6
$Z_7'=$	iZ_7	iZ_7	iZ_7	iZ_7	Z_3+iZ_7	Z_1+iZ_7	Z_6+iZ_7	Z_2+iZ_7.

Again $J^{2}_{ijk,lmn}=I_{ijk,lmn}$ (I in G_{64}).

These modular involutions may, by interchanging the roles of G_1, G_2 in **32** (17), be identified with the involutions attached in order to the following points in the basis notation:

(8) $P_{12},\ P_{34},\ P_{56},\ P_{78},\ P_{23},\ P_{45},\ P_{67},\ P_{2345}$.

If these operations are applied to L^4 it is transformed into L_4', of the same form as L^4, but with coefficients α' which are linear functions of the α's. Thus there arises the modular group of order $8!\,36$ which is generated also by eight involutions corresponding to those in (7). They are

$\alpha'=$					$\alpha-\alpha_6$	$\alpha-\alpha_4$	$\alpha-\alpha_3$	$\alpha-\alpha_5$
$\alpha_1'=$	−			−	$\alpha_1-\alpha_2-\alpha_{126}$	$\alpha_1-\alpha_7-\alpha_{147}$	$\alpha_1-\alpha_5-\alpha_{153}$	$\alpha_1-\alpha_3-\alpha_{153}$
$\alpha_2'=$		−		−	$-\alpha_1+\alpha_2-\alpha_{126}$	$\alpha_2-\alpha_3-\alpha_{423}$	$\alpha_2-\alpha_4-\alpha_{423}$	$\alpha_2-\alpha_7-\alpha_{257}$
$\alpha_3'=$			−	−	$\alpha_3-\alpha_7-\alpha_{367}$	$-\alpha_2+\alpha_3-\alpha_{423}$	$-3\alpha-\alpha_3$	$-\alpha_1+\alpha_3-\alpha_{153}$
$\alpha_4'=$		−	−		$\alpha_4-\alpha_5-\alpha_{456}$	$-3\alpha-\alpha_4$	$-\alpha_2+\alpha_4-\alpha_{423}$	$\alpha_4-\alpha_6-\alpha_{456}$
$\alpha_5'=$	−		−		$-\alpha_4+\alpha_5-\alpha_{456}$	$\alpha_5-\alpha_6-\alpha_{456}$	$-\alpha_1+\alpha_5-\alpha_{153}$	$-3\alpha-\alpha_5$
$\alpha_6'=$	−	−			$-3\alpha-\alpha_6$	$-\alpha_5+\alpha_6-\alpha_{456}$	$\alpha_6-\alpha_7-\alpha_{367}$	$-\alpha_4+\alpha_6-\alpha_{456}$
(9) $\alpha_7'=$	−	−	−	−	$-\alpha_3+\alpha_7-\alpha_{367}$	$-\alpha_1+\alpha_7-\alpha_{147}$	$-\alpha_6+\alpha_7-\alpha_{367}$	$-\alpha_2+\alpha_7-\alpha_{257}$
$\alpha_{423}'=$		−	−	−	$\alpha_{423}-\alpha_{153}-\alpha_{147}-\alpha_{257}$	$-2\alpha_2-2\alpha_3$	$-2\alpha_2-2\alpha_4$	$\alpha_{403}-\cdots$
$\alpha_{153}'=$	−		−	−	$\alpha_{153}-\alpha_{423}-\alpha_{147}-\alpha_{257}$	$\alpha_{315}-\cdots$	$-2\alpha_1-2\alpha_5$	$-2\alpha_1-2\alpha_3$
$\alpha_{126}'=$	−	−		−	$-2\alpha_1-2\alpha_2$	$\alpha_{126}-\cdots$	$\alpha_{126}-\cdots$	$\alpha_{126}-\cdots$
$\alpha_{147}'=$	−	−	−	−	$\alpha_{147}-\alpha_{423}-\alpha_{153}-\alpha_{257}$	$-2\alpha_1-2\alpha_7$	$\alpha_{147}-\cdots$	$\alpha_{147}-\cdots$
$\alpha_{257}'=$	−	−	−	−	$\alpha_{257}-\alpha_{423}-\alpha_{153}-\alpha_{147}$	$\alpha_{257}-\cdots$	$\alpha_{257}-\cdots$	$-2\alpha_2-2\alpha_7$
$\alpha_{367}'=$	−	−	−	−	$-2\alpha_3-2\alpha_7$	$\alpha_{367}-\cdots$	$-2\alpha_6-2\alpha_7$	$\alpha_{367}-\cdots$
$\alpha_{456}'=$	−	−	−	−	$-2\alpha_4-2\alpha_5$	$-2\alpha_5-2\alpha_6$	$\alpha_{456}-\cdots$	$-2\alpha_4-2\alpha_6$.

The first four of these generators yield only the changes of sign indicated. In the fifth all of the terms are given. In the last three the terms indicated by $\cdots$ are to be supplied as is done in the fifth with the group of terms $\alpha_{423}, \alpha_{153}, \alpha_{147}, \alpha_{257}$.

(10) *Under the modular group of order* 8! 36 *the modular form* α *is one of a set of* 135 *conjugates, each one invariant under a subgroup of order* $2^6 \cdot 168$, *which are permuted in the same way as the* 135 *Göpel planes in the finite geometry under* G_{NC}.

To prove this we first observe that the $Z_1, \cdots, Z_7$ attached above to the points of a finite plane (mod. 2) may be permuted in 168 ways without destroying the linearity of triads. In fact the collineation group in the finite plane is of order 168. These 168 permutations are collineations in S_7 which, with corresponding collineations on $\alpha_1, \cdots, \alpha_7$ and $\alpha_{423}, \cdots, \alpha_{456}$, leave L^4, and therefore its manifold K_3^{24} of double points, invariant. This G_{168} on the α's must be the result of a period transformation on the Z's. It is a group which permutes the seven modular involutions $J_{000,ijk}$ $(i, j, k \neq 0, 0, 0)$ in all geometrically possible ways. These seven involutions, four of which appear in the first columns of (9), are themselves merely changes of sign of the α's. They are permutable and are subject to no other relation than that their product is the identity. These seven involutions generate an abelian G_{2^6}. The entire $G_{2^6} \cdot G_{168} = G_{2^6 \cdot 168}$ (in which G_{2^6} is an invariant subgroup) is determined in the finite geometry by the Göpel plane with seven points, $\{000, ijk\}_2$. Since (cf. **28**) there are 135 conjugate Göpel planes, $G_{2^6 \cdot 168}$ is precisely the subgroup of the modular group which corresponds to an invariant Göpel space and α is an obvious invariant of this subgroup. This theorem is fundamental for the transition from modular forms to algebraic invariants of the ternary quartic. We identify similarly (Chap. III) the leading coefficient of the Kummer surface with one of a conjugate set of 15 Göpel invariants ($p = 2$).

We give finally one of the 63 cubic relations on the coefficients α of L^4. One of the fixed spaces of $I_{000,100}$ is

$Z_1 = Z_5 = Z_6 = Z_7 = 0$. The section of L^4 by this S_3 is the Kummer quartic surface **32** (22) in variables Z, Z_2, Z_3, Z_4 and coefficients α.

The cubic relation **32** (24) is now

$$(11) \qquad \alpha^3 - \alpha(\alpha_2^2 + \alpha_3^2 + \alpha_4^2 - \alpha_{423}^2) + 2\alpha_2\alpha_3\alpha_4 = 0.$$

The 63 conjugates of this may be obtained by applying the generators (9).

Some further properties of K_3^{24} particularly with reference to the 64 M_2^{12}'s along which K_3^{24} is tangent to the 4-fold linear spaces of W_3^{24} (the extension of the conics in the tropes of the Kummer surface) are found in Chap. IV. Here and throughout the admirable account of Wirtinger[76] should be consulted.

34. Algebraic and abelian functions. If a canonical system of $2p$ cuts is drawn on the Riemann surface T defined by an algebraic curve, $F(x, y) = 0$, of order q and genus p, the p normal integrals of the first kind $u = u_1, \cdots, u_p$ have simultaneous periods on each of the $2p$ cuts which coincide in form with the $2p$ periods of $\vartheta(u)$ in **18** (6) and which satisfy the convergence condition of $\vartheta(u)$ ([67] § 15). These normal integrals are

$$(1) \qquad u_i = \int \varphi_i(x, y)\, dx / F'_y \qquad (i = 1, \cdots, p)$$

where $\varphi_1, \cdots, \varphi_p$ are properly chosen canonical adjoints of F (cf. **12**). The $p(p+1)/2$ moduli a_{ij} which arise in this way from $F(x, y) = 0$ must depend only on the $3p-3$ $(p>1)$ algebraic moduli of F. The theta functions with a period scheme thus defined by an algebraic curve are subject to $(p-2)(p-3)/2$ conditions and are called *abelian* theta functions. The single condition for $p = 4$, as given by Schottky, is found in **57** (15); for values of $p > 4$ they are as yet undetermined.

The theorem of Abel is of especial importance for geometric applications. This states that if α $(x = \alpha, y = \beta)$ is a fixed point of T and $x_1, \cdots, x_n$ a variable set of n points on T which on the curve $F(x, y) = 0$ is in a fixed g_r^n then

$$\sum_{i=1}^{i=n} {}_{\alpha}^{x_i}u \equiv m \tag{2}$$

where m is constant as the set x varies in g_r^n. The equation (2) stands for the system of p equations obtained by setting $u, m = u_j, m_j$ $(j = 1, \cdots, p)$. The possible variation of m indicated by the $\equiv$ sign is due to the possible variation in the paths of integration on T from α to x_i. Another version of the theorem states that if $x_1, \cdots, x_n$ and $x_1', \cdots, x_n'$ are two sets in the same g_r^n, i. e., the zeros and poles of a rational function, then

$$\sum_{i=1}^{i=n} {}_{x_i'}^{x_i}u \equiv 0. \tag{3}$$

Conversely (3) is a sufficient condition that the two sets are equivalent. If the complete g_r^n is not special (cf. 11), the p equations (2) define uniquely the position of the remaining $n - r = p$ points of a set of g_r^n when r of the points are given; if g_r^n is special and $n - r = p - i$ then only $p - i$ of the equations (1) are independent.

Let $u(o)$ be the normal integrals of the first kind taken with fixed lower limit at $\alpha = \alpha, \beta$ and with variable upper limit as in $o = x, y$ on T or $F(x, y) = 0$. Then the function,

$$\vartheta(u(o) - e) \tag{4}$$

with *parameters* $e_1, \cdots, e_p$, regarded as a function of position of o on T is known as the *Riemannian* theta function. The moduli of ϑ are of course those defined by the u's on T. This Riemannian theta function has, for general choice of the parameters e, p zeros on T at points $\eta_1, \cdots, \eta_p$ which are connected with the parameters e by the congruence

$$e \equiv \sum_{j=1}^{j=p} u(\eta_j) + k \tag{5}$$

where $k_1, \cdots, k_p$, the so-called Riemannian constants, are independent of the parameters e and the zeros η (cf. [41] IX

§§2–7; [67] §§27–32). If for example e is an odd half period P_m for which therefore $\vartheta_m(u) = E\cdot(\vartheta(u - P_m))$ is an odd function, then the zeros of $\vartheta(u(o) - P_m)$ or of $\vartheta_m(u(o))$ are, in addition to $o = \alpha$ at which $u(o) \equiv 0$, the $p-1$ points at which $F(x, y) = 0$ is tangent to one of the $2^{p-1}(2^p - 1)$ contact canonical adjoints, $\varphi_m(x, y)$. If however e is an even half period and l_α the tangent to F at α which meets F in $q-2$ further points ζ then the zeros of the even function $\vartheta_m(u(o))$ are found at the p contacts of one of the $2^{p-1}(2^p+1)$ adjoints, $\psi_m(x, y)$, of order $q-2$ which can be passed through the points ζ to touch F again at p points. If in the odd case $\varphi_m \cdot l_\alpha$ is set equal to $\psi_m(x, y)$ then for any two half periods P_m, P_n the function $\vartheta_m^2(u(o))/\vartheta_n^2(u(o))$, a uniform function of position on T, has the same poles and zeros as $\psi_m(x, y)/\psi_n(x, y)$ whence

$$(6) \quad \vartheta_m(u(o))/\vartheta_n(u(o)) = c_{m,n}\sqrt{\psi_m(x, y)}/\sqrt{\psi_n(x, y)}.$$

Such radicals of rational functions as can be expressed as uniform functions of $u(o)$ are called *root functions*. The general function of this character is discussed by Stahl ([67] p. 228 (IV)).

From (4) and (5) there follows that in general the function

$$(7) \quad \vartheta\left(u(o) - \sum_{j=1}^{j=p} u(\eta_j) - k\right)$$

vanishes only at the p points $o = \eta_1, \cdots, \eta_p$ on T. If however the p points η are on a canonical adjoint φ then ϑ in (7) vanishes identically, i. e., for the values u determined by all points o on T. In particular, for $o = \eta_p$, a point on the adjoint φ determined by arbitrary $\eta_1, \cdots, \eta_{p-1}$, ϑ also vanishes whence

$$(8) \quad \vartheta\left(\sum_{j=1}^{j=p-1} u(\eta_j) + k\right) = 0$$

for any $p-1$ points η.

The determination of the p upper limits η, when the lower limits δ and the U are given in

$$\sum_{j=1}^{j=p} {}_{\delta_j}^{\eta_j}u_i \equiv U_i \qquad (i = 1, \cdots, p), \tag{9}$$

is known as the *inversion problem.* Since ${}_{\delta_j}^{\eta_j}u_i \equiv u_i(\eta_j) - u_i(\delta_j)$ there follows, again from (4) and (5) that the required points η are the zeros of the function

$$\vartheta\left(u(o) - \sum_{j=1}^{j=p} u(\delta_j) - U - k\right).$$

The solution is in general unique. If however the U's are so chosen that this function vanishes identically, i. e., if for given points δ the U's are such as would arise in (9) from p points η on a canonical adjoint then one or more of the points $\eta_1, \cdots, \eta_p$ of the solution can be taken arbitrarily. It is by means of this inversion problem that the p remaining points of a set of g_r^n in (2) or (3) are determined when r points of the set are given. For the properties of rational and symmetric functions of the coördinates x, y of the p points η as abelian functions of the U's we refer to Stahl ([67] § 36).

CHAPTER III

GEOMETRIC APPLICATIONS OF THE FUNCTIONS OF GENUS TWO

The theta functions of genus two are necessarily of the hyperelliptic type defined by an algebraic curve on which there is a g_1^2 (cf. 11) with $2p+2=6$ branch points. There are two standard canonical forms for the hyperelliptic curve of genus p. To obtain the first, of greater geometric interest, the curve is transformed birationally in such wise that the g_1^2 is cut out by a pencil of lines, $x_0-tx_1=0$. It is then a curve H_p^{p+2} of order $p+2$ and genus p with a p-fold point at $O(0,0,1)$ whose equation is

$$H_p^{p+2} \equiv f_p\, x_2^2 + 2 f_{p+1}\, x_2 + f_{p+2} = 0,$$

where f_j is a binary form of order j in x_0, x_1. For given value of $t=x_0/x_1$ the two further intersections of the line on O are separated by the irrationality

$$\begin{aligned} z &= \{f_{p+1}^2(t,1) - f_p(t,1)\cdot f_{p+2}(t,1)\}^{1/2} \\ &= \{a_0(t-e_1)\cdots(t-e_{2p+2})\}^{1/2} = \{(\alpha\, t)^{2p+2}\}^{1/2}. \end{aligned}$$

This latter canonical form, $z^2=(\alpha\, t)^{2p+2}$, is the one commonly employed in the study of the hyperelliptic algebraic functions and their integrals, and of related transcendental matters. The birational moduli of the curve are then the absolute projective invariants of the binary $(2p+2)$-ic, $(\alpha\, t)^{2p+2}$.

35. The figure, P_6^1, of six points on a line. For $p=2$ the fundamental binary form of order $2p+2$ is the sextic

$$(1) \qquad (\alpha\, t)^6 = a_0(t-e_1)\cdots(t-e_6).$$

The group, $G_{m,1}$ (cf. 15), is, in the absence of Cremona transformations on the line, merely the permutation group, $g_{6!}$, of

8

the six points $e_1, \cdots, e_6$. Its invariants are therefore the usual projective invariants, rational or irrational, of the binary sextic. The complete system (cf. [17] I § 3) is most conveniently determined in terms of the 15 irrational Göpel invariants

$$(2) \qquad (ij)(kl)(mn) = (e_i - e_j)(e_k - e_l)(e_m - e_n).$$

The even subgroup of $g_{6!}$ is peculiar in that it contains two distinct systems of six conjugate ikosahedral subgroups. In the first system the subgroups have the individual e's as invariants. In the second system the subgroups arise from the six essentially distinct ways in which $e_1, \cdots, e_6$ can be identified with the six diagonals of an ikosahedron. The latter system is defined by a set of six irrational invariants,

$$A, B, C, D, E, F,$$

due originally to Joubert (for references cf. [15]). We give

$$(3) \qquad \begin{aligned} A = (25)(13)(46) + (51)(42)(36) + (14)(35)(26) \\ + (43)(21)(56) + (32)(54)(16). \end{aligned}$$

Here and hereafter we avoid lists of conjugate formulae by giving merely a sample along with the generating substitutions which produce the entire set. In this case the substitutions in cycle form are

$$(4) \qquad \begin{aligned} (12)&: \quad (AD)(BE)(CF), \\ (23456)&; \quad (ADBFE). \end{aligned}$$

An odd permutation also changes the sign of $A, \cdots, F$.

The ratios of these six irrational invariants are functions of three moduli and therefore subject to two relations which are

$$(5) \quad A + B + \cdots + F = 0, \quad A^3 + B^3 + \cdots + F^3 = 0.$$

Other irrational invariants are the following:

$$(6) \qquad \begin{aligned} A + B &= 4(51)(42)(36), \\ A - B &= 4[(53)(41)(26) - (34)(25)(16)]. \end{aligned}$$

If $A - B = 0$, the pairs e_1, e_5; e_2, e_4; e_3, e_6 are pairs of an involution. Moreover (cf. [17] I p. 168):

(7) *The invariants,* $\sum A^2$, $\sum A^4$, $\sum A^6$, $(\sum A^5)^2$, *and* $\prod (A - B)$ *constitute a complete system of rational integral invariants of the binary sextic,* P_6^1. *The square of* $\prod$ *is reducible.*

The identification of this complete system with a classic system is given in ([15] p. 317).

The 15 three term relations among the Göpel invariants (cf. **30** (14), **28** (11)) are now according to (6) a consequence of the one linear relation (5). This invariant theory of P_6^1 is based entirely on the 15 discriminant conditions $e_i - e_j$ of P_6^1.

36. The figure, Q_6^3, of six points in space and its congruent figures. In space S_3 with dual coördinates $y_0, \cdots, y_3$; $\zeta_0, \cdots, \zeta_3$ six points $q_1, \cdots, q_6$ are associated (cf. **16**) with the P_6^1 of the preceding section. The points Q_6^3 are on a unique cubic norm curve C^3 and have on C^3 parameters t which are projective to P_6^1. With properly chosen factors of proportionality for the points q there is a bilinear identity in ζ, t of the form

$$(1) \qquad (q_1 \zeta) \cdot (t e_1) + \cdots + (q_6 \zeta) \cdot (t e_6) \equiv 0.$$

The determinants formed for four points q are then proportional to those formed for the two complementary points e. With ε defined as in **29** (5)

$$(2) \qquad |q_i \, q_j \, q_k \, q_l| = \varrho \, \varepsilon_{ijklmn} \cdot (e_m \, e_n).$$

The projective invariants of Q_6^3 are composed of such determinants and are proportional to corresponding invariants of P_6^1.

We pass then at once to a study of the set Q_6^3 under regular Cremona transformation. If the cubic transformation A_{1234} has F-points at $q_1, \cdots, q_4$ and inverse F-points at $q_1', \cdots, q_4'$ and ordinary corresponding pairs, q_5, q_5' and q_6, q_6' then Q_6^3 and $Q_6'^3$ are congruent under A_{1234} (**15**). A surface of order γ_0 with multiplicity γ_i at q_i is transformed by A_{1234} into a surface of order γ_0' with multiplicity γ_i' at q_i' where (**15** (4))

$$(3)\quad A_{1234}: \begin{array}{ll} \gamma_i' = \gamma_i + L_{1234} & (L_{1234} = 2\gamma_0 - \gamma_1 - \cdots - \gamma_4), \\ \gamma_j' = \gamma_j & (i = 0, 1, \cdots, 4;\ j = 5, 6). \end{array}$$

The linear group, $g_{6,3}$, with integer coefficients is generated by this element A_{1234}, and by permutations of $\gamma_1, \cdots, \gamma_6$ which will be written in cycle form. The group has the invariant forms

$$(4)\quad L = 4\gamma_0 - \gamma_1 - \cdots - \gamma_6, \quad Q = 2\gamma_0^2 - \gamma_1^2 - \cdots - \gamma_6^2.$$

By combining generators A_{ijkl} only three types of regular transformation with six or fewer F-points are obtained ([17] II p. 363), namely

$$(5)\quad T^3: \begin{array}{c|ccc} & & 4 & 2 \\ \hline & 3 & -1 & 0 \\ 4 & 2 & 0,-1 & 0 \\ 2 & 0 & 0 & 1,0 \end{array}; \qquad T^5: \begin{array}{c|ccc} & & 2 & 4 \\ \hline & 5 & -2 & -1 \\ 2 & 4 & -2,-1 & -1 \\ 4 & 2 & -1 & -1,0 \end{array};$$

$$T^7: \begin{array}{c|cc} & & 6 \\ \hline & 7 & -2 \\ 6 & 4 & -2,-1 \end{array}.$$

The numbers without an array are the numbers of F-points (direct at the top and inverse at the left) of like multiplicities. Of the columns within the array the first gives the order and multiplicities of the transform of a plane section, and the others the same data for the P-surfaces which correspond respectively to directions about the F-points. The notation i, j within an array represents a square matrix with principal diagonal elements i and other elements j. A precise definition of the three types is

$$(6)\quad \begin{array}{l} T^3 = A_{1234}; \qquad T^5 = A_{1234}(56)\cdot A_{1256}(34); \\ T^7 = A_{1234}(56)\cdot A_{1256}(34)\cdot A_{3456}(12). \end{array}$$

The symmetric type T^7 has the explicit expression:

$$(7)\quad T\colon \gamma_0' = -\gamma_0 + 2L, \quad \gamma_i' = -\gamma_i + L \qquad (i = 1, \cdots, 6).$$

Corresponding to the ways in which the groups of F-points can be selected from Q_6^3 there is one type T^1 (a collineation), 15 types T^3, 15 types T^5, and one type T^7. The 6! permutations of the γ's combined with each of these types yields the group $g_{6,3}$ of order 6! 32.

The surface of order γ_0 and multiplicities γ_i may sometimes be represented more conveniently by the linear polar of the value system γ with respect to the invariant quadratic form Q in (4). Thus the form

$$(8) \qquad 2c_0\gamma_0 - c_1\gamma_1 - \cdots - c_6\gamma_6$$

represents a surface of order c_0 and multiplicities c_i. In particular the invariance of L indicates that the web of quadrics on Q_6^3 passes into the web on the congruent set $Q_6'^3$. Also the 32 types of Cremona webs noted above are represented by the forms:

$$(9) \qquad \begin{aligned} T^1 &\equiv 2\gamma_0; \qquad T^3 \equiv 6\gamma_0 - 2\gamma_1 - \cdots - 2\gamma_4, \\ T^5 &\equiv 10\gamma_0 - 4\gamma_1 - 4\gamma_2 - 2\gamma_3 - \cdots - 2\gamma_6; \\ T^7 &\equiv 14\gamma_0 - 4\gamma_1 - \cdots - 4\gamma_6. \end{aligned}$$

The P-surfaces of Q_6^3, read off from the colums of the arrays (5), are 32 in number and are represented by

$$(10) \qquad \begin{aligned} P(i)^0 &\equiv \gamma_i; \qquad P(ijk)^1 \equiv 2\gamma_0 - \gamma_i - \gamma_j - \gamma_k; \\ P(i^2jklmn)^2 &\equiv 4\gamma_0 - 2\gamma_i - \gamma_j - \cdots - \gamma_n. \end{aligned}$$

These divide into 16 pairs, each pair a quadric on Q_6^3,

$$(11) \qquad \begin{aligned} P_i &= P(i)^0 \cdot P(i^2jklmn); \\ P_{ijk} &= P_{lmn} = P(ijk)^1 \cdot P(lmn)^1, \end{aligned}$$

which are permuted as entities under $g_{6,3}$.

The equations (7) show that T^7, as a collineation on the γ's, is a harmonic perspectivity with linear space L of fixed points and center at the pole of L as to Q. It is then an invariant element of $g_{6,3}$ which with the identity makes up an invariant subgroup h_2 of $g_{6,3}$. The same equations (7) show that the quadrics on Q_6^3 are transformed into those on $Q_6'^3$ in such wise that each of the 16 quadrics (11) passes into the like quadric on $Q_6'^3$. From this it might be inferred that the two sets Q_6^3, $Q_6'^3$ congruent under T^7 are projective. The same inference is an immediate result of the defining property, noted in **15**, of the cubic transformation A_{1234}; namely that if two ordered sets Q_6^3, $Q_6'^3$ are congruent under A_{1234}, they are projective under $A_{1234}(56)$. For T^7 is expressed in (6) as a product of three elements of this latter type. The sets Q_6^3, $Q_6'^3$ congruent under T^7 may then be taken as superposed and T^7 is then an involution I^7 which leaves every quadric on Q_6^3 unaltered. Also the net of quadrics on y is invariant whence I^7 transforms y into y' where Q_6^3, y, y' are the eight base points of a net. If y' coincides with y in some direction, one quadric of the net on y must have a node at y; conversely if a quadric has a node at y, the eighth base point y' of the net on Q_6^3, y is at y. Hence the locus of fixed points of I^7 is the Weddle quartic surface, $W(y)$, the jacobian of the web of quadrics on Q_6^3, the locus of nodes of quadrics of the web i. e. the locus of points y from which Q_6^3 projects into a set R_6^2 on a conic. From the Clebsch transference principle the equation of this locus is

$$(12) \quad W(y) = \begin{vmatrix} |135\,y|\,|425\,y| & |145\,y|\,|235\,y| \\ |136\,y|\,|426\,y| & |146\,y|\,|236\,y| \end{vmatrix} = 0.$$

The equation (12) shows that $W(y)$ contains the line $\overline{q_1 q_3}$ and by reason of symmetry all 15 lines $\overline{q_i q_j}$. The tangent plane of $W(y)$ at q_i must vanish since it can not contain the five lines $\overline{q_i q_j}$. Hence $W(y)$ has a node at q_i with tangent cone $P(i^2jklmn)^2$. Also, as a nodal locus, $W(y)$ must contain the norm cubic C^3 on Q_6^3 and the ten double lines of the pairs of planes $P(ijk)^1$, $P(lmn)^1$.

The involution, $A_{1234}(56) = (56)A_{1234}$, shares with I^7 the property that congruence of Q_6^3, $Q_6'^3$ under it implies projectivity in the identical order. When Q_6^3, $Q_6'^3$ coincide, $A_{1234}(56)$ is the cubic Cremona involution with superposed

F-points at $q_1, \cdots, q_4$ which interchanges q_5 and q_6. As elements of $g_{6,3}$ the 15 involutions of this type satisfy the relations,

$$A_{1234}(56) \cdot A_{1235}(46) = A_{1235}(46) \cdot A_{1234}(56) = A_{1236}(45),$$

$$(13) \quad A_{1234}(56) \cdot A_{1256}(34) = A_{1256}(34) \cdot A_{1234}(56)$$
$$= I^7 \cdot A_{3456}(12) = A_{3456}(12) \cdot I^7.$$

These are checked most readily by noting that any two equal products have the same order and the same effect upon the six P-loci, $P(i)^0$.

The 32 elements 1, I^7, $A_{ijkl}(mn)$, $I^7 A_{ijkl}(mn)$ constitute according to (13) and (6) an abelian subgroup of $g_{6,3}$ for whose elements congruence implies projectivity. This subgroup, h_{32}, is therefore an invariant subgroup of $g_{6,3}$. When the congruent sets are brought into coincidence there results:

(14) *The six nodes Q_6^3 of the Weddle surface $W(y)$ define an abelian Cremona group h_{32} in space which leaves $W(y)$ unaltered. On $W(y)$ the element I^7 of h_{32} is the identical transformation and the pair of elements, $T^3 = A_{1234}(56)$ and $T^5 = I^7 \cdot A_{1234}(56)$, has the same effect. The resulting h_{16} on the points of $W(y)$ is isomorphic with the group $G_{2^{2p}}$ (cf.* **32**) *of additive half periods* ($p = 2$).

If indeed $A_{1234}(56)$ is identified with $u' = u + P_{56}$ where P_{56} is a half period in the basis notation, the multiplicative relations (13) are isomorphic with the half period relations, $P_{56} + P_{46} = P_{45}$, $P_{56} + P_{34} = P_{12}$.

The elements of the invariant subgroup, h_{32}, of $g_{6,3}$ account for each type of Cremona transformation. The factor group $h_{6!}$ of h_{32} with respect to $g_{6,3}$ is therefore isomorphic with the permutation group of $\gamma_1, \cdots, \gamma_6$ or of the points $q_1, \cdots, q_6$. This factor group can be represented as the Moore[47] cross-ratio Cremona group, $G_{6,3}$, in S_3. With ordered $q_1, \cdots, q_6$ the first five points are taken at an ordered basis B in S_3 and the sixth at y. If the points are taken in permuted order π, say $q_{i_1}, \cdots, q_{i_6}$, and if the first five are then transformed linearly into ordered B, the sixth is transformed into the point $y' = \pi(y)$. The 6! points y' so obtained will each with B constitute a $Q_6'^3$ which is projective in some order π to $Q_6^3 = B, y$. A simple algebraic apparatus for the representation of this group is the coördinate system $y_1, \cdots, y_5$ where

$$y_1 + \cdots + y_5 = 0. \tag{15}$$

The base B then has points with coördinates:

$$q_1 = -4, 1, 1, 1, 1; \cdots; q_5 = 1, 1, 1, 1, -4.$$

If the parameters of Q_6^3 on C^3 are the $e_1, \cdots, e_6$ of **35** and if this binary sextic be transformed linearly in such wise that e_6 becomes ∞ and that the sum of the transforms of $e_1, \cdots, e_5$ is zero then these five transforms are the coördinates $y_1, \cdots, y_5$ of the point which with B forms a Q_6^3 associated to P_6^1 (cf. [15]). The Göpel invariants (**35** (2)) then yield Göpel covariants of $G_{6,3}$ which are quadrics on B, namely

$$(12)(34)(56) = (y_1 - y_2)(y_3 - y_4). \tag{16}$$

The six irrational invariants $A, \cdots, F$ of P_6^1 yield irrational covariants $A(y), \cdots, F(y)$ of $G_{6,3}$ where

$$A(y) = (y_2 - y_5)(y_1 - y_3) + \cdots + (y_3 - y_2)(y_5 - y_4);\ \text{etc.} \tag{17}$$

The six quadrics $A(y), \cdots, F(y)$ on B, subject to the relations

$$A(y) + \cdots + F(y) \equiv 0, \qquad A^3(y) + \cdots + F^3(y) \equiv 0, \tag{18}$$

define the linear system of ∞^4 quadrics on B. The linear relation is therefore to be expected. An interpretation of

$$A^2(y) \cdot A(y') + \cdots + F^2(y) \cdot F(y') = 0 \tag{19}$$

is furnished by polarizing the cubic identity to get

$$A^2(y) \cdot A(y, y') + \cdots + F^2(y) \cdot F(y, y') \equiv 0.$$

This shows that for given y, the quadric (19) in variables y' has a node at y. Hence (cf. [15] § 2).

(20) *The equation* (19) *for given* y *and variable* y' *is the quadric cone with node at* y *and on* B; *for given* y' *and variable* y *it is the Weddle quartic surface with nodes at the* $Q_6^3 = B, y'$.

We have thus identified the invariant h_{32} of the linear group, $g_{6,3}$, as well as its factor group $h_{6!}$ with Cremona groups in S_3. There is however no Cremona group in S_3 which is simply isomorphic with $g_{6,3}$ because of the projectivity of congruent sets Q_6^3.

The behavior of curves under A_{1234} leads to the dual form of $g_{6,3}$. A curve of order c_0 with multiplicities c_i at q_i becomes under A_{1234} a curve of order c_0' with multiplicities c_i' at q_i' where

$$
(21) \qquad \begin{gathered} c_0' = c_0 + 2M_{1234}, \quad c_i' = c_i + M_{1234}, \quad c_j' = c_j \\ (i = 1, \cdots, 4; \quad j = 5, 6; \quad M_{1234} = c_0 - c_1 - \cdots - c_4). \end{gathered}
$$

This substitution is the transposed substitution of the involution A_{1234} in (3), and therefore its dual, with invariant forms

$$
(22) \qquad 2c_0 - c_1 - \cdots - c_6, \quad c_0^2 - 2c_1^2 - \cdots - 2c_6^2.
$$

The situation is expressed most simply by observing that a curve of order c_0 and multiplicities c_i is transformed under regular Cremona transformation as the form

$$
c_0\gamma_0 - c_1\gamma_1 - \cdots - c_6\gamma_6
$$

is transformed under $g_{6,3}$. The F-curves of the second kind (cf. [52] p. 198) for the transformations T with F-points in Q_6^3 are sixteen in number and are represented by the forms:

$$
(23) \qquad f = 3\gamma_0 - \gamma_1 - \cdots - \gamma_6, \quad f_{ij} = \gamma_0 - \gamma_i - \gamma_j.
$$

They are then respectively C^3 and the 15 lines $\overline{q_i q_j}$. They are transformed among themselves by the elements of $g_{6,3}$ with a change of sign if a particular curve is an F-curve of the corresponding Cremona transformation. For example A_{1234} transforms f_{12} into $-f_{34}$, f_{15} into f_{15}, and f_{56} into f. The occurence of $-f_{34}$ indicates that f_{12} is an F-curve of the second kind of A_{1234}.

We shall have occasion in the next chapter to consider the reduction in the order of the transform of a curve due to incidence with these F-curves of the second kind.

37. Schottky's parametric expression of the Weddle quartic surface. An elegant parametric expression of $W(y)$ in terms of theta functions of genus two has been given by Schottky[61]. We reproduce this in part. The eight theta products of second order and like characteristic (P_{56} in the basis notation) are (cf. 30 (5)):

$$\vartheta_1(u)\,\vartheta_{156}(u), \quad \vartheta_2(u)\,\vartheta_{256}(u), \quad \vartheta_3(u)\,\vartheta_{356}(u), \quad \vartheta_4(u)\,\vartheta_{456}(u);$$
$$\vartheta_{125}(u)\,\vartheta_{345}(u), \quad \vartheta_{135}(u)\,\vartheta_{245}(u), \quad \vartheta_{145}(u)\,\vartheta_{235}(u), \quad \vartheta_5(u)\,\vartheta_6(u);$$

of which the first four are odd, the last four even. In either set of four any three are linearly related (cf. 30 (8)). Hence the 20 odd functions,

$$(1) \qquad F_{ijk} = \vartheta_i(u)\,\vartheta_j(u)\,\vartheta_k(u)\,\vartheta_{ijk}(u),$$

are such that, in any set of four like

$$F_{156},\ F_{256},\ F_{356},\ F_{456},$$

any three are linearly related. Thus all 20 can be expressed linearly in terms of the four,

$$F_{456},\ F_{356},\ F_{346},\ F_{345};$$

moreover each F_{ijk} which contains a subscript 6 can be expressed linearly in terms of $F_{456}, F_{356}, F_{346}$. Only four of the twenty are linearly independent and these four may be equated to independent linear functions of $y = y_0, \cdots, y_3$. Each F is then equated to a plane in $S_3(y)$. Since the planes $F_{456}, F_{356}, F_{346}$ meet in a point, say q_6, each F with subscript 6 is a plane on q_6. Thus from the symmetry of (1) there exists a set Q_6^3 in S_3 such that F_{ijk} is equated to the plane on q_i, q_j, q_k. If four independent equations of this kind are solved for the coördinates y in terms of the four F_{ijk} then for variable $\pm u$ the point y runs over a surface. An equation of this surface is

$$F_{135}\,F_{425}\,F_{146}\,F_{236} - F_{136}\,F_{426}\,F_{145}\,F_{235} = 0.$$

This is identically satisfied by (1) due to $\vartheta_{ijk}(u) = \vartheta_{lmn}(u)$. Since F_{ijk} is a constant times the determinant $|ijky|$ of the coördinates of the four points q_i, q_j, q_k, y, this equation may be written

$$
(2) \qquad \begin{aligned} &|135y|\,|425y|\,|146y|\,|236y| \\ &- c\,|136y|\,|426y|\,|145y|\,|235y| = 0. \end{aligned}
$$

It may also have the alternative form

$$
\begin{aligned} &|135y|\,|425y|\,|126y|\,|436y| \\ &- c'\,|136y|\,|426y|\,|125y|\,|435y| = 0. \end{aligned}
$$

On subtracting these two and using the identity

$$
|146y|\,|236y| - |126y|\,|436y| = |136y|\,|246y|,
$$

there results, after factoring out $|136y|\,|246y|$,

$$
|135y|\,|425y| = -c\,|145y|\,|235y| + c'\,|125y|\,|435y|
$$

whence $c = c' = 1$. Thus the surface (2) is the Weddle surface $W(y)$ (cf. 36 (12)).

Schottky goes on to show that the theta squares are proportional to quadratic functions of the F's. In the second set of four functions above $\vartheta_5(u)\,\vartheta_6(u)$ is linear in $\vartheta_{135}(u)\,\vartheta_{245}(u)$, and $\vartheta_{145}(u)\,\vartheta_{235}(u)$. Multiplying all three by $\vartheta_1(u) \cdots \vartheta_4(u)\,\vartheta_5^2(u)$ and setting

$$
(3) \qquad \vartheta_1(u) \cdot \vartheta_2(u) \cdot \cdots \cdot \vartheta_6(u) = \Pi
$$

the linear relation becomes

$$
\Pi \cdot \vartheta_5^2(u) = a F_{135} F_{245} + b F_{145} F_{235}.
$$

On the right there is a quadric with node at q_5 and simple points at $q_1, \cdots, q_4$. Due to the symmetry on the left this quadric must pass through q_6 also. If then we set

$$
(4) \qquad \Pi \cdot \vartheta_5^2(u) = G_5,
$$

G_5 is the quadric cone $P(5^2\,12346)^2 \cdot P(5)^0$ (36 (11)). Again if we set

(5) $$G_{123} = G_{456} = F_{123}\,F_{456}$$

then

(6) $$\Pi \cdot \vartheta_{123}^2(u) = G_{123}$$

where G_{123} is the pair of planes $P(123)^1 \cdot P(456)^1$. In (4) and (6) the theta squares are expressed as quadrics on the nodes Q_6^3 of $W(y)$. But the theta squares are themselves linear in the coördinates $Z(u)$ of a point on the Kummer surface (32). Hence K is the map of the Weddle surface by the web of quadrics on its nodes Q_6^3.

In this rational transformation from space $S_3(y)$ to space $S_3(Z)$ in which planes in $S_3(Z)$ correspond to quadrics on Q_6^3 in $S_3(y)$, the net of planes on Z corresponds to a net of quadrics on Q_6^3 which contains Q_6^3 and a further pair of the involution I^7. The transformation is then 2 to 1 in general but becomes birational on $W(y)$, $K(Z)$ since on $W(y)$ the members of a pair of I^7 coalesce. If in the quartic equation $K(Z) = 0$ the coördinates Z are replaced by their values as quadrics on Q_6^3, the square of $W(y)$ must be obtained. Hence

(7) *The square of the jacobian, $W(y)$, of a web of quadrics on six points is a quartic polynomial in four quadrics of the web. This quartic polynomial is the equation of a Kummer surface birationally equivalent to $W(y)$.*

The f-curve, C^3, of Q_6^3 is on a net of quadrics of the web. It corresponds therefore to a point on K. Since the cones $G_1, \cdots, G_6$ in (4) all contain C^3, and the corresponding theta squares all vanish for $u = 0$, this f-curve is determined on $W(y)$ by $u = 0$ and corresponds to the node $u = 0$ on K. Similarly the 15 f-curves $\overline{q_i\,q_j}$ on $W(y)$ correspond to the 15 nodes $u = P_{ij}$ on K. It is clear also from (6) that the conics in the even tropes of K correspond on $W(y)$ to the double lines of the pairs of planes on Q_6^3 and again from (4) that the conics in the odd tropes correspond to the directions on $W(y)$ at the nodes.

38. Cremona theory of the hyperelliptic curve H_p^{p+2}. The hyperelliptic plane curve, H_p^{p+2}, of order $p+2$ and genus p with p-fold point at $O(0, 0, 1)$ and equation

$$(1) \qquad H_p^{p+2} \equiv f_p\, x_2^2 + 2 f_{p+1}\, x_2 + f_{p+2} = 0,$$

has $3p-1$ absolute projective constants. Indeed the $(p+2) \times (p+5)/2$ constants in the general $(p+2)$-ic are reduced by the $p(p+1)/2$ conditions for the p-fold point at O and the six constants in a collineation C which leaves O fixed. The forms f_p, f_{p+1}, f_{p+2} contain $3p+6$ coefficients one of which is a factor of proportionality and six of which may be removed by C. The curve H_p^{p+2} is birationally equivalent to

$$(2) \quad z^2 = (\alpha\, t)^{2p+2} = f_p f_{p+2} - f_{p+1}^2 \quad (t_0 : t_1 = x_0 : x_1),$$

whose moduli are the $2p-1$ independent cross-ratios of the binary $(2p+2)$-ic. Hence

(3) *There are ∞^p curves H_p^{p+2} all of which are projectively distinct but birationally equivalent.*

Before proving that these projectively distinct types are equivalent under Cremona transformation we consider the Cremona transformations T under which H_p^{p+2} is invariant. The curve has a unique g_1^2 cut out by lines t on O and the coincidences of g_1^2 occur at the $2p+2$ *branch points* $r_1, \cdots, r_{2p+2}$ on H_p at which the *branch lines*, or tangents to H_p from O, touch. If H_p is invariant under T, the g_1^2 is also invariant, and therefore the $2p+2$ branch lines also. For $p \geqq 2$ the branch lines are in general self projective in their pencil only in the identical order whence T leaves every line t on O unaltered. Then T effects on a particular line t a projectivity π which either (a) interchanges the two points of H_p on T or (b) leaves each of the two points of H_p unaltered. In case (a) π is involutorial with two distinct fixed points whose locus for variable t is a curve H_q with equation,

$$(4) \qquad H_q^{q+2} \equiv g_q\, x_2^2 + 2 g_{q+1}\, x_2 + g_{q+2} = 0.$$

Since the pairs cut out on t by H_p, H_q are harmonic,

$$(5) \qquad f_p\, g_{q+2} - 2 f_{p+1}\, g_{q+1} + f_{p+2}\, g_q \equiv 0.$$

As thus defined by H_q, $T \equiv IH_q$ is involutorial. The branch points of H_p are fixed points of IH_q and therefore are on H_q; from the symmetry of (5) the branch points of H_q are on H_p. The $2q+2$ branch points of H_q are the simple F-points of IH_q whose P-curves are the lines t on them. The point O is an F-point of order $q+1$ of IH_q whose P-curve,

$$(6) \qquad L_q \equiv g_q\, x_2 + g_{q+1} = 0,$$

is of order $q+1$ with q-fold point at O and simple points at the $2q+2$ branch points. Transformations of this type were discovered by de Jonquières (cf. [52] p. 81; [35] p. 98). The $2p+2+2q+2$ intersections of H_p and H_q outside O are determined from the eliminant of (1) and (4) with respect to x_2, namely

$$4(f_p f_{p+2} - f_{p+1}^2)(g_q\, g_{q+2} - g_{q+1}^2) \\ - (f_p\, g_{q+2} - 2 f_{p+1}\, g_{q+1} + f_{p+2}\, g_q)^2 = 0.$$

Hence

(7) *The two curves, H_p (1) and H_q (4), with common multiple point O, are each on the branch points of the other if* (5) *is satisfied. Then each curve is invariant under the Jonquières involution for which the other is the locus of fixed points. The involutions IH_p and IH_q are permutable and form with the identity and their product, IH_{p+q+1} with fixed curve* (8), *a four group.*

The equation of H_{p+q+1}, the locus of the common harmonic pair of (1) and (4) is

$$(8) \qquad H_{p+q+1} \equiv \begin{vmatrix} f_p & f_{p+1} & f_{p+2} \\ g_q & g_{q+1} & g_{q+2} \\ 1 & -x_2 & x_2^2 \end{vmatrix} = 0.$$

For special relative values of the coefficients of the binary forms f, g in x_0, x_1, linear factors in x_0, x_1 may separate out and the order of the product IH_{p+q+1} be correspondingly reduced.

For given H_p the curves H_q which satisfy (5) lie in a linear system and cut out on H_p a g^{2q+2}_{2q+2-p}. The existence of this system for $q \geqq (p-2)/2$ furnishes conditions on O and the $2p+2$ branch points r. This set, R^2_{2p+3}, which in general has $4p-2$ absolute projective constants, has in the present case only the $3p-1$ inherent in H_p and therefore is subject to $p-1$ conditions. If p is odd $(p=2k+1)$ there is a pencil of curves H_k with k-fold point O and simple points r. Of these simple base points k are determined by the others and the $2k=p-1$ conditions thus obtained. If however p is even $(p=2k+2)$ there is a unique curve H_k and a net H_{k+1} on R^2_{2p+3}. If O and $3k+5$ of the points r are given, the web of curves H_{k+1} on O and these points (which contains a pencil made up of H_k and an arbitrary line t) cuts H_k in a g^{k+1}_1. If then a further point r' is given on H_k the k remaining points r are determined as the intersections of H_k with a proper H_{k+1} of the web on r'. Thus the set is subject to $2k+1=p-1$ conditions. Hence (cf. [3]).

(9) *The $p-1$ conditions on the planar set of points R^2_{2p+3} consisting of the $2p+2$ branch points $r_1, \cdots, r_{2p+2}$ of H^{p+2}_p and of $r_{2p+3}=O$ are: when $p=2k+1$, that there be a pencil of curves H_k on R^2_{2p+3} with k-fold point at O; and when $p=2k+2$ that an H_k and an H_{k+1} with multiple point at O meet in R^2_{2p+3}.*

There are *special* hyperelliptic curves H_p for which curves H_q $(q<(p-2)/2)$ satisfying (5) will exist. For example H_{p+q+1} in (8) is such a special curve if $q \neq p$ when $p+q$ is even.

Returning to the case (b) above in which H_p is the locus of fixed points of T, the effect of T upon t is determined when the correspondent of O on t is located. The locus of these correspondents for variable t is a rational curve of type L_q in (6) and T is a Jonquières transformation of order $q+2$. Since the q directions at O are self corresponding they must coincide with those of H_p at O and L_q has the form

$$(10) \qquad L_q = f_p\, g_{q-p}\, x_2 + g_{q+1} = 0.$$

The $2q+2$ simple F-points of T are the intersections of L_q and H_p outside the $qp+p$ absorbed at O. For particular choices of L_q, T is periodic; e. g. if $L_q = f_p x_2 + f_{p+1} = 0$, T is the involution $I H_p$. Hence

(11) *The infinite discontinuous group of Jonquières transformations which leaves H_p unaltered has an invariant abelian subgroup of index two each element of which has H_p as a curve of fixed points and is determined by a curve L_q in* (10). *The remaining elements all are involutorial of type $I H_q$ in* (7).

Let H_p and H_p' be two projectively distinct curves which are equivalent under a birational transformation B'. Since B' followed by a properly chosen collineation will superpose O' and O as well as the two sets of $2p+2$ branch lines, this situation will be assumed. Then B' transforms a point of H_p on a line t into a point of H_p' on this same line. Let $\alpha_1, \cdots, \alpha_p$ be the points on H_p which pass into the p-ad of points on H_p' at O. Through the points α pass a curve L_q, of sufficiently high order, to meet H_p outside O in $2q+2$ further points β. Let the line t cut H_p in γ_1, γ_2; H_p' in γ_1', γ_2'; and L_q in δ. On t there is a projectivity which sends $\gamma_1, \gamma_2, \delta$ into γ_1', γ_2', O. For variable t these projectivities define a Jonquières transformation J_q' of the plane, which is of order $q+2$ since L_q passes into directions at O. The projectivity becomes illusory only for lines t on the $2q+2$ points β. These points and O are the F-points of J_q'. Any other choice of rational curve $L_{q'}$ on $\alpha_1, \cdots, \alpha_p$ would lead to a $J_{q'}'$ such that $J_q' \cdot J_{q'}'^{-1}$ would leave H_p invariant point by point. Thus $J_{q'}'$ is the product of J_q' and an element of the invariant subgroup of (10). Hence

(12) *Two curves, H_p and H_p', which are birationally equivalent are equivalent under Jonquières transformation of the plane. The ∞^p projectively distinct curves H_p' which are birationally equivalent to H_p are determined by the ∞^p p-ads $\alpha_1, \cdots, \alpha_p$ on H_p.*

Two points which are paired in g_1^2 will be called "superposed points"; and the superposed points of a k-ad of points will form the "superposed k-ad." Since under IH_q in (7) the p-ad at O and its superposed p-ad on H_p are interchanged, any two superposed p-ads on H_p determine the same projective type H'_p.

The planar set R^2_{2p+3} made up of the branch points $r_2, \cdots, r_{2p+2}$ of H_p and of O is projectively a special set subject to the $p-1$ conditions implied by its situation with respect to H_p. It is therefore natural to develop the theory of sets congruent to it under Cremona transformation with particular reference to those transformations which leave the form of these conditions unaltered, i. e. which convert H_p into H'_p. These are the Jonquières transformations generated by A_{O12} and permutations of $r_1, \cdots, r_{2p+2}$. With superposed sets the elements

(13) $$I_{12} \equiv A_{O12} \cdot (12) = (12) \cdot A_{O12}$$

are involutorial and satisfy the relations

(14) $$\begin{aligned} I_{12}\, I_{13} &= I_{13}\, I_{12} = I_{23}, \\ I_{12}\, I_{34} &= I_{34}\, I_{12} = I_{13}\, I_{24} = \cdots = A_{1234}, \quad \text{etc.} \end{aligned}$$

Including the identity the

$$\binom{2p+2}{0} + \binom{2p+2}{2} + \cdots + \binom{2p+2}{2p+2} = 2^{2p+1}$$

elements $I_{1,2,\cdots,2k}$ constitute an abelian $g_{2^{2p+1}}$. The element $I_{1,2,\cdots,2p+2}$ is IH_p under which the set R^2_{2p+3} is congruent to itself. Hence

(15) *Under Jonquières transformation the planar set R^2_{2p+3} defined by H_p is congruent to 2^{2p} projectively distinct sets. The two sets congruent to R^2_{2p+3} under $I_{1,2,\cdots,2k}$ and $I_{2k+1,\cdots,2p+2}$ are projective.*

Under such transformation the F-curves of the set, i. e. the curves which correspond to directions about the points, divide into two conjugate sets. The first set contains 2^{2p+1} members; one, the directions at O, and the others, the curves of type $L_k(r_1, \cdots, r_{2k+2})$ $(k = 1, \cdots, p)$. The 2^{2p+1} divide into the 2^{2p} pairs, $L_k(r_1, \cdots, r_{2k+2})$ and $L_{p-k-1}(r_{2k+3}, \cdots, r_{2p+2})$.

(16) *Each curve of the pair* L_k, L_{p-k-1} *meets* H_p *in the same* p*-ad of points outside* R^2_{2p+3}. *The* 2^{2p} p*-ads thus defined on* H_p *represent as in* (12) *the* 2^{2p} *projectively distinct curves* H'_p *whose sets* R'^2_{2p+3} *are congruent to* R^2_{2p+3} *under Jonquières transformation.*

For, L_k, L_{p-k-1} are interchanged by IH_p and the intersections of either with H_p outside R^2_{2p+3} are on both. A particular pair, L_{-1}, L_p, is the directions at O and the unique curve L_p.

The second conjugate set of F-curves divide into $2p+2$ pairs each pair consisting of (a) directions at r_i and (b) the line Or_i.

The arithmetic group, $j_{2p+3,2}$, attached to these Jonquières transformations has the order $(2p+2)!\ 2^{2p+1}$. The group has the invariant involutorial element defined by IH_p. The group also has the invariant abelian subgroup $g_{2^{2p+1}}$ mentioned above. The factor group of order $(2p+2)!$ appears as the permutation group of the conjugate set of $2p+2$ pairs of F-curves in which A_{O12} effects the transposition (12).

The discriminant conditions of the set R^2_{2p+3} also divide into two conjugate sets under Jonquières transformation. In the first conjugate set there are $(2p+2)(2p+1)/2$ pairs $\delta_{ij}=0$, $\delta(O\,r_i\,r_j)^1=0$, indicating respectively that r_i, r_j coalesce, and that O, r_i, r_j are on a line. In either case H_p acquires a node at $r_i=r_j$ and the genus p is reduced. These are, of course, the discriminant conditions of the binary $(2p+2)$-ic in (2).

The second conjugate set arises from the condition that r_i coincides with O, i. e. that one branch of H_p at O has a flexpoint at O. The conjugates of this are of the form $\delta(O^k r_1,\cdots,r_{2k+3})^{k+1}=0$ $(k=0,\cdots,p-1)$ which represents the condition that there exists a curve L_k on $2k+3$ rather than $2k+2$ of the branch points.

(17) *The set* R^2_{2p+3} *has under Jonquières transformation a conjugate set of* 2^{2p+1} *discriminant conditions which divide into* 2^{2p} *pairs*

$$\delta(O^k r_1,\cdots,r_{2k+3})^{k+1}=0,$$
$$\delta(O^{p-k-2} r_{2k+4},\cdots,r_{2p+2})^{p-k-1}=0,$$

either of which implies the other.

In fact under IH_p the one condition passes into the other. The extreme case $\delta(O^{p-1}\, r_1, \cdots, r_{2p+1})^p = 0$ pairs with the coincidence of r_{2p+2} and O.

39. Transcendental theory of H_p^{p+2}. Application to the Weddle surface ($p = 2$). The points x of H_p^{p+2} are in one-to-one correspondence with the points z, t of the Riemann surface F defined by 38 (2). To the branch points $r_1, \cdots, r_{2p+2}$ of H_p there correspond the branch points $e_1, \cdots, e_{2p+2}$ of F; and to the p-ad of points of H_p which coalesce at O there corresponds p distinct points on F. It is convenient to denote the point z, t on F by its corresponding point x on H_p.

Let p points $x_1, \cdots, x_p$ be selected on F as well as a path of integration to each from the fixed branch point r_1. If v_i is one of the p normal integrals of the first kind and we set

$$(1) \qquad {}_{r_1}^{x_1}v_i + {}_{r_1}^{x_2}v_i + \cdots + {}_{r_1}^{x_p}v_i = u_i \qquad (i = 1, \cdots, p)$$

then for given $u = u_1, \cdots, u_p$ there is determined in general a unique p-ad, $x_1, \cdots, x_p$, the solution of the inversion problem (cf. 34), on F or on the curve H_p. The superposed p-ad is determined by $-u$ since in each term of (1) the integrands and limits change sign on F with z.

The theorem of Abel (34) states that if $x_1, \cdots, x_{p+2}$ are the points of intersection of a line with H_p then

$$(2) \qquad {}_{r_1}^{x_1}v_i + \cdots + {}_{r_1}^{x_{p+2}}v_i = m_i \qquad (i = 1, \cdots, p)$$

where the m_i are constant for a variable line section. If the line section is on O then $x_1, \cdots, x_p$ is the p-ad at O and the other two points are in g_1^2 whence ${}_{r_1}^{x_{p+1}}v_i + {}_{r_1}^{x_{p+2}}v_i = 0$. Hence

(3) *If in* (1) $u_i = m_i$ *when* $x_1, \cdots, x_p$ *are at* O *on* H_p *then the* $p+2$ *intersections of any line with* H_p *are subject to the relations* (2). *Also the* $\nu = (n-k)\,p + 2\,n$ *intersections with* H_p *outside of* O, $x_1, \cdots, x_\nu$, *of any curve of order* n *with* k*-fold point at* O *are subject to the* p *relations*

(4) $$\int_{r_1}^{x_1} v_i + \cdots + \int_{r_1}^{x_\nu} v_i \equiv (n-k)\, m_i \quad (i = 1, \cdots, p).$$

If $k = n-1$ these are just sufficient to determine as in (1) the remaining p intersections when $2n$ are given to determine the curve.

It is proved by Krazer ([41] p. 448) that the values of the integrals $\int_{r_1}^{r_j} v \; (j = 2, \cdots, 2p+2)$ are congruent to certain half periods which, being mutually azygetic and subject to no other relation than that their sum is congruent to zero, may be denoted by P_{1j} in the basis notation. This transfer to the basis notation is, more precisely, the following:

(5) $$\begin{pmatrix} \varepsilon_1 & \varepsilon_2 & \cdots & \varepsilon_p \\ \varepsilon_1' & \varepsilon_2' & \cdots & \varepsilon_p' \end{pmatrix}_2 = \begin{pmatrix} P_{23} & P_{2345} & P_{234567} & \cdots & P_{2,\cdots,2p+1} \\ P_{34} & P_{56} & P_{78} & \cdots & P_{2p+1,2p+2} \end{pmatrix}.$$

If $x_1, \cdots, x_p$ are the points of H_p on a line through two points r_i, r_j then from (1) and (4) this p-ad is determined by

(6) $$u \equiv m + P_{ij}.$$

And in general

(7) *The p-ad of points cut out on H_p by the pair of curves L_k, L_{p-k-1} of* 38 (16) *is determined as in* (1) *by*

$$u \equiv m + P_{1,\cdots,2k+2} \equiv m + P_{2k+3,\cdots,2p+2}.$$

The 2^{2p} projectively distinct types H_p' with sets $R_{2p+3}'^2$ congruent to R_{2p+3}^2 under Jonquières transformation have p-ads at O' which arise from the p-ad at O on H_p by the operations of the group of additive half periods.

The following construction for the p-ad superposed to that at O on H_p is a consequence of Abel's theorem.

(8) *The linear system of curves H_{p-2} on $r_1, \cdots, r_{2p+2}$ cuts H_p in a g_{p-2}^{2p-2}. The unique curve L_{p-2} on a set of g_{p-2}^{2p-2} cuts H_p in the p-ad superposed to the p-ad at O.*

For, Abel's sum formed for the $2p+2$ points r_i is $P_{11} + P_{12} + \cdots + P_{1,2p+2} \equiv 0$. Hence according to (3) a set $x_1, \cdots, x_{2p-2}$ of g_{p-2}^{2p-2} is defined by $\int_{r_1}^{x_1} v + \cdots + \int_{r_1}^{x_{2p-2}} v \equiv 2m$. Also the additional p-ad n on L_{p-2} is defined by $2m + n \equiv m$; or $n \equiv -m$.

If H_p is birationally given, projective properties of it which are not also invariant under birational transformation are expressed by conditions on the parameters m of the point O. If for example m itself is a half period and congruent to $-m$, the p-ad at O coincides with its superposed p-ad and H_p has a flex on each branch at O. In this case the condition is directly on the parameters m. We proceed to find others expressed by the vanishing of theta functions of m.

Let the branch point r_1 move up to O to produce a flex at O. The remaining $p-1$ points $x_2, \cdots, x_p$ at O are still arbitrary and $\overset{x_2}{\underset{r_1}{v}} + \cdots + \overset{x_p}{\underset{r_1}{v}} \equiv m$. Then ([41] p. 456 VI; $s=1$) $\vartheta(m+k)=0$ where ([41] p. 451 (31)) $k = \begin{pmatrix} 1 & 1 & 1 & 1 & \cdots & 1 \\ 1 & 0 & 1 & 0 & \cdots & \eta \end{pmatrix}$ with $\eta = 0, 1$ according as p is even or odd. By comparison with (5), $k = P_{357\cdots, 2p+1}$ (p even); $k = P_{1357\cdots, 2p+1}$ (p odd). There is still a choice for the designation of the original even theta function $\vartheta[0, 0]_2(u)$ and we set

$$\vartheta[0, 0]_2(u) = \vartheta_{1357\cdots, 2p+1}(u). \tag{9}$$

Then $\vartheta(m+k) = 0$ when in the basis notation $\vartheta_1(m) = 0$ (p even), or $\vartheta(m) = 0$ (p odd). Hence

(10) *If the basis notation is introduced as in* (5) *and* (9), *the condition on the parameters m of O on H_p that r_1 coincide with O is $\vartheta_1(m) = 0$ or $\vartheta(m) = 0$ according as p is even or odd.*

This condition is one of the conjugate set of discriminant conditions described in **38** (17). The conjugates are obtained by carrying out the parallel transformations, I_{ij} (**38** (13)), and $m' \equiv m + P_{ij}$ (cf. (6)). The result is

(11) *The conjugate set of 2^{2p} discriminant conditions* **38** (17) *is given by the vanishing of the 2^{2p} odd and even thetas for the parameters m of the p-ad O on H_p; more precisely, if p is even, $\delta(O^k r_i r_j r_k \cdots)^{k+1} = 0$ when $\vartheta_{ijk\cdots}(m) = 0$; if p is odd $\delta(O^k r_1 r_i r_j r_k \cdots)^{k+1} = 0$ if $\vartheta_{ijk\cdots}(m) = 0$, while $\delta(O^k r_i r_j r_k \cdots)^{k+1} = 0$ if $\vartheta_{1ijk\cdots}(m) = 0$.*

When $p = 2$ the set R_7^2 of six branch points $r_1, \cdots, r_6$ and node $O = r_7$ of H_2^4 is subject to the single condition

(38 (9)) that $r_1, \cdots, r_6$ are on a conic K. Under $I H_2^4$, K is projected into itself from O and therefore meets H_2^4 at the pair of contacts of tangents to K from O. The line joining these contacts meets H_2^4 again in the duad superposed to the node (cf. (8)).

The planar set R_7^2 determines projectively its associated set Q_7^3 in space. On the norm cubic curve C^3 through $q_1, \cdots, q_6$ these six points have parameters projective to those of the six lines from r_7 to $r_1, \cdots, r_6$ (cf. 16c). Hence for all sets R_7^2 defined by curves $H_2'^4$ birationally equivalent to H_2^4 the points $q_1, \cdots, q_6$ may be fixed on C^3. If $q_1, \cdots, q_6$ are projected from q_7 into a planar set Q_6^2 this set is associated to $r_1, \cdots, r_6$ (16b). Since $r_1, \cdots, r_6$ are on a conic, Q_6^2 is likewise on a conic (16c). Hence q_7 is a point y on the Weddle surface $W(y)$ with nodes at $q_1, \cdots, q_6$. Thus the ∞^2 projectively distinct curves $H_2'^4$ which are birationally equivalent and therefore determine the same $q_1, \cdots, q_6$ are represented by the ∞^2 points y on $W(y)$.

If, in R_7^2, r_i, r_j, r_k are on a line, i. e. $\vartheta_{ijk}(m) = 0$ (cf. (11)), then, in Q_7^3, q_l, q_m, q_n, y are on a plane (16 (9)); if r_i and $O = r_7$ coincide, i. e. $\vartheta_i(m) = 0$, then q_i and $y = q_7$ coincide. Hence (cf. 37 (4), (6)) the parameters $u \equiv m$ of the nodal pair on H_2^4 are the parameters u in Schottky's parametric equation of $W(y)$. The $2^{2p} = 16$ sets $R_7'^2$ congruent to R_7^2 under Jonquières transformation (cf. 38 (15)) determine on $W(y)$ 16 points y which form a conjugate set under the half period group, $u' \equiv u + P_{ij}$ ((9), 36 (4)). Indeed the association of R_7^2 and Q_7^3 is unaltered when the transformations (12) A_{012} and (12) A_{3456} are applied to these respective sets (16 (8)).

If four points $r_1, \cdots, r_4$ are on a line, r_5, r_6 are flex points at the node and $u \equiv m \equiv P_{56}$; y is then on the line $q_5\, q_6$. If $u \equiv m \equiv 0$ the nodal pair is a pair of g_1^2. This is the indeterminate case of the inversion problem and can occur only when H_2^4 is a doubly covered conic whose double points are the sets of g_1^2. Then R_7^2 is on a conic with O at any point of the conic; the associated Q_7^3 is on C^3 with y at any point of C^3.

The proportionality of the determinants $|ijk|$ and $|lmny|$ ($y = q_7$) was noted above. It persists for $|ijx|$ ($x = O$) and $|klmn|$. For, if $|klmn|$ vanishes, then (ij) for the sextic $(at)^6$ also vanishes (cf. 36 (2)); similarly if r_i, r_j, O are on a line, two branch points coalesce to form an additional node and (ij) for the sextic of branch lines vanishes. These proportionalities are used in 42 for the interpretation of the theta relations. An extension of this application to the Weddle surface to values $p > 2$ has been indicated by the author [23].

40. The figure, R_6^2, of six points in a plane. Occasion frequently arises to make use of the set R_6^2, usually as part of a larger set, and some of its properties will be developed here. The determinants formed from the coördinates of three of the points r, or of two points r and a variable point x, are denoted by $|ijk|$ or $|ijx|$ respectively ($i, j, k = 1, \cdots, 6$). The invariants of the binary set P_6^1 (cf. 35) are expressed so simply in terms of the Göpel invariants that a natural point of departure for R_6^2 is the system of Göpel covariants

$$(1) \qquad |ijx|\,|klx|\,|mnx|.$$

According to the Clebsch principle of transference these satisfy the same relations as the corresponding binary invariants. If then a set of six covariants $a, \cdots, f$ is defined as in 35 (3):

$$(2) \qquad a = |25x|\,|13x|\,|46x| + \cdots + |32x|\,|54x|\,|16x|,$$

with conjugates derived from the parallel substitutions 35 (4), there follows

$$(3) \qquad a + b = 4\,|51x|\,|42x|\,|36x|,$$

and furthermore

$$(4) \qquad a + b + \cdots + f \equiv 0, \qquad a^3 + b^3 + \cdots + f^3 \equiv 0.$$

The ratios of $a, \cdots, f$ subject to (4) define projectively the pencil of lines from x to R_6^2.

The covariants (1), and therefore $a, \cdots, f$ as well, are cubic curves on R_6^2. Only four of these are linearly independent and there must be a second linear relation connecting $a, \cdots, f$. By proper combination with the first, the second may be taken to be

$$(5) \quad \bar{a}\, a + \bar{b}\, b + \cdots + \bar{f}\, f = 0, \quad \bar{a} + \bar{b} + \cdots + \bar{f} = 0.$$

These values $\bar{a}, \cdots, \bar{f}$ must be linear invariants of R_6^2. For if R_6^2 is given there can be only ∞^2 line pencils from variable x to R_6^2. Hence (5) must express the condition on the invariants $a, \cdots, f$ of this pencil that it may exist. If on the other hand values $a, \cdots, f$ are given and five points $r_1, \cdots, r_5$ of R_6^2 also are given, x is uniquely determined and r_6 must lie on the sixth line of the pencil. As linear invariants, $\bar{a}, \cdots, \bar{f}$ must be expressible in terms of the ten linear invariants

$$(6) \qquad |ijk|\,|lmn|.$$

The explicit form of these expressions and some of their algebraic consequences may be stated as follows ([17] I § 4):
(7) *If* $|ij, kl, mn| = |ikl|\,|jmn| - |imn|\,|jkl|$ *is the determinant of the coördinates of the three lines* $|ijx|$, $|klx|$, $|mnx|$ *then*

$$6\bar{a} \equiv |15, 24, 36| + |14, 35, 26| + |12, 43, 56| + |23, 45, 16| + |13, 52, 46|.$$

Under parallel odd substitution, **35** (4), $\bar{a}, \cdots, \bar{f}$ *do not change sign. Further typical relations are*

$$(8) \qquad \begin{aligned} \bar{a} - \bar{b} &= |15, 24, 36|, \\ \bar{d} + \bar{e} + \bar{f} &= -|123|\,|456|. \end{aligned}$$

The condition that R_6^2 be on a conic is an alternating invariant whose sign is fixed by setting

$$(9) \qquad d_2 = \begin{vmatrix} |341|\,|561| & |531|\,|461| \\ |342|\,|562| & |532|\,|462| \end{vmatrix}.$$

The complete system of invariants of R_6^2 is as follows (cf. [17] I § 5):

(10) *If* $a_2, \cdots, a_6$ *are the elementary symmetric functions of* $\bar{a}, \cdots, \bar{f}$, *a complete system of rational projective invariants of* R_6^2 *consists of* a_2, a_3, a_4, a_5, a_6 *and* $d_2\sqrt{d}$ *where* $d_2^2 = a_2^2 - 4a_4$ *and* $\sqrt{d} = \Pi(\bar{a} - \bar{b})$.

The linear system of ∞^3 cubic curves on R_6^2 maps the plane upon a cubic surface M_2^3 in S_3. If the system is given by $a, \cdots, f$ the equation of M_2^3 appears in Cremona's hexahedral form (4), the sum of six cubes. The 45 tritangent planes of M_2^3, as planar cubics, comprise 15 of the type $a + d = 0$ and 15 pairs whose equations are

$$\begin{aligned} &\overline{ad}\,(a+d)^2 + \overline{be}\,(b+e)^2 + \overline{cf}\,(c+f)^2 = 0, \\ &(\overline{ad} = a_2 + 2\bar{a}^2 + 2\bar{a}\,\bar{d} + 2\bar{d}^2). \end{aligned} \tag{11}$$

This pair can be factored in three ways one of which is

$$(\overline{be} \mp d_2)(b+e) - (\overline{cf} \pm d_2)(c+f) = 0. \tag{12}$$

The three ways are equivalent due to the relation

$$-d_2^2 = \overline{be}\,\overline{cf} + \overline{cf}\,\overline{ad} + \overline{ad}\,\overline{be}. \tag{13}$$

There are 72 sets $R_6'^2$ congruent to R_6^2 under Cremona transformation. In fact the only types of transformation with six or fewer F-points in R_6^2 are the collineation, A_{123}, $A_{123}A_{145}$, $A_{123}A_{456}$, and $A_{123}A_{456}A_{123}$ of orders 1, 2, 3, 4, 5 respectively and numbering, according to the choice of the F-points in R_6^2, 1, 20, 30, 20, 1 respectively. The F-curves of the set are 27 in number: the 6 sets of directions at points r_i, the 15 lines $r_i r_j$, and the 6 conics on r_i, r_j, r_k, r_l, r_m. These 27 curves map into the 27 lines on M_2^3. For given R_6^2 the directions at the points r_i map into a line-six on M_2^3, i. e., six skew lines. For each of the 72 congruent sets there is on M_2^3 such a line-six and a pair of sets congruent under the quintic transformation (two associated six-points) determine a pair of line-sixes which make up a double six on M_2^3.

An analysis of the projective conditions implied by congruence is given in ([17] II § 2).

The arithmetic group $g_{6,2}$ (cf. **7**), determined by the permutations of $\gamma_1, \cdots, \gamma_6$ and the 72 types of congruence noted above, has the order $6!\,72$ and is isomorphic with the permutation group of the 27 lines on M_2^3 ([17] II § 3). It has an invariant subgroup of index two consisting of elements of determinant $+1$. The Cremona group $G_{6,2}$ in space Σ_4, isomorphic to $g_{6,2}$, is discussed in ([17] III) as an essential element in the determination of the 27 lines on M_2^3.

The complete projective system of R_6^2 given in (10) is open to the objection that one of the invariants, $d_2 \sqrt{d}$ has factors which have different projective meanings. For, $d_2 = 0$ implies that R_6^2 is on a conic, and $\sqrt{d} = 0$ implies that R_6^2 is made up of two perspective triangles. This situation is due to the choice of the group under which invariance is required. If the group is reduced to the alternating $g_{6!/2}$, a complete system consists of $a_2, \cdots, a_6, d_2$ and $\sqrt{d}$; if it is enlarged to a $g_{6!2}$ by allowing the passage to the associated set $R_6'^2$ of R_6^2, the system consists merely of $a_2, \cdots, a_6$.

Of greater interest however is the complete system of R_6^2 under congruent transformation. This is the complete system of the cubic surface M_2^3. Its members can be built up out of the 36 discriminant conditions of the set R_6^2. For they are likewise projective invariants and as such expressible in terms of the $|ijk|$ which are discriminant conditions. The 36 conditions comprise 15 of type $\delta_{ij} = 0$ which implies a coincidence of r_i, r_j; 20 of type $\delta_{ijk} = 0$ which implies that r_i, r_j, r_k are collinear; and $\delta = d_2 = 0$ which implies that R_6^2 is on a conic. Each condition is associated with a double-six and, when satisfied, requires that M_2^3 have a node and that the two line-sixes of the double-six coalesce into the six lines of M_2^3 on the node.

A coincidence of two points is not to be regarded as the identity of the two (two conditions) but rather as the coalescence of the two in some direction (only one condition because the direction has one degree of freedom). Thus a coincidence is peculiar in that it can not be expressed by a single condition on the coördinates of R_6^2. Other discriminant conditions when expressed in the coördinates of R_6^2 are satisfied when certain coincidences occur; thus $|ijk| = 0$ is satisfied by $\delta_{ij} = 0$, $\delta_{ik} = 0$, $\delta_{jk} = 0$; and $d_2 = 0$ is satisfied by $\delta_{ij} = 0$. The proper procedure is illustrated by the two types of irrational invariants of R_6^2:

$$(14) \quad d_2 \, |123| \, |456|, \quad |134| \, |234| \, |356| \, |456| \, |512| \, |612|.$$

Each is of degree three in each point and their ratio λ is therefore an absolute projective invariant. Each vanishes at least once for any coincidence. But the first vanishes twice for the coincidences $\delta_{12}, \delta_{13}, \delta_{23}, \delta_{45}, \delta_{46}, \delta_{56}$ and the second vanishes twice for the coincidences $\delta_{12}, \delta_{34}, \delta_{56}$. The ratio λ has then simple zeros when any one of the nine discriminant conditions, $\delta_{12}, \delta_{13}, \delta_{23}, \delta_{45}, \delta_{46}, \delta_{56}, \delta, \delta_{123}, \delta_{456}$ vanishes; and simple poles when any one of the nine, δ_{12}, $\delta_{34}, \delta_{56}, \delta_{134}, \delta_{234}, \delta_{356}, \delta_{456}, \delta_{512}, \delta_{612}$ vanishes. The 10 invariants of the first type (14) and the 30 invariants of the second type (14) are conjugate members in the simplest linear system of dimension 10 of irrational invariants of R_6^2 under Cremona transformation (cf. [17] III § 3). The 40 conjugate irrational invariants are permuted under Cremona transformation just as the corresponding 40 products of nine discriminant conditions under the permutation group of the lines.

We return to the case when R_6^2 is on a conic K ($d_2 = 0$). The ternary projective invariant theory of R_6^2 can then be reduced to a binary theory. For with

$$(15) \qquad x_0 = t^2, \quad x_1 = t, \quad x_2 = 1$$

as a parametric equation of K, and with $t = e_1, \cdots, e_6$ as the parameters of R_6^2 on K the ternary determinants are expressed in terms of the binary determinants, $(ij) = e_i - e_j$, by

$$(16) \qquad |ijk| = (ij)\,(ik)\,(jk).$$

In this particular case there is an interesting connection between the linear invariants $\bar{a}, \cdots, \bar{f}$ of R_6^2 and the linear invariants $A, \cdots, F$ of the parameters P_6^1 $(e_1, \cdots, e_6)$. Also when x is on K

$$(17) \qquad A : B : \cdots : F = a : b : \cdots : f,$$

since $|ijx|\,|klx|\,|mnx| = (ij)\,(kl)\,(mn)\cdot(it)\,(jt)\cdots(nt)$.

For given P_6^1, the value system $A, \cdots, F$ subject to 35 (5) determines a point A on a cubic spread M_3^3 in S_4, the ∞^3 points of M_3^3 representing the ∞^3 projectively distinct binary sextics. If two roots of the sextic coincide, say $e_5 = e_6$, the point A covers the plane π_{56} whose equations are $A+D = B+F = C+E = 0$ (35 (6)). If three roots e_4, e_5, e_6, or the complementary three e_1, e_2, e_3, coincide, the point A is fixed at $1, 1, 1, -1, -1, -1$ which evidently is a node, $n_{123} = n_{456}$, on M_3^3. Hence

(18) *The Göpel invariants* 35 (2) *map the system of ∞^3 projectively distinct binary sextics upon an M_3^3 in S_4 with ten nodes, $n_{ijk} = n_{lmn}$, the maps of sextics with a triple root. M_3^3 contains* 15 *planes π_{56} each a map of sextics with a double root. The plane π_{ij} and node n_{ijk} are incident. M_3^3 is also the map of an $S_3(y)$ by quadrics on five points $q_1, \cdots, q_5$* (cf. 36 (18)).

In the latter mapping directions at q_1 map into the plane π_{16}; the plane $q_1 q_2 q_3$ maps into the plane π_{45}; and the line $q_1 q_2$ maps into the node n_{345}.

For given R_6^2 on a conic, the linear invariants $\bar{a}, \cdots \bar{f}$ subject to (5) behave contragrediently to $a, \cdots, f$, and therefore (cf. (17)) to $A, \cdots, F$ also. Hence these invariants map such sets R_6^2 on K upon the S_3's, $\bar{a}$, in S_4. Since $\bar{d}_2^2 = a_2^2 - 4a_4$ (cf. (10)), the locus of $\bar{a}$ is a quartic envelope, E_3^4. When $e_5 = e_6$ then (cf. (8)), $\bar{a} = \bar{d}$, $\bar{b} = \bar{f}$, $\bar{c} = \bar{e}$ and the S_3, $\bar{a}$, is on the plane π_{56}. It is easy to verify from the equation of E_3^4 that $\bar{a}$ is then a double S_3 on E_3^4. But M_3^3, having 10 nodes, is an envelope of class four and an S_3 on one of its 15 planes is a double S_3 of this envelope. Hence

(19) *The linear invariants of* R_6^2 *on a conic map such sets upon the* S_3*'s in* S_4 *which lie on the quartic envelope* E_3^4, *the reciprocal of* M_3^3.

One may verify algebraically that point, A, and S_3, $\overline{a}$, are point of M_3^3 and tangent space at the point. This situation gives rise to equations which express the $\overline{a}$ as quadratic polynomials in the A as well as equations which express the $A\sqrt{\Delta}$ ($\sqrt{\Delta} = \Pi(e_i - e_j)$) as cubic polynomials in the $\overline{a}$ (cf. 15 (26), (32), (33)).

Given point A on M_3^3 and tangent space $\overline{a}$ at A; the polar quadric of A cuts M_3^3 in a locus of points A' which is the map from S_3 of the Weddle quartic surface $W(y)$ (cf. 36 (19), (20)), the point y mapping into A'. The tangent space $\overline{a}'$ at A' passes through A. On the other hand the point section at A of the reciprocal E_3^4, i. e. the S_3's, $\overline{a}'$, of E_3^4 which are on A and therefore have contacts A', has a section by an arbitrary Σ_3 not on A which is a quartic envelope with 16 tropes, and therefore a Kummer quartic envelope K^4. The tropes of K^4 are the sections by Σ_3 of the tangent S_3, $\overline{a}$, and the 15 S_3's which join A to the planes π_{ij}. Thus point y on the Weddle which maps into point A' on M_3^3 is birationally related to the plane of K^4 in which Σ_3 cuts the S_3, $\overline{a}'$, tangent to M_3^3 at A'.

41. Theory of the Weddle and Kummer quartic surfaces in binary notation. The special coördinate system (cf. 17) set up by a cubic norm curve, C^3, in S_3 leads to interesting forms of the Weddle and Kummer surface. We recall that the coefficients, $-a_3$, $3a_2$, $-3a_1$, a_0, of the binary cubic,

$$(at)^3 = (a_0 t_0 + a_1 t_1)^3$$
$$\equiv a_0 t_0^3 + 3a_1 t_0^2 t_1 + 3a_2 t_0 t_1^2 + a_3 t_1^3,$$

are taken as the coördinates of a point in space S_3. If in particular $(at)^3$ is an actual and not merely a symbolic cube the point lies on C^3. The reader will observe the double use of a_0, a_1 on the one side as binary symbols and on the other as actual coefficients.

With equivalent symbols,

$$f = (at)^3 \equiv (bt)^3,$$

the Hessian is $(ab)^2(at)(bt)^* = (pt)\cdot(qt)$, and the cubic can be expressed as $f = \lambda(pt)^3 + \mu(qt)^3$; i. e. the point represented by f is on the chord of C^3 which cuts C^3 in points p, q whose parameters are the roots of the Hessian. Thus for given t in

$$(1) \qquad (ab)^2(at)(bt) = 0,$$

we have the equation of a quadric (since the coefficients of f enter to the second degree) which is the locus of points on a bisecant of C^3 through t, i. e. the quadric cone on C^3 with vertex at t. For variable t, (1) is the equation of the quadratic system of cones on C^3.

Let $(\beta t)^2 = \beta_0 t_0^2 + 2\beta_1 t_0 t_1 + \beta_2 t_1^2$ be any quadratic form. Then

$$(2) \qquad (ab)^2(a\beta)(b\beta) = 0$$

is again the equation of a quadric, the locus of bisecants of C^3 which meet C^3 in pairs of points with parameters apolar to $(\beta t)^2$. For variable β, (2) is the equation of the net of quadrics on C^3.

The Weddle surface is determined when its six nodes Q_6^3 on C^3 are given; let these points have parameters determined by

$$(3) \qquad (\alpha t)^6 = \alpha_0 t_0^6 + 6\alpha_1 t_0^5 t_1 + \cdots + \alpha_6 t_1^6 = 0.$$

Then

$$(4) \qquad (\alpha a)^3(\alpha b)^3 = 0$$

is a quadric which cuts C^3 in these six nodes. For if $(at)^3$ is a perfect cube, i. e. $(at)^3 \equiv (bt)^3 \equiv (t'\,t)^3 \equiv (t_0' t_1 - t_1' t_0)^3$ then $(\alpha a)^3(\alpha b)^3 = (\alpha t')^3(\alpha t')^3 = (\alpha t')^6$. Hence the quadric (4) does not contain C^3 and therefore the web (cf. Baker [2] p. 56)

$$(5) \qquad \beta_3(\alpha a)^3(\alpha b)^3 + (ab)^2(a\beta)(b\beta) = 0$$

* It should be noted that the interpretation of the symbol (mn) depends upon m, n. For cogredient symbols, $(\alpha\beta) = \alpha_0\beta_1 - \alpha_1\beta_0$; for cogredient variables, $(st) = s_0 t_1 - s_1 t_0$; for symbols α and variables t, contragredient to each other, $(\alpha t) = \alpha_0 t_0 + \alpha_1 t_1$.

of quadrics with parameters $\beta_3, \beta_0, \beta_1, \beta_2$ is the web on Q_6^3. On account of the symmetry of (5) in the equivalent symbols of f the polarized form of the web is obtained by assuming $(at)^3 \neq (bt)^3$.

If we assume that the polarized quadric (5) has a node at the point a the coefficients of b_0, b_1, b_2, b_3 must vanish giving rise to four equations bilinear in the a's and the β's. If from these the β's are eliminated, the equation of the Weddle surface W determined by Q_6^3 is obtained. If however the a's are eliminated, the equation in planar coördinates β of the Kummer surface K is obtained since plane sections of K correspond to adjoint quadrics of W (cf. 37).

The expanded form of (5) is

$$\begin{aligned}\beta_3\{&a_0(b_0\alpha_6 - 3b_1\alpha_5 + 3b_2\alpha_4 - b_3\alpha_3)\\ &\qquad - 3a_1(b_0\alpha_5 - 3b_1\alpha_4 + 3b_2\alpha_3 - b_3\alpha_2)\\ &+ 3a_2(b_0\alpha_4 - 3b_1\alpha_3 + 3b_2\alpha_2 - b_3\alpha_1)\\ &\qquad - a_3(b_0\alpha_3 - 3b_1\alpha_2 + 3b_2\alpha_1 - b_3\alpha_0)\}\\ &+ \beta_0\{a_1b_3 - 2a_2b_2 + a_3b_1\}\\ &\qquad + \beta_1\{-a_0b_3 + a_1b_2 + a_2b_1 - a_3b_0\}\\ &+ \beta_2\{a_0b_2 - 2a_1b_1 + a_2b_0\}.\end{aligned}$$

The eliminations mentioned yield the following determinant forms of W and K (Baker[2] p. 65, p. 56):

$$(6)\qquad W = \begin{vmatrix} a_0\alpha_6 - 3a_1\alpha_5 + 3a_2\alpha_4 - a_3\alpha_3 & 0 & -a_3 & a_2 \\ 3(-a_0\alpha_5 + 3a_1\alpha_4 - 3a_2\alpha_3 + a_3\alpha_2) & a_3 & a_2 & -2a_1 \\ 3(a_0\alpha_4 - 3a_1\alpha_3 + 3a_2\alpha_2 - a_3\alpha_1) & -2a_2 & a_1 & a_0 \\ -a_0\alpha_3 + 3a_1\alpha_2 - 3a_2\alpha_1 + a_3\alpha_0 & a_1 & -a_0 & 0 \end{vmatrix} = 0;$$

$$(7)\qquad K = \begin{vmatrix} \beta_3\alpha_6 & -3\beta_3\alpha_5 & 3\beta_3\alpha_4 + \beta_2 & -\beta_3\alpha_3 - \beta_1 \\ -3\beta_3\alpha_5 & 9\beta_3\alpha_4 - 2\beta_2 & -9\beta_3\alpha_3 + \beta_1 & 3\beta_3\alpha_2 + \beta_0 \\ 3\beta_3\alpha_4 + \beta_2 & -9\beta_3\alpha_3 + \beta_1 & 9\beta_3\alpha_2 - 2\beta_0 & -3\beta_3\alpha_1 \\ -\beta_3\alpha_3 - \beta_1 & 3\beta_3\alpha_2 + \beta_0 & -3\beta_3\alpha_1 & \beta_3\alpha_0 \end{vmatrix} = 0.$$

From the mode of derivation W is a simultaneous invariant of the cubic $(a\,t)^3$ and the sextic $(\alpha\,t)^6$, of degrees four and one respectively. It must therefore be the bilinear invariant of $(\alpha\,t)^6$ and that sextic which is the product of $f = (a\,t)^3$ and its cubic covariant $f' = (a'a'')^2\,(a'a''')\,(a''t)\,(a'''t)^2$, i. e.

$$(8) \qquad W = (a\alpha)^3\,(a''\alpha)\,(a'''\alpha)^2\,(a'\,a'')^2\,(a'a''').$$

Since the relation of f and f' is mutual, the corresponding points in S_3 are partners in a Cremona involution of the third order (whose pairs are on chords of C^3 and harmonic to the crossings) under which W is invariant.

Similarly if K is arranged according to powers of β_3, say

$$(9) \qquad K = K_0\,\beta_3^4 + 4\,K_1\,\beta_3^3 + 6\,K_2\,\beta_3^2 + 4\,K_3\,\beta_3 + K_4,$$

the coefficients of the various powers are simultaneous invariants of the quadratic $(\beta\,t)^2$ and the sextic $(\alpha\,t)^6$. From the degrees in the coefficients as well as, to some extent, their form, we may conclude at once that, to within numerical multiples, K_0 is the catalecticant of $(\alpha\,t)^6$; K_1 is the bilinear invariant of $(\beta\,t)^2$ and the quadratic covariant $(\alpha\,\alpha')^4\,(\alpha\,\alpha'')^2 \times(\alpha'\alpha'')^2\,(\alpha''t)^2$; K_2 is a linear combination of the bilinear invariant of $[(\beta\,t)^2]^2$ and $(\alpha\,\alpha')^4\,(\alpha\,t)^2\,(\alpha'\,t)^2$, and of the product $(\beta\,\beta')^2\cdot(\alpha\,\alpha')^6$; K_3 is the bilinear invariant of $[(\beta\,t)^2]^3$ and $(\alpha\,t)^6$; and K_4 is $[(\beta\,\beta')^2]^2$. From the form of K_4 it is clear that $\beta_3 = 0$ is the planar equation of the node $0, 0, 0, 1$ of K. This is the node $u = 0$ in **37** which corresponds on W to the cubic curve C^3 since the quadric (5) contains C^3 when $\beta_3 = 0$.

A section of the Weddle by a quadric on the nodes is determined when β_3 and the coefficients, β_0, β_1, β_2, of $(\beta\,t)^2$ are given. The corresponding section of the Kummer is by a plane with coefficients β_0, β_1, β_2, β_3. In particular, if

$$(10) \quad (\alpha\,t)^6 = (t_1\,t)\cdot(t_2\,t)\cdot\cdots\cdot(t_6\,t) \qquad [(t_i\,t) = t_{i0}\,t_1 - t_{i1}\,t_0],$$

the tropes of the Kummer corresponding to the six cones on C^3 with respective nodes at Q_6^3 are given by

$$(\beta t)^2,\ \beta_3 = (t_i t)^2,\ 0 \qquad (i = 1, \cdots, 6). \tag{11}$$

We seek then the further values $(\beta t)^2$, β_3 which define the ten remaining tropes of K associated with the ten pairs of planes on Q_6^3. If then

$$\begin{gathered}(\alpha t)^6 = (c t)^3 \cdot (\gamma t)^3, \\ (c t)^3 = (t_1 t)(t_2 t)(t_3 t), \qquad (\gamma t)^3 = (t_4 t)(t_5 t)(t_6 t),\end{gathered} \tag{12}$$

the equation of such a pair of planes is

$$D_{123} = D_{456} = (a c)^3 \cdot (a' \gamma)^3 = 0. \tag{13}$$

In order to express this in the standard form (5) the product $(c t)^3 \cdot (\gamma \tau)^3$ is written in a Clebsch-Gordan expansion ([31] II p. 86); i. e.

$$\begin{aligned}(c t)^3 \cdot (\gamma \tau)^3 &= \{(c t)^3 (\gamma t)^3\}_{\tau^3} + 3\{(c\gamma)(c t)^2 (\gamma t)^2\}_{\tau^2} \cdot (t\tau)/2 \\ &\quad + 9\{(c\gamma)^2 (c t)(\gamma t)\}_{\tau} \cdot (t\tau)^2/10 + \{(c\gamma)^3\} \cdot (t\tau)^3/4.\end{aligned}$$

If in this t, τ are replaced by contragredient symbols a, a' respectively there results

$$\begin{aligned}D_{123} &= (c a)^3 (\gamma a')^3 = (\alpha a)^3 (\alpha a')^3 \\ &+ 3[(c\gamma)(c t)^2 (\gamma t)^2,\ (a a')(a t)^2 (a' t)^2]^1/2 \\ &+ 9[(c\gamma)^2 (c t)(\gamma t),\ (a a')^2 (a t)(a' t)]^2/10 \\ &+ (c\gamma)^3 \cdot (a a')^3/4,\end{aligned}$$

where $[f, \varphi]^k$ is the k-th transvectant of f and φ. Since $(a a')(a t)^2 (a' t)^2 \equiv 0$ and $(a a')^3 = 0$,

$$\begin{aligned}D_{123} &= (c a)^3 (\gamma a')^3 \\ &= (\alpha a)^3 (\alpha a')^3 + 9[(a a')^2 (a t)(a' t),\ (c\gamma)^2 (c t)(\gamma t)]/10.\end{aligned}$$

By comparison with (5) we find that

(14) *The ten even tropes $D_{123} = D_{456}$ of the Kummer surface are given by*

$$(\beta t)^2,\ \beta_3 = 9(\beta_{123} t)^2,\ 10$$

where

$$(\beta_{123} t)^2 = (\beta_{456} t)^2 = [(t_1 t)(t_2 t)(t_3 t),\ (t_4 t)(t_5 t)(t_6 t)]^2.$$

The ten quadratics $(\beta_{123}\, t)^2$ are linear irrational covariants of $(\alpha\, t)^6$. It may be proved that only nine such covariants are linearly independent with respect to numerical coefficients and that all may be expressed in terms of the ten given in (14) which themselves are related as in

$$\Sigma_{10}\, (\beta_{ijk}\, t)^2 = 0. \tag{15}$$

In order to find the nodes of the Kummer surface it is convenient to use as point coördinates b_0, b_1, b_2 (the coefficients of the quadratic $(b\, t)^2$), and b_3 with an incidence condition with respect to the above planar coördinates of the form

$$(b\, \beta)^2 + b_3\, \beta_3 = 0. \tag{16}$$

In terms of the usual coördinates y, η this is equivalent to

$$\begin{aligned} \eta_0 : \eta_1 : \eta_2 : \eta_3 &= \beta_0 : \quad \beta_1 : \beta_2 : \beta_3; \\ y_0 : y_1 : y_2 : y_3 &= b_2 : -2\, b_1 : b_0 : b_3. \end{aligned}$$

The node P_{ij} is on the three tropes $D_i = (t_i\, t)^2, 0$; $D_j = (t_j\, t)^2, 0$ (cf. (11)); and $D_{ijk} = 9\, (\beta_{ijk}\, t)^2, 10$. It is easily verified that $P_{ij} = 10\, (t_i\, t)\, (t_j\, t), \; -9\, [(t_i\, t)\, (t_j\, t), \; (\beta_{ijk}\, t)^2]^2$. The result should be symmetric in k, l, m, n and by a direct calculation of the transvectants one verifies that

$$c_{ij} = [(t_i\, t)^2\, (t_j\, t)^2, \; (t_k\, t) \cdots (t_n\, t)]^4 = 3\, [(t_i\, t)\, (t_j\, t), \; (\beta_{ijk}\, t)^2]^2. \tag{17}$$

Thus

(18) *The fifteen nodes P_{ij} other than $u = 0$ $(0, 0, 0, 1)$ of the Kummer surface are given by*

$$(b\, t)^2, \; b_3 = 10\, (t_i\, t)\, (t_j\, t), \; -3\, c_{ij}.$$

Since $(\beta_{ijk}\, t)^2$ can be determined from its bilinear invariants with respect to the three independent quadratics $(t_i\, t)\, (t_j\, t)$, $(t_i\, t)\, (t_k\, t)$, $(t_j\, t)\, (t_k\, t)$, there follows from (17) that

$$\begin{aligned} 3\, (\beta_{ijk}\, t)^2 &= 3\, (\beta_{lmn}\, t)^2 \\ &= \{(jk)\, c_{jk}\, (t_i\, t)^2 + (ki)\, c_{ki}\, (t_j\, t)^2 + (ij)\, c_{ij}\, (t_k\, t)^2\} / (ijk) \\ &= \{(mn)\, c_{mn}\, (t_l\, t)^2 + (nl)\, c_{nl}\, (t_m\, t)^2 \\ &\qquad + (lm)\, c_{lm}\, (t_n\, t)^2\} / (lmn) \end{aligned} \tag{19}$$

where $(lmn \cdots) = (t_l\, t_m)\, (t_l\, t_n)\, (t_m\, t_n) \cdots$.

Parametric equations of the Kummer and Weddle surface are easily obtained in the representations used here. If $(at)^3$ is a point on W such that the bisecant to C^3 meets C^3 in $t=r$, $t=s$ then $(at)^3 = \lambda(rt)^3 + \mu(st)^3$. The cubic covariant of $(at)^3$ is then $\lambda(rt)^3 - \mu(st)^3$. The product of the two is apolar to $(\alpha t)^6$ as in (8) if $\lambda^2(\alpha r)^6 - \mu^2(\alpha s)^6 = 0$, i. e., if $\lambda : \mu = \sqrt{(\alpha s)^6} : \sqrt{(\alpha r)^6}$. If then we take the underlying relation of genus two in the form

$$(20) \qquad S: \quad z^2 = (\alpha t)^6, \quad z_r = \sqrt{(\alpha r)^6},$$

$(at)^3$ represents a point on W if $(at)^3 = z_s(rt)^3 + z_r(st)^3$. It arises from the pair of points r, z_r; s, z_s (or the superposed pair $r, -z_r$; $s, -z_s$) on S. The second way of pairing these four points on S into r, z_r; $s, -z_s$ and $r, -z_r$; s, z_s yields the second point on W on the bisecant to C^3 through the first point.

If in the equation (5) of the quadric on Q_6^3 we set $(at)^3 = (bt)^3 = z_s(rt)^3 + z_r(st)^3$ it becomes the equation in planar coördinates $\beta_0, \beta_1, \beta_2, \beta_3$ of the corresponding point on the Kummer surface. The result is

$$\beta_3 [z_r z_s + (\alpha r)^3 (\alpha s)^3] + (rs)^2 (\beta r)(\beta s) = 0.$$

By comparison with the incidence relation (16) we find that

(21) *The parametric equation of W referred to C^3 in terms of two points, r, z_r and s, z_s, of S in* (20) *is given by* $(at)^3 \equiv z_s(rt)^3 + z_r(st)^3$; *that of K is*

$$(bt)^2, b_3 \equiv (rs)^2 (rt)(st), \; z_r z_s + (\alpha r)^3 (\alpha s)^3.$$

If $r = s$, the point of K is the node $u = 0$ with coördinates $0, 0, 0, 1$; if $(rt)(st) = (t_1 t)(t_2 t)$, the point of K is the node given in (18) since $(\alpha t_1)^3 (\alpha t_2)^3 = -3c_{ij}(t_1 t_2)^2/10$. This parametric equation of K differs from that given by Hudson ([36] p. 19) in that the coördinate plane opposite the node $u = 0$ is not a particular even trope but rather the sum of the ten even tropes for which, according to (15), $(\beta t)^2 \equiv 0$ in $(\beta t)^2, \beta_3$. This particular plane section of K

corresponds to that quadric on Q_6^3, $(\alpha a)^3 (\alpha b)^3 = 0$, which has the projective definition that it is apolar to the net of quadric envelopes on the planes of C^3.

The planar parametric equation of K in terms of (rt), (st) is derived as follows. For $(at)^3$ in (5) a point of W, $(\beta t)^2$, β_3 are to be determined so that (5) is satisfied by all values of b. Thus on replacing symbols b by variables t to obtain $\beta_3 (\alpha a)^3 (\alpha t)^3 - (a\beta)(at)^2(\beta t) \equiv 0$, and then replacing $(at)^3$ by $z_s (rt)^3 + z_r (st)^3$, the β's are to be determined so that

$$\beta_3 [z_s (\alpha r)^3 + z_r (\alpha s)^3] (\alpha t)^3 \\ + (\beta t) [z_s (rt)^2 (\beta r) + z_r (st)^2 (\beta s)] \equiv 0.$$

If $(\beta t)^2$ is expressed as

$$(22) \qquad (\beta t)^2 = k_0 (rt)^2 + 2 k_1 (rt)(st) + k_2 (st)^2;$$

and if (αt) is expressed in terms of (rt), (st) from $(rs)(\alpha t) = (st)(\alpha r) - (rt)(\alpha s)$, the identity is satisfied when

$$k_0 = -3 z_s [z_s \cdot (\alpha r)^5 (\alpha s) + z_r \cdot (\alpha r)^2 (\alpha s)^4],$$

$$(23)\ \beta_3 = z_r z_s (rs)^4; \quad k_1 = z_r z_s [z_r z_s + (\alpha r)^3 (\alpha s)^3],$$

$$k_2 = -3 z_r [z_s \cdot (\alpha r)^4 (\alpha s)^2 + z_r \cdot (\alpha r)(\alpha s)^5].$$

Hence

(24) *In terms of the binary parameters* r, s *and the irrationality* z_r, z_s *the parametric expression of the envelope* K *is given by the* $(\beta t)^2$, β_3 *defined in* (22), (23).

In order to connect the algebraic parameters r, s of a point on W or K with the hyperelliptic parameters u_1, u_2 we pursue the article of Schottky [61] begun in **37**. We first observe that, as a consequence of **18** B the function

$$\vartheta[g', h'](u) \cdot \partial/\partial u_1\, \vartheta[g, h](u) - \vartheta[g, h](u)\, \partial/\partial u_1\, \vartheta[g', h'](u)$$

is a theta function of the second order and characteristic $[g+g', h+h']$. The theta relation,

$$(25) \qquad \alpha_1 \vartheta_1 \vartheta_{234} + \alpha_2 \vartheta_2 \vartheta_{314} + \alpha_3 \vartheta_3 \vartheta_{124} \equiv 0,$$

has the derivative

$$\textstyle\sum_i \alpha_i [\vartheta_i\, \partial/\partial u_1\, \vartheta_{jk4} + \vartheta_{jk4}\, \partial/\partial u_1\, \vartheta_i] \equiv 0 \quad (i, j, k = 1, 2, 3).$$

Setting $\sum_i \alpha_i [\vartheta_i \partial/\partial u_1 \vartheta_{jk4} - \vartheta_{jk4} \partial/\partial u_1 \vartheta_i] \equiv 2 R_1$ we observe that R_1 is an even function of the second order and characteristic P_{56}. It is therefore a linear combination of $\vartheta_5 \vartheta_6$ and $\vartheta_{125} \vartheta_{126}$. Since R_1 and $\vartheta_5 \vartheta_6$ vanish for $u = 0$ while $\vartheta_{125} \vartheta_{126}$ does not, R_1 must be $a_1 \vartheta_5 \vartheta_6$. With a similar argument for derivatives with respect to u_2 we find that

$$(26) \quad \begin{aligned} &\alpha_1 \vartheta_1 \, d\vartheta_{234} + \alpha_2 \vartheta_2 \, d\vartheta_{314} + \alpha_3 \vartheta_3 \, d\vartheta_{124} \\ &\qquad = (a_1 \, du_1 + a_2 \, du_2) \, \vartheta_5 \vartheta_6, \\ &\alpha_1 \vartheta_{234} \, d\vartheta_1 + \alpha_2 \vartheta_{314} \, d\vartheta_2 + \alpha_3 \vartheta_{124} \, d\vartheta_3 \\ &\qquad = -(a_1 \, du_1 + a_2 \, du_2) \, \vartheta_5 \vartheta_6. \end{aligned}$$

If (25), (26) are multiplied by $\Pi \vartheta_1 \vartheta_2 \vartheta_3 \vartheta_4$ and the planes F_{ijk} and quadric cones G_i introduced as in **37** (1), (3), (4) then

$$(27) \quad \begin{aligned} &\alpha_1 G_1 F_{234} + \alpha_2 G_2 F_{134} + \alpha_3 G_3 F_{124} \equiv 0, \\ &\alpha_1 F_{234} \, dG_1 + \alpha_2 F_{134} \, dG_2 + \alpha_3 F_{124} \, dG_3 \\ &\qquad = -2\Pi^2 (a_1 \, du_1 + a_2 \, du_2). \end{aligned}$$

Let

$$(28) \quad P = G_1 \cdot G_2 \cdot \cdots \cdot G_6 = \Pi^8.$$

Then $\Pi^2 = P^{1/4}$ and $\Pi^4 = P^{1/2} = G_1 G_2 G_3 F_{456}/F_{123}$. Hence on W there is a linear function, $v = -(a_1 u_1 + a_2 u_2)$, of the arguments u_1, u_2 whose differential can be expressed in the form

$$(29) \quad \begin{aligned} dv &= (\alpha_1 F_{234} \, dG_1 + \alpha_2 F_{134} \, dG_2 + \alpha_3 F_{124} \, dG_3)/P^{1/4} \\ &= -(\alpha_1 G_1 \, dF_{234} + \alpha_2 G_2 \, dF_{134} + \alpha_3 G_3 \, dF_{124})/P^{1/4} \end{aligned}$$

where

$$\alpha_1 G_1 F_{234} + \alpha_2 G_2 F_{134} + \alpha_3 G_3 F_{124} \equiv 0.$$

Because of the unsymmetric character of this expression with respect to the nodes 1, 2, 3, 4 of W there must be a second differential of the same sort and hence du_1 and du_2 admit of such expression. Schottky then closes with the remark that if y_0, y_1, y_2, y_3 are planes in S_3 and Q_0, Q_1, Q_2, Q_3 quadrics on Q_6^3, the 16 cubic surfaces $y_i Q_j$ on Q_6^3 must be connected by two linear relations $F_i(y_0, \cdots, y_3; Q_0, \cdots, Q_3) \equiv 0$

$(i = 1, 2)$ (cf. Baker[2] p. 78). Thus on $W = 0$, for proper choice of the numerical factors in F_i,

$$(30) \quad du_i = F_i(dy_0, \cdots, dy_3;\ Q_0, \cdots, Q_3)/P^{1/4} \qquad (i = 1, 2).$$

If now the point y in S_3 is determined by the coefficients of the binary cubic $(a\,t)^3$ with hessian $(h\,t)^2$, the two relations, $F_i = 0$, just mentioned are a consequence of the fact that the cubic has no linear covariant, and therefore the polar $(h\,a)^2\,(a\,\sigma)$ vanishes identically with respect to σ_0, σ_1. For here the coefficients of $(h\,t)^2$ are quadrics not merely on Q_6^3 but on C^3. Hence

$$(31) \qquad dv = (h\,d\,a)^2\,(d\,a\,\sigma)/(h\,t_1)^2 \cdot (h\,t_2)^2 \cdot \cdots \cdot (h\,t_6)^2$$

since $(h\,t_i)^2$ is a constant multiple of G_i in P and the constant can be incorporated with σ_0, σ_1. Taking

$$(a\,t)^3 = (r\,t)^3/z_r + (s\,t)^3/z_s, \qquad (h\,t)^2 = 2\,(r\,s)^2\,(r\,t)\,(s\,t)/z_r\,z_s$$

in order to ensure that $W = 0$ and also that the coefficients are of degree zero in the parameters $r_0 : r_1$ and $s_0 : s_1$, and recalling that

$$z_r^2 = (\alpha\,r)^6, \qquad 2\,z_r\,dz_r = 6\,(\alpha\,r)^5\,(\alpha\,d\,r),$$

we find that

$$\begin{aligned} (d\,a\,t)^3 &= -3\,(\alpha\,r)^5\,(\alpha\,t)\cdot(r\,t)^2\,(r\,d\,r)/z_r\cdot(\alpha\,r)^6 \\ &\qquad -3\,(\alpha\,s)^5\,(\alpha\,t)\cdot(s\,t)^2\,(s\,d\,s)/z_s\cdot(\alpha\,s)^6, \\ (h\,d\,a)^2\,(d\,a\,\sigma) &= 2\,(r\,s)^3\,[-(r\,\sigma)\,(r\,d\,r)/z_r + (s\,\sigma)\,(s\,d\,s)/z_s]/z_r\,z_s, \\ (h\,t_1)^2 \cdot \cdots \cdot (h\,t_6)^2 &= 2^6\,(r\,s)^{12}/z_r^4\,z_s^4. \end{aligned}$$

On substituting these values in (31) and incorporating the factor $1/2^{1/2}$ with σ_0, σ_1 there results

$$(32) \qquad dv = -(r\,\sigma)\,(r\,d\,r)/z_r + (s\,\sigma)\,(s\,d\,s)/z_s.$$

The general integral of the first kind with parameters σ_0, σ_1, for $z^2 = (\alpha\,t)^6$ is $\int(-\sigma_0 + \sigma_1\,t)\,d\,t/z_t$, or, in homogeneous form with $t = t_0 : t_1$, is $\int (\sigma t)\,(t\,d\,t)/z_t$. Hence, $d\,v = \alpha\,d\,u_1 + \beta\,d\,u_2$ where

$$u_1 = \int^{r,\,-z_r} d\,v_1 + \int^{s,\,z_s} d\,v_1,$$

$$u_2 = \int^{r,\,-z_r} d\,v_2 + \int^{s,\,z_s} d\,v_2,$$

v_1 and v_2 being the normal integrals of the first kind.

42. Theta relations as projective relations. In the foregoing sections the theta functions of genus two have been connected with the projective figures: P_6^1 (cf. 35), six points on a line; Q_6^3, y (cf. 36), six points in S_3, and y a point on the Weddle surface: and R_6^2, O (cf. 39, 40) six points in the plane on a conic with a seventh point O in general position. The coördinates of the points in such sets are related by a variety of determinant identities. The ratio of two terms in such an identity can be expressed in terms of double ratios determined by the projective figures. On the other hand (cf. 30) the theta functions are subject to a variety of relations which essentially are identical with the projective relations which are consequences of the determinant identities. We give here the formulae by which the transition from one type of relation to the other can be effected.

Beginning with the modular relations (30 V, VI) let $P_6^2 = p_1, \cdots, p_6$ be a planar six-point with determinants $|ijk| = |p_i, p_j, p_k|$; and let

$$\text{(1)} \qquad \vartheta_{ijk}^4 = \vartheta_{lmn}^4 = \varepsilon_{ijk}\,\varepsilon_{lmn}\,|\,i\,j\,k\,|\;|\,l\,m\,n\,|.$$

The relations V then show that these zero values of the even thetas define P_6^2 projectively (cf. 40 (6)) since the determinant products satisfy the same linear relations as the ϑ_{ijk}^4 namely:

$$(-1)^l\,\varepsilon_{ijk}\,\varepsilon_{lmn}\,|\,i\,j\,k\,|\;|\,l\,m\,n\,| + (-1)^k\,\varepsilon_{ijl}\,\varepsilon_{kmn}\,|\,i\,j\,l\,|\;|\,k\,m\,n\,|$$
$$+ (-1)^j\,\varepsilon_{ikl}\,\varepsilon_{jmn}\,|\,i\,k\,l\,|\;|\,j\,m\,n\,| + (-1)^i\,\varepsilon^{jkl}\,\varepsilon_{imn}\,|\,j\,k\,l\,|\;|\,i\,m\,n\,| = 0.$$

Since the modular functions depend on three moduli and P_6^2 depends on four absolute constants the six points are subject to some relation. On substituting from (1) in **30** VI and comparing the result with **16** (5) it is clear that the set P_6^2 is self-associated and therefore on a conic. Thus of the 15 relations V only five are independent, and of the 15 additional relations VI one is sufficient to imply to others. If the conic on P_6^2 is taken in the normal form

$$x_0 : x_1 : x_2 = t^2 : t : 1,$$

and if the point p_i has the parameter $t = e_i$, then $|ijk| = (ijk)$ (cf. **30** (9), (10) (14)). The relations VI then reduce to the binary identities **30** VI° and the relations V to the determinant identities **30** V°. It should be emphasized however that the P_6^2 thus defined on a conic is not projective to the set R_6^2 mentioned above. The distinction between the two is brought out later (cf. p. 155).

In the discussion of the set Q_6^3, y it is more convenient to use the theta relations **30** I°, $\cdots$, IX° as revised by Schottky. Since Q_6^3 is associated with P_6^1 determined by $t = e_i$ $(i = 1, \cdots, 6)$ the quaternary determinants are proportional to the binary differences and we set

$$(2) \qquad |ijkl| = \varepsilon_{ijklmn}\,(m\,n).$$

The quaternary identities then are satisfied by virtue of the binary identities. In harmony with Schottky's definition (**37** (1)) of F_{ijk}, the section of the Weddle surface by the plane on q_i, q_j, q_k, we set

$$(3) \qquad |ijky| = \varepsilon_{ijklmn}\,(l\,m\,n)\,\sigma_i(u)\,\sigma_j(u)\,\sigma_k(u)\,\sigma_{ijk}(u)$$

where as before

$$(l\,m\,n) = (l\,m)\,(l\,n)\,(m\,n) = (e_l - e_m)\,(e_l - e_n)\,(e_m - e_n).$$

The relations I°, VII°, VIII° are then consequences respectively of the determinant identities:

$$\begin{aligned}&|234y||156y| - |134y||256y|\\ &\qquad + |124y||356y| - |123y||456y| = 0,\\ &|2356||156y| - |1356||256y| + |1256||356y| = 0,\\ &|126y||346y| - |136y||246y| + |236y||146y| = 0.\end{aligned}$$

The other relations involve the odd functions whose squares are to be identified with the cones G_i with vertex at q_i and on C^3. Expressions for these cones are obtained from IX°. Taking IX° in the form

$$\sigma_{125}(u)\sigma_{134}(u) - \sigma_{135}(u)\sigma_{124}(u) = \pm(23)(45)\sigma_1(u)\sigma_6(u)$$

and multiplying by $\sigma_1^2\sigma_2\sigma_3\sigma_4\sigma_5$ to introduce the determinants, the cone G_1 is obtained, and in conformity with this result we set in general

$$\begin{aligned}G_i &= |ikmn||ijln||ikly||ijmy|\\ &\qquad - |ijmn||ikln||ikmy||ijly|\\ &= (jklmn)\cdot \Pi \cdot \sigma_1^2(u).\end{aligned} \tag{4}$$

The apolary invariant, $g_{i,jk}$, of the pair of points q_j, q_k with respect to the cone G_i is

$$g_{i,jk} = \varepsilon_{jklmn}(jk)(lmn). \tag{a}$$

The relation connecting four of these cones is therefore

$$\begin{aligned}&\varepsilon_{jklmn}(imn)G_i + \varepsilon_{iklmn}(jmn)G_j\\ &\qquad + \varepsilon_{ijlmn}(kmn)G_k + \varepsilon_{ijkmn}(lmn)G_l = 0.\end{aligned} \tag{b}$$

This is a projective quaternary relation if $(imn) = (im)(in)(mn), \cdots$ are modified by using (2). If in the projective relation (b) the G_i are replaced by $\sigma_i^2(u)$ from (4), the theta relation II° is obtained in the form

$$(jkl)\sigma_i^2(u) - (ikl)\sigma_j^2(u) + (ijl)\sigma_k^2(u) - (ijk)\sigma_l^2(u) = 0. \tag{c}$$

Projective relations connecting the pairs of planes on Q_6^3 and the cones G_i may be similarly established and similarly translated to obtain the theta relations III° and IV°. Thus the transference formulae (2), (3), (4) suffice to establish the identity of the projective relations with the theta relations.

We consider now the set R_7^2 which consists of the six branch points R_6^2 and the node $O = r_7$ of the hyperelliptic curve H_2^4. It was noted in **39** that this set R_7^2 is associated with the set $Q_7^3 = Q_6^3, y$ determined by a point y on W whence

(d) $|ijkl| = \varepsilon_{ijklmn7}\,|mn7|, \quad |ijky| = \varepsilon_{ijk7lmn}\,|lmn|.$

Comparing this with (2) we can set

$$\text{(e)} \qquad |mn7| = (mn)$$

which states merely that the lines from r_7 to $r_1, \cdots, r_6$ are projective to the fundamental binary sextic. Then according to (1)

(5) $\vartheta_{ijk}^4 = \vartheta_{lmn}^4 = \varepsilon_{ijk}\,\varepsilon_{lmn}\,|ij7|\,|ik7|\,|jk7|\,|lm7|\,|ln7|\,|mn7|.$

From (3) and (d) there follows

$$|ijk|\,|lmn| = (lmn)\,(ijk)\cdot \Pi \cdot \sigma_{ijk}^2(u).$$

If Π be deleted from the squares of the $\sigma(u)$'s as a factor of proportionality and if the σ's be replaced by the ϑ's [cf. **30** (10), (11) and (5) above] then

$$\text{(6)} \qquad \vartheta_{ijk}^2\,\vartheta_{ijk}^2(u) = \varepsilon_{ijk}\,\varepsilon_{lmn}\,|ijk|\,|lmn|.$$

Again if (d) be applied to G_i in (4) the factors ε cancel and G_i becomes after deleting Π and passing to $\vartheta_i^2(u)$ (cf. **30** (10), (11))

$$\text{(7)} \qquad \begin{aligned} H_i &= |jl7|\,|km7|\,|jmn|\,|kln| - |kl7|\,|jm7|\,|jln|\,|kmn| \\ &= (jklmn)^{1/2}\cdot \vartheta_i^2(u). \end{aligned}$$

In this $(jklmn) = (jk)\cdots(mn)$ is to be evaluated for the plane from (e). Hence by virtue of (5), (6), (7) the theta relations **30** I, $\cdots$, IX are satisfied by the planar set R_7^2. The coördinates of the six branch points R_6^2 can be rationally obtained in terms of the irrational invariants $|ijk|\,|lmn|$ (cf. [17] I p. 197). When these are located the double ratios of the six lines $r_7 - r_1, \cdots, r_6$ can be rationally obtained from (5) (cf. [15]) and $r_7 = O$ thereby is rationally determined. Thus the branch points and node of H_2^4 are expressed rationally in terms of the theta and modular functions. For variation of u the projectively distinct but birationally equivalent types of H_2^4 are obtained. The ∞^2 sets of six branch points are those sets on a conic for which there exists a point r_7 with given fundamental sextic (cf. the interpretation of 40 (5)).

It is possible also to satisfy the theta relations by equating the theta squares to binary quadratics since in 41 the coördinates of the nodes and tropes of the Kummer surface are given in terms of the coefficients of such quadratics.

CHAPTER IV

GEOMETRIC APPLICATIONS OF THE FUNCTIONS OF GENUS THREE

The theta functions of genus three are defined by a period scheme which is determined by the normal integrals of the first kind attached to an algebraic curve of genus three. The canonical curve of genus three is a planar quartic curve whose 28 double tangents are associated with the 28 odd theta functions of first order. This double tangent configuration is determined by an Aronhold set of seven double tangents or dually by an Aronhold set of seven points, P_7^2. The 36 even theta functions of the first order are associated with systems of contact cubics (cf. **14**). By proper mapping the contacts become the plane sections of a space sextic, the locus of nodes of the net of quadrics on a self-associated set of points, Q_8^3. The purpose of this chapter is to obtain parametric expressions for the coördinates of the sets of points, P_7^2 and Q_8^3, in terms of theta modular functions and to show that the transition to congruent sets under Cremona transformation is due to period transformation of the functions. With respect to such transformation the irrational Göpel invariants of the quartic curve play a fundamental part. For later geometric study of the functions of genus four Cayley's dianodal surface is important. Schottky has given an interesting parametric equation of this surface which may be utilized to study the nature of the section of the generalized Kummer surface $K_p(p=3)$ by one of its 64 contact sections, i. e., the extension of the conic on a trope of the Kummer surface.

43. The figure, P_7^2, of seven coplanar points. The types of Cremona transformations under which a set $P_7^2 = p_1, \cdots, p_7$ may be congruent to a set $Q_7^2 = q_1, \cdots, q_7$ are listed in **6** (10). These types divide into pairs C_0, D_8; A_2, D_7; B_3, D_6; C_4, D_5

and C_5, D_4. In any pair either type is the product of the other and the symmetrical type D_8. With respect to D_8 we prove that

(1) *If two sets P_7^2, Q_7^2 are congruent under D_8 with directions at p_i corresponding to the principal curve $P(q_i^2 q_j \cdots q_o)^3$ and vice versa then P_7^2, Q_7^2 are projective in the natural order.*

Let $K(p)$ be an elliptic cubic on P_7^2 with canonical elliptic parameter u (i. e., $u_1 + u_2 + u_3 \equiv 0$ is the collinear condition) and with parameters $u_1, \cdots, u_7$ for P_7^2. Then $K(p)$ is transformed by D_8 into a cubic $K(q)$ on Q_7^2 and $x(u)$ on $K(p)$ is transformed into $x'(u)$ on $K(q)$. Since the direction at q_i on $K(q)$ arises from the intersection with $K(p)$ outside P_7^2 of the P-curve of q_i, the parameter of q_i on $K(q)$ is $-u_i - \sigma$ $(\sigma = u_1 + \cdots + u_7)$. Three points u, v, w of $K(q)$ are on a line if u, v, w on $K(p)$ are on an octavic with triple points at P_7^2, i. e., if $u + v + w + 3\sigma \equiv 0$. To restore the canonical parameter on $K(q)$ let $u' = u + \sigma$ and then Q_7^2 has canonical parameters $-u_1', \cdots, -u_7'$. Since $K(p)$ and $K(q)$ are birationally equivalent there is a collineation which transforms $K(p)$ into $K(q)$ with $x(u)$ passing into $x'(u')$. Also there is a collineation which leaves $K(q)$ unaltered and sends the point u' into $-u'$. Hence Q_7^2 on $K(q)$ is projective to P_7^2 on $K(p)$. The argument used here is essentially the same as that by which the generator of the group $e_{m,k}$ (cf. 15 (5)) was obtained.

The number of sets Q_7^2 congruent in some order to P_7^2 is the number of ways in which Cremona transformations of the above types can be selected with F-points in P_7^2, i. e., 2 for C_0, D_8; $2\binom{7}{3}$ for A_2, D_7; and $2\binom{7}{1}\binom{6}{2}$, $2\binom{7}{1}\binom{6}{3}$, $2\binom{7}{1}$ for the remaining pairs, or 2. 288 in all. Since congruence under D_8 implies projectivity only 288 are projectively distinct. Hence

(2) *There are* 7! 288 *projectively distinct ordered sets congruent to a given set P_7^2. These ordered sets map as in* 7 (1) *into* 7! 288 *points in Σ_6 conjugate under the Cremona group $G_{7,2}$ in Σ_6. The linear group $g_{7,2}$* (cf. 6 (2)) *is in*

2 to 1 isomorphism with $G_{7,2}$ with the invariant element of $g_{7,2}$ arising from D_8 corresponding to the identity in $G_{7,2}$.

The set P_7^2 has 56 P-curves: 7 sets of directions $P(i)^0$; 21 lines $P(ij)^1$; 21 conics $P(ijklm)^2$; and 7 cubics $P(i^2jklmno)^3$. They divide into 28 pairs each pair making up a cubic of the net K on P_7^2. We set $D_{i8} = P(i)^0 \cdot P(i^2jklmno)^3$ and $D_{ij} = P(ij)^1 \cdot P(klmno)^2$. Since these comprise all the degenerate cubics of the net on P_7^2, this division into pairs is invariant under transformation to congruent sets.

The set P_7^2 has 63 discriminant conditions: 21 of type $\delta_{ij} = \delta_{klmno8} = 0$ which express the coincidence of p_i, p_j in some direction; 35 of type $\delta_{ijk8} = \delta_{lmno} = 0$ which express that p_i, p_j, p_k are collinear; and 7 of type $\delta_{i8} = \delta_{jklmno} = 0$ which express that the six points other than p_i are on a conic. We proceed to identify the 28 pairs D_{ij} $(i, j = 1, \cdots, 8)$ of P-curves with the double points of a general quartic envelope E^4 and the 63 discriminant conditions $\delta = 0$ with the irrational factors of the discriminant of E^4.

Since Q_7^2, congruent to P_7^2 under D_8, is projective to P_7^2 the sets Q_7^2 and P_7^2 may be taken as superposed and D_8 then becomes a transformation I. The square of I is the identity since, under I^2, $P(i)^0$ is invariant. The involutorial transformation I carries the net K of cubics on P_7^2 into itself and interchanges the parts of a degenerate cubic D_{ij}. Within this net K, either every cubic is invariant or else there is a pencil of invariant cubics and one invariant cubic not in the pencil. Since 27 of the invariant cubics D_{ij} can not lie in a pencil, every cubic in K is invariant. The pencil of invariant cubics on x must pass through the partner x' of x under I. Thus I is the Geiser involution (cf. [52] p. 122) of pairs which with P_7^2 make up the base points of a pencil of cubics.

If x' coincides with x along some direction, the pencil on x has contact at x and some member of the pencil has a node at x. The converse is true also, and the locus of fixed points of I is the jacobian, J^6, of the net K, the locus of nodes of the net, a sextic of genus three with nodes at P_7^2. The directions at p_i on the P-curve $P(i^2jklmno)^3$ are self-corresponding and are the directions at p_i on J^6 which passes also through the intersections of $P(ij)^1$ and $P(klmno)^2$. A line ξ is transformed by I into an octavic which meets ξ in the six points common to ξ and J^6 and in the one further pair of corresponding points x, x' on ξ. As ξ revolves about a point y the pair of I on ξ runs over a curve which passes once through y for $\xi = yy'$ and once through p_i for $\xi = yp_i$. The curve is therefore a cubic of the net K, say $K(y)$, which has an equation of the form $(\alpha x)^2 \, |\alpha xy| = 0$. Thus the pairs of I form the principal coincidence of a connex $(1, 2)$, $(\alpha x)^2 (a\xi) = 0$, $(x\xi) = 0$ (cf. [50] II 1 p. 411). The four tangents from y to $K(y)$ are the lines ξ on y on which the pair x, x' coincide at x on J^6 whence the locus of lines ξ on which the involutorial pair is coincident is a quartic envelope E^4 of class four and genus three birationally equivalent to J^6. Two of these four tangents from y coincide only when $K(y)$ has a node and the four coincide into two pairs only when $K(y)$ is binodal and therefore degenerate. Thus the 28 double points of E^4 arise from the points y attached to the 28 cubics D_{ij}. On D_{i8} the lines ξ joining pairs of I are on $y = p_i$, i. e., E^4 has nodes at P_7^2. On D_{12} the node y of E^4 is constructed as follows. If x is the meet of $P(12)^1$ and $P(34)^1$, its partner x' is the meet of the conics $P(34567)^2$, $P(12567)^2$ outside p_5, p_6, p_7 which is rationally known. The line ξ on x, x' meets $P(34567)^2$ in $y = d_{12}$, a node of E^4. Hence the 21 remaining nodes of E^4 are each rationally known and linearly constructible when the seven nodes P_7^2 are given. Since P_7^2 has 6 absolute projective constants, E^4 must also have 6 absolute constants and be a projectively general envelope. Otherwise E^4 would arise from infinitely many sets P_7^2 and have infinitely many nodes. It will appear in the next section that any three of the nodes in P_7^2 are azygetic (cf. 14) and thus that P_7^2 is an Aronhold set of 7 nodes.

If Q_7^2 is congruent to P_7^2 under Cremona transformation T other than I, the net of cubics K on P_7^2 is transformed into the net K' on Q_7^2 and the nodal locus J^6 of K into the nodal locus J'^6 of K'. Since J^6 and J'^6 are thus birationally equivalent, the envelopes E^4, E'^4 defined by P_7^2, Q_7^2 are also birationally equivalent and therefore (as canonical curves of genus three) are projective. The collineation which carries E'^4 into E^4 carries the Aronhold set of nodes Q_7^2 of E'^4 into an Aronhold set of E^4 which is different from P_7^2 since P_7^2, Q_7^2 are not projective. Hence

(3) *The* 288 *sets* P_7^2 *congruent in some order to a given set are projective to the* 288 *Aronhold sets of seven nodes of the quartic envelope* E^4. *The Cremona group* $G_{7,2}$ *in* Σ_6 (cf. (2)) *is the Galois group in the problem of the determination of the* 28 *double tangents of a planar quartic curve* (*the dual of* E^4).

In particular under the quadratic transformation A_{ijk} the nodes d_{l8}, d_{m8}, d_{n8}, d_{o8} of E'^4 arise from the like named nodes of E^4 but the nodes d_{i8}, d_{j8}, d_{k8} of E'^4 arise from the nodes d_{jk}, d_{ik}, d_{ij} respectively of E^4. The collineation mentioned carries the points $q_i, \cdots, q_o$ of Q_7^2 into the nodes $d_{jk}, d_{ik}, d_{ij}, d_{l8}, \cdots, d_{o8}$ of E^4. Since the three inverse F-points of a quadratic transformation are each rationally known when the three direct F-points and four corresponding pairs are given there follows:

(4) *The geometric relations among the* 28 *nodes of a general quartic envelope are consequences of the theorem that* (*with given nodes* $d_{i8}, \cdots, d_{o8}$) *the quadratic transformation with fixed points* $d_{l8}, \cdots, d_{o8}$ *and F-points* d_{i8}, d_{j8}, d_{k8} *has inverse F-points* d_{jk}, d_{ik}, d_{ij}.

Some of the congruence properties of the nodes (cf. [17] II pp. 357–9) are derived by Conner [26] by projection from one of the 8 base points of a net of quadrics.

If the discriminant condition $\delta_{78} = 0$ is satisfied the conic, $P(12 \cdots 6)^2$, factors twice from the octavics which correspond to line sections under I, and once from J^6. The involution I is then of the fourth order with a curve H_2^4 of fixed points with node at p_7 and branch points at $p_1, \cdots, p_6$. The envelope E^4 is the doubly covered pencil of lines on p_7

with branch lines on $p_1, \cdots, p_6$ (cf. [35] pp. 125–6). If $\delta_{567} = 0$, the line $P(567)^1$ factors once from the octavics and once from J^6. Then E^4 has $P(567)^1$ as a double line. If $\delta_{67} = 0$, i. e. if p_7 coincides with p_6 along the line ξ, J^6 has a triple point at p_6 with ξ as one tangent and E^4 has ξ as a double tangent. Thus the discriminant factors of P_7^2 are the conditions that the genus of E^4 shall be reduced and they therefore are factors of the discriminant of E^4. These special cases are perhaps more conveniently studied by mapping the pairs of I upon the points of a plane $S(y)$ by means of the net K (cf. [50] p. 412 (16)). J^6 maps upon a quartic curve C^4 projective to E^4 and the cubics D_{ij} upon the double tangents of C^4.

In order to identify the group $G_{7,2}$ with the group of period transformations, reduced mod. 2, let the discriminant factors δ_{ij}, δ_{ijkl} be identified with the half periods P_{ij}, P_{ijkl} respectively (**23** (5)); and the pairs of P-curves D_{ij} with the odd functions ϑ_{ij} (cf. **25** (1)). The group $G(2)$ is generated by a conjugate set of involutions I_{ij}, I_{ijkl} (**22** (10)) and the effect of I_{ij} is merely to interchange subscripts i and j (**25** (4)). We identify I_{ij} $(i, j = 1, \cdots, 7)$ with the transposition of p_i, p_j and I_{ijk8} with the quadratic transformation A_{ijk}. The planar permutations thus effected on δ_{ij}, δ_{ijkl} and on D_{ij} are precisely those effected by the corresponding involutions of $G(2)$ upon P_{ij}, P_{ijkl} and upon ϑ_{ij} whence $G_{7,2}$ and $G(2)$ are simply isomorphic.

44. The figure, Q_8^3, of eight self-associated points in space. If eight points in space, Q_8^3, are self-associated, the bilinear identity (**16** (2)) connecting the set with its associated set becomes an identity connecting the squares of the eight points whence quadrics on seven of the points pass also through the eighth and the eight are the base points of a net of quadrics. This net and its relation to the planar quartic curve has been discussed in **14**. We consider here more particularly the congruence properties of Q_8^3.

According to **16** (b) for $r = 1$, $s = 0$, the projection of $q_1, \cdots, q_7$ from q_8 is a planar set P_7^2 whose associated set in space is $Q_7^3 = q_1, \cdots, q_7$. If Q_7^3 is congruent to $Q_7'^3$ under the regular cubic transformation A_{ijkl} $(i, j, k, l = 1, \cdots, 7)$ the net of quadrics on Q_7^3 is transformed into the net on $Q_7'^3$, whence the eighth base point q_8 determined by Q_7^3 is transformed into the eighth base point q_8' determined by $Q_7'^3$. Under regular transformation in space Q_7^3 is congruent to 288 projectively distinct sets $Q_7'^3$. For according to 16 (8) the association of the planar set P_7^2 and the spatial set Q_7^3 is unaltered when the two sets are replaced by their congruent sets under respectively A_{ijk} and A_{lmno}. Thus to each type of Cremona transformation in the plane with 7 or less F-points there corresponds in space a type of regular Cremona transformation with 7 or less F-points. To the types with 6 or less F-points given in 36 (5) we add those with 7 F-points:

$$
T'^7:\ \begin{array}{c|cccc}
 & & 1 & 3 & 3 \\ \hline
 & 7 & -3 & -2 & -1 \\
1 & 6 & -2 & -2 & -1 \\
3 & 4 & -2 & -1 & 0,-1 \\
3 & 2 & -1 & 0,-1 & 0
\end{array}\ ; \qquad
T^9:\ \begin{array}{c|cccc}
 & & 3 & 3 & 1 \\ \hline
 & 9 & -3 & -2 & -1 \\
3 & 6 & -2 & -2,-1 & -1 \\
3 & 4 & -2,-1 & -1 & 0 \\
1 & 2 & -1 & 0 & 0
\end{array}\ ;
$$

$$
(1)\quad T'^9:\ \begin{array}{c|ccc}
 & & 1 & 6 \\ \hline
 & 9 & -4 & -2 \\
1 & 8 & -3 & -2 \\
6 & 4 & -2 & 0,-1
\end{array}\ ; \qquad
T^{11}:\ \begin{array}{c|cccc}
 & & 1 & 4 & 2 \\ \hline
 & 11 & -4 & -3 & -2 \\
1 & 8 & -3 & -2 & -2 \\
4 & 6 & -2 & -1,-2 & -1 \\
2 & 4 & -2 & -1 & -1,0
\end{array}\ ;
$$

$$
T^{13}:\ \begin{array}{c|ccc}
 & & 3 & 4 \\ \hline
 & 13 & -4 & -3 \\
3 & 8 & -3,-2 & -2 \\
4 & 6 & -2 & -2,-1
\end{array}\ ; \qquad
T^{15}:\ \begin{array}{c|cc}
 & & 7 \\ \hline
 & 15 & -4 \\
7 & 8 & -3,-2
\end{array}\ .
$$

As for the planar case P_7^2 there are here 288.2 possible types of congruence. Since in the plane congruence under D_8 implies projectivity (cf. 43 (1)) then in space congruence under some T must imply projectivity. From symmetry this T must be T^{15}. A direct verification follows from the product formulae:

$$
(2)\quad \begin{aligned}
D_8 &= A_{567} \cdot A_{234} \cdot A_{156} \cdot A_{127} \cdot A_{347} \cdot (127)\,(34)\,(56); \\
T^{15} &= A_{1234} \cdot A_{1567} \cdot A_{2347} \cdot A_{3456} \cdot A_{1256} \cdot (127)\,(34)\,(56).
\end{aligned}
$$

These formulae are checked by observing that, for example, under the product D_8 the P-curve $P(1)^0$ passes successively into $P(1)^0, P(1)^0, P(56)^1, P(12567)^2, P(12345677^2)^3, P(1^2234567)^3$. This check for each $P(i)^0$ is sufficient. Hence

(3) *If two sets Q_7^3, $Q_7'^3$ are congruent under T^{15} with directions at q_i corresponding to the principal surface $P(q_i'^3 q_j'^2 \cdots q_o'^2)^4$ and vice versa then Q_7^3, $Q_7'^3$ are projective in the natural order. The* 7! 288 *projectively distinct ordered sets $Q_7'^3$ congruent to a given set Q_7^3 map, as in* **7** (1) *into* 7! 288 *points in Σ_6 conjugate under the Cremona group $G_{7,3} = G_{7,2}$. The linear group $g_{7,3}$, simply isomorphic with $g_{7,2}$, is in* 2 *to* 1 *isomorphism with $G_{7,3}$.*

If Q_7^3, $Q_7'^3$, congruent under T^{15} and therefore projective, are superposed, the square of T^{15} is the identity and T^{15} is an involutorial transformation I^{15}. The set Q_7^3 has 2.63 P-surfaces which pair off under I^{15} into the 63 pairs:

$$(4) \qquad \begin{aligned} \Delta_{i8} &= \Delta_{jklmno} = P(i)^0 \cdot P(i^3 j^2 \cdots o^2)^4; \\ \Delta_{ijk8} &= \Delta_{lmno} = P(ijk)^1 \cdot P(ijkl^2 \cdots o^2)^3; \\ \Delta_{ij} &= \Delta_{klmno8} = P(i^2 k \cdots o)^2 \cdot P(j^2 k \cdots o)^2. \end{aligned}$$

These pairs comprise all degenerate quartic surfaces with nodes at Q_7^3 and the pairing is invariant under transformation to $Q_7'^3$. The set Q_7^3 defines also 28 rational curves which play the part of F-curves of the second kind for the transformations (1) and **36** (5). They comprise the 21 lines and 7 cubic curves:

$$(5) \qquad F_{ij} = F(ij)^1; \qquad F_{i8} = F(jk \cdots o)^3.$$

The involution I^{15} carries the net of quadrics on Q_7^3 into itself whence q_8 is a fixed point of I^{15}. The linear transformation effected by I^{15} within this net is the identity. For it is easily verified from the product expression (2) that each of the curves F_{ij} is invariant under I^{15} whence the pencil of quadrics on each curve F_{ij} is invariant. The common member of any two of these pencils must be invariant and therefore every quadric of the net is invariant. Each of the ∞^2 elliptic quartic curves C^4 on Q_7^3, and therefore on $Q_8^3 = Q_7^3, q_8$, as the base curve of a pencil in the net, is invariant under I^{15}. If A_0, A_1, A_2 are three quadrics which define the net, then

$$\sum a_{ij} A_i A_j = 0 \qquad (i, j = 0, 1, 2) \tag{6}$$

is a system (∞^5) of *syzygetic* 8-nodal quartic surfaces with nodes at Q_8^3. Each surface of this system is invariant. Moreover Δ_{ij} in (4) is a quartic surface invariant under I^{15} with nodes at Q_7^3 but not on q_8. Hence the linear system (∞^6) of all quartic surfaces with nodes at Q_7^3 is

$$a B + \sum a_{ij} A_i A_j = 0 \tag{7}$$

where B is any surface of the system not on q_8. Each surface of the system (7) is invariant under I^{15}. If then C^4 is an elliptic quartic curve on Q_8^3 with canonical elliptic parameter u and with $u_1, \cdots, u_8$ $(\sigma = u_1 + \cdots + u_8 \equiv 0)$ as parameters of Q_8^3, the involution on C^4 cut out by the system (7), and therefore by I^{15}, is $u + u' \equiv 2u_8$. The fixed points, for which $2u \equiv 2u_8$, are $u_8 = q_8$, or $u_8 + \omega_1/2$, $u_8 + \omega_2/2$, $u_8 + (\omega_1 + \omega_2)/2$. If r_8 is one of the last three, we define the eight points $R_8^3 = Q_7^3, r_8$ to be a *half-period set of eight points.* The sum of the canonical elliptic parameters of such a half period set on the unique C^4 through it is a proper elliptic half period. All surfaces (7) on r_8 touch C^4 at r_8 whence there is a system (∞^3) of surfaces (7) with a node at r_8. If A_0, A_1 are two quadrics of the net on r_8, this system has the form

$$a B + a_{00} A_0^2 + 2 a_{01} A_0 A_1 + a_{11} A_1^2 = 0. \tag{8}$$

The members of this system for which $a \neq 0$ are *azygetic* 8-nodal quartic surfaces. We observe that quartic surfaces with eight given nodes apparently lie in a system (∞^2). But the C^4 on the eight nodes must contain the eight points, either as a half period set (one condition on R_8^3), or as a self-associated set (three conditions on Q_8^3). Thus the conditions on the nodes relieve the conditions on the surface and the system is either (∞^3) as in (8) or (∞^5) as in (6). If B has nodes at R_8^3, the system (∞^3), $\lambda B + \lambda_0 A_0^2 + \lambda_1 A_1^2 + \lambda_2 A_2^2$, has at least one member in common with (8) whence r_8 is on the Jacobian $J(B, A_0, A_1, A_2)$. This sextic surface is Cayley's "dianode" (cf. [11] p. 35). As the locus of the eighth node of azygetic quartic surfaces with nodes at Q_7^3 it must be transformed by A_{ijkl} into the dianode determined by the congruent set and hence must have triple points at Q_7^3. As the locus of fixed points of I^{15} other than q_8 it must have the same tangent cone at q_i $(i = 1, \cdots, 7)$ as the P-surface $P(i^3 j^2 k^2 \cdots o^2)^4$. Since the matrix of J determined by A_0, A_1, A_2 vanishes on the nodal locus $C^6(y)$ of the net (cf. 14), the dianode contains $C^6(y)$. An elliptic quartic C^4 on Q_7^3 is on four nodal quadrics with nodes on $C^6(y)$. The three reflections in the pairs of opposite edges of the tetrahedron of these nodes carries q_8 into the three points in which C^4 cuts the dianode outside Q_7^3. If C^4 is nodal with node on $C^6(y)$ it cuts the dianode in one further point, and if binodal is contained in part at least on the dianome. Thus the dianode contains the 28 F-curves (5). A parametric expression for the dianode is given in 48.

The relation of the set Q_7^3 to the double tangent problem is precisely similar to that of its associated set P_7^2 (cf. [30] § 7). This becomes evident under projection from q_8. The ∞^2 curves C^4 project into the net of cubic curves on P_7^2 and the nodal locus $C^6(y)$ projects into J^6. The 28 degenerate curves C^4 project into the 28 degenerate cubics on P_7^2 each associated with a double point of the quartic envelope E^4. From this point of view a full discussion has been given by Conner[26].

The eight sets of seven points found in the self-associated Q_8^3 are congruent Q_7^3's. For, under the transformation T^7 in 36 (5), in its involutorial form I^7, the set $q_1, \cdots, q_6, q_7$ is congruent to $q_1, \cdots, q_6, q_8$. The 288 projectively distinct sets Q_7^3 congruent to a given set are then presumably found in 36 congruent self-associated sets Q_8^3. This is a consequence of the theorem:

(9) *The sets* $Q_8'^3$ *and* $Q_8''^3$ *congruent to the self-associated set* Q_8^3 *under* A_{ijkl} *and* A_{mnop} *respectively are projective to each other.*

For, according to **15** (5) the self-associated set Q_8^3 on an elliptic C^4 with canonical parameters $u_1, \cdots, u_8$ is congruent under A_{1234} to a set $Q_8'^3$ on the same curve with parameters

$$u_i' \equiv u_i - \sigma_{1234}/2, \qquad u_j' \equiv u_j + \sigma_{1234}/2$$
$$(i = 1, \cdots, 4;\ j = 5, \cdots, 8;\ \sigma_{1234} = u_1 + \cdots + u_4).$$

Under A_{5678} the set $Q_8''^3$ is projective to

$$u_i'' \equiv u_i + \sigma_{5678}/2, \qquad u_j'' = u_j - \sigma_{5678}/2.$$

Since $\sigma_{1234} + \sigma_{5678} \equiv 0$, $\sigma_{1234}/2 \equiv -\sigma_{5678}/2 + \omega/2$. Hence $Q_8'^3$ and $Q_8''^3$ on C^4 are projective under the collineation on C^4 whose parametric equation is $u' \equiv u + \omega/2$.

Thus in transforming Q_8^3 into projectively distinct congruent sets $Q_8'^3$ it is sufficient to replace A_{ijk8} by A_{lmno}. Only such sets $Q_8'^3$ will be obtained as arise from congruent Q_7^3's and the 288 congruent Q_7^3's are distributed, as noted earlier, into 36 congruent Q_8^3's. Hence

(10) *The Cremona group* $G_{8,3}$ *in* Σ_9 *determined by a general set* Q_8^3 *is infinite and discontinuous. In* Σ_9 *there is a manifold* M_6, *the map of self-associated* Q_8^3*'s, which is invariant under* $G_{8,3}$ *and upon which the infinite* $G_{8,3}$ *appears as a finite group* $f_{8!36}$. *This* $f_{8!36}$ *is the factor group of an infinite invariant subgroup of* $G_{8,3}$ *which leaves each point of* M_6 *unaltered.*

The constitution of this infinite invariant subgroup of $G_{8,3}$ is determined in ([17] II (44) p. 377). The factor group $f_{8!36}$ is evidently the double tangent group. The manifold M_6 is rational. For, if Q_7^3 is represented by a point in Σ_6, the eighth base point q_8 is rationally known from the involution I^7 determined by Q_6^3 and thus Q_8^3, and the point on M_6 which corresponds to the point on Σ_6, is determined.

The transformation from Q_8^3 to congruent $Q_8'^3$ carries the net of quadrics on Q_8^3 into the net on $Q_8'^3$ and the nodal locus $C^6(y)$ of the one net into the nodal locus $C'^6(y)$ of the other. Plane sections of each nodal locus correspond to contacts of azygetic systems of contact cubics of the birationally equivalent planar quartic. The 36 azygetic systems of contacts are therefore represented on $C^6(y)$ itself by first the plane sections, and second the sections by the homaloidal web $(i^2 j^2 k^2 l^2)^3$ which coincide with those of the web $(m^2 n^2 o^2 p^2)^3$. Hence though the number of regular homaloidal webs determined by Q_8^3 is infinite they cut out on $C^6(y)$ only 36 g_3^6's. The section of $C^6(y)$ by the plane $P(234)^1$ corresponds to the section of the quartic by the azygetic triad of double tangents, d_{34}, d_{24}, d_{23}. Under A_{1234} this triad corresponds to d_{12}, d_{13}, d_{14} which therefore is also azygetic. Thus the seven nodes $d_{18}, \cdots, d_{78}$ of the quartic envelope E^4 attached to P_7^2 are an azygetic set.

The special character of the self-associated Q_8^3 is reflected in the fact that it possesses only a finite number of discriminant conditions. These are first the 28 of type $\delta_{ij}=0$ which expresses coincidence of q_i, q_j; and second the 35 of type $\delta_{ijkl}=\delta_{mnop}=0$ which express that four points are coplanar. For, if $q_i, \cdots, q_l$ are coplanar, one quadric on Q_8^3 must contain this plane as a factor and $q_m, \cdots, q_p$ are on the plane determined by the other factor. Also δ_{1234} is transformed by A_{4567} into the condition that the quadric with node at q_4 is on $q_1, q_2, q_3, q_5, q_6, q_7$, while $\delta_{5678}=\delta_{1234}$ is transformed into δ_{48}; etc. Thus further discriminant conditions coalesce with the 63 mentioned.

45. Theta relations ($p=3$). In the case $p=3$ with 28 odd functions $\vartheta_{ij}=\vartheta_{ij}(u)$ and 36 even functions $\vartheta_{ijkl} =\vartheta_{mnop}=\vartheta_{ijkl}(u)$, we set (cf. **28** (1), (6)) $\vartheta_{ijkl}(0)=c_{ijkl}$ and also attach to each odd function ϑ_{ij} a constant c_{ij}. Later the c's are replaced by the 63 constants $e_{ij}, e_{ijkl}=e_{mnop}$. This transition to constants e leads to simpler results for such products of constants c as have a greater number of distinct factors. A few relations of this type will suffice for the applications in **46**.

We begin therefore with products of like characteristic P_{78} rather than with the theta squares of characteristic zero. Of the 32 products of this sort only 16 have factors of like parity namely

(1) $p_i = \vartheta_{i7}\,\vartheta_{i8},\ \ p_{ijk} = p_{lmn} = \vartheta_{ijk7}\,\vartheta_{ijk8}\ (i, j, \cdots = 1, \cdots, 6).$

These products behave like the 16 theta squares for $p = 2$ in the following respects: they are even; only four are linearly independent (cf. **20** (9)); the first six vanish for $u = 0$ and the last ten do not. Moreover under addition of the half periods formed with indices $1, \cdots, 6$ their behavior is the same as that of the theta squares ($p = 2$). But these properties just mentioned were sufficient to establish the relations **30** I, $\cdots$, VI. From VI there is then obtained

$$\text{(A)} \qquad \Sigma_\alpha \pm c_{\alpha 457}\, c_{\alpha 458}\, c_{\alpha 467}\, c_{\alpha 468} = 0 \quad (\alpha = 1, 2, 3).$$

Let u_{ij} denote the initial term (linear in the variables u_1, u_2, u_3) of the development of $\vartheta_{ij}(u)$; and let

$$\text{(2)} \qquad u_{ij} = c_{ij}\, v_{ij}.$$

In the relation similar to that of **30** II namely

$$\text{(3)} \qquad \Sigma_\alpha \pm c_{\alpha 567}\, c_{\alpha 568}\, \vartheta_{\alpha 7}\, \vartheta_{\alpha 8} = 0 \quad (\alpha = 1, \cdots, 4)$$

let u be increased by the half period P_{58}. Then

$$\Sigma_\alpha \pm c_{\alpha 567}\, c_{\alpha 568}\, \vartheta_{\alpha 578}\, \vartheta_{\alpha 5} = 0.$$

The initial terms then yield

$$\text{(B)} \qquad \Sigma_\alpha \pm c_{\alpha 567}\, c_{\alpha 568}\, c_{\alpha 578}\, c_{\alpha 5}\, v_{\alpha 5} \equiv 0 \quad (\alpha = 1, \cdots, 4).$$

We prove also the further relation connecting these initial terms,

$$\text{(C)} \quad \Sigma_\alpha \pm (c_{\alpha 467}\, c_{\alpha 468}\, c_{\alpha 567}\, c_{\alpha 568}\, c_{\alpha 457}\, c_{\alpha 458}\, c_{\alpha 7}\, c_{\alpha 8}\, v_{\alpha 7}\, v_{\alpha 8})^{1/2} = 0 \quad (\alpha = 1, 2, 3).$$

An algebraic lemma is necessary (cf. [63] p. 256):

(4) *If* $2n$ *linear homogeneous functions* $x_1, \cdots, x_{2n}$ *of* n *variables satisfy an identical relation* $g_1 x_1^2 + \cdots + g_{2n} x_{2n}^2 \equiv 0,$

and if $A_1 x_1 + \cdots + A_{n+1} x_{n+1} \equiv 0$ *and* $B_1 x_1 + \cdots$ $\cdots + B_n x_n + B_{n+2} x_{n+2} \equiv 0$ *then* $A_1^2/g_1 + \cdots + A_{n+1}^2/g_{n+1}$ $= 0$ *and* $A_1 B_1/g_1 + \cdots + A_n B_n/g_n = 0$.

For, the given identity states that the quadric

$$Q = g_1 x_1^2 + \cdots + g_{n+1} x_{n+1}^2 \equiv -(g_{n+2} x_{n+2}^2 + \cdots + g_{2n} x_{2n}^2) = 0$$

in the underlying space S_{n-1} has a node at $x_{n+2} = \cdots = x_{2n} = 0$. But Q in S_{n-1} has a node if Q, as a quadric in S_n with variables $x_1, \cdots, x_{n+1}$, touches the S_{n-1}, $A_1 x_1 + \cdots + A_{n+1} x_{n+1} = 0$. The dual form of Q in S_n is $\xi_1^2/g_1 + \cdots + \xi_{n+1}^2/g_{n+1} = 0$ and it touches $\xi = A$ if $A_1^2/g_1 + \cdots + A_{n+1}^2/g_n = 0$. Secondly the quadric

$$Q' = g_1 x_1^2 + \cdots + g_n x_n^2 \equiv -(g_{n+1} x_{n+1}^2 + \cdots + g_{2n} x_{2n}^2)$$

in S_{n-1} has $x_{n+1}, \cdots, x_{2n}$ as a self-polar n-edron. The pair of S_{n-2}'s, $x_{n+1} = 0$, $x_{n+2} = 0$, is then apolar to Q'. The dual form of Q' is $\xi_1^2/g_1 + \cdots + \xi_n^2/g_n = 0$ and the apolarity condition of the pair, $-A_{n+1} x_{n+1}$, $-B_{n+2} x_{n+2}$, or $A_1 x_1 + \cdots + A_n x_n$, $B_1 x_1 + \cdots + B_n x_n$, is $A_1 B_1/g_1 + \cdots$ $\cdots + A_n B_n/g_n = 0$.

On the transcendental side let the theta functions be of the abelian type derived from an algebraic curve $F(x, y)$ with contact canonical adjoints $\varphi(x, y)$ or, for $p = 3$, $\varphi_{ij}(x, y)$ $(i, j = 1, \cdots, 8)$. Then (cf. [63] pp. 254–6; [67] pp. 271–2) for proper choice of the constant factors in the φ_{ij} the initial terms u_{ij} of the odd functions ϑ_{ij} satisfy the same linear relations as the φ_{ij}. Moreover for the restricted range of values u_1, u_2, u_3 which arise from the normal integrals of the first kind taken between limits x, y; x', y' the odd theta functions can be represented by

$$\vartheta_{\alpha\beta} = k \,\{ \varphi_{\alpha\beta}(x, y) \cdot \varphi_{\alpha\beta}(x', y') \}^{1/2}. \tag{5}$$

If then

$$w_\alpha = u_{\alpha 7}\, u_{\alpha 8} \tag{6}$$

there follows from (3) and (5) that

$$\Sigma_\alpha \pm p_{\alpha 56}(0)\,(w_\alpha w'_\alpha)^{1/2} = 0 \quad (\alpha = 1, \cdots, 4). \tag{7}$$

If x', y' is chosen, first so that $w'_4 = 0$, and second equal to x, y then

$$(8) \qquad \begin{aligned} \Sigma_\alpha K_\alpha (w_\alpha)^{1/2} &= 0 \qquad (\alpha = 1, 2, 3); \\ \Sigma_\alpha \pm p_{\alpha 56}(0)(w_\alpha^{1/2})^2 &= 0 \quad (\alpha = 1, \cdots, 4). \end{aligned}$$

Thus only two of the four functions $w_\alpha^{1/2}$ are linearly independent and according to (4)

$$(9) \qquad \Sigma_\alpha \pm K_\alpha^2 / p_{\alpha 56}(0) = 0 \qquad (\alpha = 1, 2, 3).$$

The same constants K_α appear in (8) when the indices 56 are replaced by 45 or 46 whence by virtue of the relations (A) the K_α's have values

$$(10) \qquad \varrho K_\alpha^2 = p_{\alpha 45}(0) \cdot p_{\alpha 46}(0) \cdot p_{\alpha 56}(0).$$

On setting these values in (8) and applying (1), (2) the desired relation C is obtained. Evidently the relation C expresses the linear relation connecting three members of a contact g_1^4 (cf. [67] pp. 244–5).

We now replace the constants c in relations (A), (B), (C) by constants e from the equations **28** (1). For this a lemma from the finite geometry is useful. If G is a linear system of half periods $P_\varkappa$, or their squared null spaces, and if G is enlarged by adding a quadric Q_α (α and $\varkappa$ properly chosen sets of subscripts) then the quadrics in the enlarged linear system are all of the form $Q_{\alpha\varkappa}$ as $\varkappa$ varies in G, including also $\varkappa = 0$.

(11) *If P_λ is syzygetic with G (i. e. on all the null-spaces of points of G) then P_λ is on all or none of the quadrics $Q_{\alpha\varkappa}$; if P_λ is not syzygetic with G then P_λ is on just half of the quadrics $Q_{\alpha\varkappa}$.*

For if, in $Q_\alpha + Q_{\alpha\varkappa} + P_\varkappa = 0$, α is fixed and $\varkappa$ varies over G and if P_λ is on every null space $P_\varkappa$ then it is on both of Q_α, $Q_{\alpha\varkappa}$ or on neither. Hence P_λ is on all or none of the $Q_{\alpha\varkappa}$ according as it is on or not on Q_α. In the second case let $P_\varkappa$ be one of the points of G whose null space is not on P_λ. Then P_λ is on just one of the pair Q_α, $Q_{\alpha\varkappa}$ and for variable α all the quadrics divide into such pairs.

The space G for each of the three terms in (A) is P_{56}, P_{78}, P_{5678}. The points P_λ syzygetic with G are those in G, the six points P_{ij} $(i,j=1,\cdots,4)$, and the six further points on the line joining these to P_{56}. Any other point P_λ is on only half of the quadrics attached to the four factors and the $(e_\lambda)^{1/2}$ factors out of each product in the sum and divides out of the relation. The 15 points noted are to be tested out on one factor of each product. Thus c_{1457} corresponds to Q_{1457} which is not on P_{14}, P_{23}, P_{1456}, P_{2356}. Hence the first term yields e_{14}, e_{23}, e_{1456}, e_{2356} and the relation reads:

$$\text{(A}'\text{)} \quad e_{14}\, e_{23}\, e_{1456}\, e_{2356} \pm e_{24}\, e_{13}\, e_{2456}\, e_{1356} \pm e_{34}\, e_{12}\, e_{3456}\, e_{1256} = 0.$$

For (B) the space G is P_{78}, P_{68}, P_{67}; and the syzygetic space, an S_3, contains the 15 points P_{678i}, P_{ij} $(i,j=1,\cdots,5)$. These 15 points tested on the factor c_{15} of the first term, i. e. on Q_{15}, yield $e_{15}\, e_{1678}\, e_{5678}\, e_{23}\, e_{24}\, e_{34}$ and, after factoring out e_{5678}, the relation becomes

$$\text{(B}'\text{)} \qquad \Sigma_1^4 \pm e_{15}\, e_{23}\, e_{24}\, e_{34}\, e_{1678}\, v_{15} = 0.$$

In (C) the space G is the plane of P_{78}, P_{45}, P_{56}; its syzygetic space is the plane with points P_{12}, P_{13}, P_{23}, P_{1456}, P_{2456}, P_{3456}, P_{78}. Testing these on Q_{18} we find that

$$\text{(C}'\text{)} \qquad \Sigma_1^3 \pm e_{23}\, e_{1456}\, (v_{17}\, v_{18})^{1/2} = 0.$$

From these relations others are obtained by transformation of the periods. Such transformations are generated by the involutions attached to the particular half periods, I_{ij} and I_{ijkl}, the first of which amounts merely to a transposition of the indices. The validity of the transformed relation is due to the fact that its proof will exactly parallel that of the original relation. Thus relation (A$'$) is defined by the Göpel line, P_{56}, P_{78}, P_{5678}. There is however a Göpel line of type P_{1278}, P_{3478}, P_{5678} which arises from the other by the transformation $I_{3467}\, I_{67}$. With reference to symmetry in 8 indices there are then two types of relations (A$'$) which we denote by (A · 1) and (A · 2). Furthermore with reference to eventual application to P_7^2 and isolated index 8 there is a second type included in (A · 1) whose Göpel line is P_{45}, P_{67}, P_{4567}. These two are indicated by (A·1·1) and (A·1·2) respectively. The three subtypes are:

(A.1.1) $\qquad \Sigma_1^3 e_{14}\, e_{23}\, e_{1456}\, e_{2356} = 0;$

(A.1.2) $\qquad \Sigma_1^3 e_{18}\, e_{23}\, e_{1678}\, e_{1458} = 0;$

(A.2.1) $\quad e_{12}\, e_{34}\, e_{56}\, e_{78} + \Sigma_1^2 \pm e_{2468}\, e_{2358}\, e_{1458}\, e_{1368} = 0.$

For, the factors e and factors v permute under transformation like the points P and quadrics O respectively.

The relation (B′) involves the initial terms of four azygetic odd thetas. That involving $v_{18}, v_{28}, v_{38}, v_{48}$ is transformed into three new types by $I_{1238}, I_{1235}, I_{1258}$. The respective sets of four are:

(1) $v_{18}, v_{28}, v_{38}, v_{48}$; $\qquad$ (2) $v_{23}, v_{31}, v_{12}, v_{48}$;

(3) $v_{18}, v_{28}, v_{38}, v_{67}$; $\qquad$ (4) $v_{25}, v_{15}, v_{38}, v_{48}$.

The corresponding relations are

(B.1) $\Sigma_1^4 \pm e_{18}\, e_{23}\, e_{24}\, e_{34}\, e_{2348}\, v_{18} = 0;$

(B.2) $\Sigma_1^3 \pm e_{18}\, e_{14}\, e_{23}\, e_{1348}\, e_{1248}\, v_{23} \pm e_{12}\, e_{13}\, e_{23}\, e_{1234}\, e_{1238}\, v_{48} = 0;$

(B.3) $\Sigma_1^3 \pm e_{23}\, e_{2348}\, e_{2358}\, e_{1345}\, e_{1245}\, v_{18} \pm e_{23}\, e_{31}\, e_{12}\, e_{48}\, e_{58}\, v_{67} = 0;$

(B.4) $\begin{aligned} &\Sigma_1^2 \pm e_{28}\, e_{34}\, e_{2358}\, e_{2458}\, e_{1348}\, v_{15} \\ &\qquad + \Sigma_3^4 \pm e_{12}\, e_{45}\, e_{1235}\, e_{2458}\, e_{1458}\, v_{38} = 0. \end{aligned}$

The isolation of index 8 yields 13 subtypes as follows:

(B.1.1) $v_{18}, v_{28}, v_{38}, v_{48}$; $\qquad$ (B.3.1) $v_{18}, v_{28}, v_{38}, v_{67}$;

(B.1.2) $v_{15}, v_{25}, v_{35}, v_{45}$; $\qquad$ (B.3.2) $v_{14}, v_{24}, v_{34}, v_{78}$;

(B.1.3) $v_{15}, v_{16}, v_{17}, v_{18}$; $\qquad$ (B.3.3) $v_{78}, v_{71}, v_{72}, v_{34}$;

(B.2.1) $v_{23}, v_{31}, v_{12}, v_{48}$; $\qquad$ (B.3.4) $v_{12}, v_{13}, v_{14}, v_{56}$;

(B.2.2) $v_{23}, v_{31}, v_{12}, v_{45}$; $\qquad$ (B.4.1) $v_{15}, v_{25}, v_{38}, v_{48}$;

(B.2.3) $v_{78}, v_{68}, v_{67}, v_{12}$; $\qquad$ (B.4.2) $v_{15}, v_{25}, v_{67}, v_{87}$;

(B.4.3) $v_{15}, v_{25}, v_{36}, v_{46}$.

The relations (B.i.j) are derived by obvious permutations from the relations (B.i) above. The thirteen projective relations derived from them in the next section are all distinct.

Similarly the relation (C) on the three pairs of odd functions drawn from a Steiner complex yields three types of relations:

$$(\text{C.1}) \qquad \Sigma_1^3 \pm e_{23}\, e_{1456}\, (v_{17}\, v_{18})^{1/2} = 0;$$

$$(\text{C.2}) \qquad \Sigma_1^3 \pm e_{23}\, e_{14}\, (v_{14}\, v_{23})^{1/2} = 0;$$

$$(\text{C.3}) \qquad \Sigma_1^2 \pm e_{2356}\, e_{2378}\, (v_{13}\, v_{24})^{1/2} \pm e_{12}\, e_{34}\, (v_{56}\, v_{78})^{1/2} = 0.$$

From these the following subtypes arise by the permutations indicated:

(C.1.1) $v_{17}\, v_{18},\ v_{27}\, v_{28},\ v_{37}\, v_{38}$;

(C.1.2) $v_{16}\, v_{17},\ v_{26}\, v_{27},\ v_{36}\, v_{37}$;

(C.1.3) $v_{61}\, v_{62},\ v_{71}\, v_{72},\ v_{81}\, v_{82}$;

(C.2.1) $v_{12}\, v_{34},\ v_{13}\, v_{24},\ v_{14}\, v_{23}$;

(C.2.2) $v_{56}\, v_{78},\ v_{57}\, v_{68},\ v_{58}\, v_{67}$;

(C.3.1) $v_{13}\, v_{24},\ v_{23}\, v_{14},\ v_{56}\, v_{78}$;

(C.3.2) $v_{12}\, v_{34},\ v_{56}\, v_{78},\ v_{57}\, v_{68}$.

46. Theta relations as projective and Cremonian relations on P_7^2 and Q_8^3. In translating the theta relations of the preceding section into projective relations certain types of products of the e_{ij} and the v_{ij} recur for which the following abbreviations are employed:

$$(1) \qquad \begin{aligned} (ijkl\,\cdots\cdots) &= e_{ij}\, e_{ik}\, e_{il}\, e_{jk}\, e_{jl}\, e_{kl}\cdots, \\ (ij\cdots;\, kl\cdots) &= e_{ik}\, e_{il}\, e_{jk}\, e_{jl}\cdots\cdots, \\ v_{ijkl}\cdots\cdots &= v_{ij}\, v_{ik}\, v_{il}\, v_{jk}\, v_{jl}\, v_{kl}\cdots, \\ v_{ij\cdots;\,kl\cdots}\cdots &= v_{ik}\, v_{il}\, v_{jk}\, v_{jl}\cdots\cdots. \end{aligned}$$

For applications to P_7^2 the following notation is employed:

$$(2) \qquad \begin{aligned} D_{ijk} &= (ijk)\, e_{ijk8}; & D_i &= e_{i8}\, (jklmno); \\ \xi_i &= v_{i8}\, e_{i8}\, (1234567)^2; & \xi_{ij} &= v_{ij}\, e_{ij}\, (klmno). \end{aligned}$$

If then the relations (A.1.1), (A.1.2), (A.2.1) of **45** are multiplied respectively by (7; 1234), (123; 4567) (4567) $e_{45}\, e_{67}$, (12; 3456) (34; 56), they become

(A.1.1) $D_{237}\, D_{147} \pm D_{247}\, D_{137} \pm D_{347}\, D_{127} = 0,$
(A.1.2) $D_1\, D_{145}\, D_{167} \pm D_2\, D_{245}\, D_{267} \pm D_3\, D_{345}\, D_{367} = 0,$
(A.2.1) $D_7 \pm D_{246}\, D_{235}\, D_{145}\, D_{136} \pm D_{146}\, D_{135}\, D_{245}\, D_{236} = 0.$

The relation (A.1.1) then shows that the D_{ijk} are proportional to the determinants formed from the coördinates of seven points, P_7^2, in the plane. The ratio of any two terms in the relation as written is a double ratio of the four lines from p_7 to $p_1, \cdots, p_4$. The terms of this relation arise from the terms in 45 (A) merely by inclusion or exclusion of certain factors so that the double ratios of four-lines determined by P_7^2 are expressed in terms of the zero values of the even thetas in the form

(3) $\pm c_{1458}\, c_{1468}\, c_{2358}\, c_{2368} / c_{2458}\, c_{2468}\, c_{1358}\, c_{1368}.$

The relation (A.2.1) shows that D_i is the expression of degree two in the six points $p_j, \cdots, p_o$ whose vanishing requires that the six be on a conic. With the invariants D_{ijk} and D_i of P_7^2 thus defined the relation (A.1.2) is a syzygy which can be verified projectively.

The relations (B.i.j) now become

(B.1.1) $\Sigma_1^4 \pm D_{234}\, \xi_1 = 0,$
(B.1.2) $\Sigma_1^4 \pm D_{167}\, \xi_{15} = 0,$
(B.1.3) $\Sigma_5^7 \pm D_{167}\, D_6\, D_7\, \xi_{15} \pm D_{234}\, \xi_1 = 0,$
(B.2.1) $\Sigma_1^3 \pm D_{124}\, D_{134}\, D_1\, D_4\, \xi_{23} \pm D_{567}\, D_{123}\, \xi_4 = 0,$
(B.2.2) $\Sigma_1^3 \pm D_{267}\, D_{367}\, \xi_{23} \pm D_{467}\, D_{567}\, \xi_{45} = 0,$
(B.2.3) $\Sigma_6^7 \pm D_{127}\, D_{345}\, \xi_6 \pm D_{127}\, D_{126}\, D_1\, D_2\, \xi_{67}$
$\pm D_{167}\, D_{267}\, D_6\, D_7\, \xi_{12} = 0,$
(B.3.1) $\Sigma_1^3 \pm D_{234}\, D_{235}\, D_{267}\, D_{367}\, D_2\, D_3\, \xi_1$
$\pm D_1\, D_2\, D_3\, D_4\, D_5\, \xi_{67} = 0,$
(B.3.2) $\Sigma_1^3 \pm D_{157}\, D_{167}\, D_{247}\, D_{347}\, D_7\, \xi_{14} \pm \xi_7 = 0,$
(B.3.3) $\Sigma_1^2 \pm D_{257}\, D_{267}\, D_{156}\, D_{347}\, D_2\, D_7\, \xi_{17} \pm D_1\, D_2\, D_7\, \xi_{34}$
$\pm D_{346}\, D_{345}\, D_{256}\, D_{156}\, \xi_7 = 0,$
(B.3.4) $\Sigma_2^4 \pm D_{256}\, D_{237}\, D_{247}\, D_{134}\, \xi_{12} \pm D_1\, \xi_{56} = 0,$

$$(B.4.1)\quad \Sigma_1^2 \pm D_{234}\, D_{135}\, D_{145}\, D_1\, D_3\, D_4\, \xi_{25} + \Sigma_3^4 \pm D_{467}\, D_{145}\, D_{245}\, D_4\, \xi_3 = 0,$$

$$(B.4.2)\quad \Sigma_1^2 \pm D_{157}\, D_{267}\, D_{234}\, D_6\, D_7\, \xi_{25} \pm D_{157}\, D_{257}\, D_{347}\, D_5\, D_7\, \xi_{67} \pm D_{134}\, D_{254}\, D_{125}\, \xi_7 = 0,$$

$$(B.4.3)\quad \Sigma_1^2 \pm D_{157}\, D_{247}\, D_{237}\, \xi_{25} + \Sigma_3^4 \pm D_{147}\, D_{247}\, D_{367}\, \xi_{46} = 0.$$

These are verified most easily by replacing the D's and ξ's from (2) and removing the common factor to obtain **45** (B.1), $\cdots$, (B.4).

The ξ_i are proportional to v_{i8} which in turn are proportional to the initial terms in u_1, u_2, u_3 in the expansion of the odd thetas. We have seen that these initial terms are themselves proportional to the 28 contact canonical adjoints of the curve of genus three, i. e., to the double points of a quartic envelope E^4. If then u_1, u_2, u_3 are interpreted as line coördinates in a plane the $\xi_i = 0$ are the equations of seven azygetic double points of E^4. According to (B.1.1) these seven points are P_7^2 and the E^4 is that attached to P_7^2 as in **43**. For the envelope is rationally and uniquely known when seven azygetic nodes are given. The 21 $\xi_{ij} = 0$ are then the remaining nodes of E^4. They are related projectively to the seven given nodes and to each other as in (B.i.j).

A similar procedure applied to the relations (C) yields:

$$(C.1.1)\quad \Sigma_1^3 \pm D_{237}\, (D_2\, D_3\, \xi_1\, \xi_{17})^{1/2} = 0,$$

$$(C.1.2)\quad \Sigma_1^3 \pm D_{145}\, (\xi_{16}\, \xi_{17})^{1/2} = 0,$$

$$(C.1.3)\quad \Sigma_6^7 \pm D_7\, D_{127}\, (D_1\, D_2\, \xi_{16}\, \xi_{26})^{1/2} \pm D_{345}\, (\xi_1\, \xi_2)^{1/2} = 0,$$

$$(C.2.1)\quad \Sigma_1^3 \pm (\xi_{23}\, \xi_{14})^{1/2} = 0,$$

$$(C.2.2)\quad \Sigma_5^7 \pm (D_5\, \xi_5\, \xi_{67})^{1/2} = 0,$$

$$(C.3.1)\quad \Sigma_1^2 \pm D_{147}\, D_{237}\, (D_7\, \xi_{13}\, \xi_{24})^{1/2} \pm (\xi_{56}\, \xi_7)^{1/2} = 0,$$

$$(C.3.2)\quad \Sigma_6^7 \pm D_{126}\, D_{346}\, (D_6\, \xi_7\, \xi_{56})^{1/2} \pm D_5\, (D_6\, D_7\, \xi_{12}\, \xi_{34})^{1/2} = 0.$$

From these by rationalization many forms of the equation of E^4 are obtained but none which involve only the arbitrarily given points of P_7^2. We observe that the double point ξ_{ij}

plays the role of a contravariant of P_7^2 of degree -4 in p_i, p_j and of degree -3 in the coördinates of the remaining points. All of the theta relations (A), (B), (C) have now been translated into relations which are purely projective.

The important theorem that the various Aronhold sets of seven nodes are congruent under Cremona transformation may be observed in these relations. It is sufficient to see that the quadratic transformation Q defined by the two F-triangles, ξ_1, ξ_2, ξ_3 and $\xi_{23}, \xi_{31}, \xi_{12}$, and by the fixed point ξ_4, will also have the fixed point ξ_5. Under Q the pencil of lines on ξ_1 becomes the pencil on ξ_{23} in such wise that $a = \xi_1 \xi_2$, $b = \xi_1 \xi_3$, $c = \xi_1 \xi_4$ correspond respectively to $a' = \xi_{23} \xi_{12}$, $b' = \xi_{23} \xi_{31}$, $c' = \xi_{23} \xi_4$. If in this projectivity the line $d = \xi_1 \xi_5$ corresponds to the line $d' = \xi_{23} \xi_5$, i. e., if the double ratios $(ab; cd)$ and $(a'b'; c'd')$ are equal, and if similar equal double ratios exist for another pair of F-points, then ξ_5 is a fixed point. But $(ab; cd) = (a'b'; c'd')$ is expressed in terms of the coördinates by

$$D_{124} D_{135} / D_{125} D_{134} = D_{23,31,5} D_{23,12,4} / D_{23,31,4} D_{23,12,5},$$

and from (B.2.1) it appears that

$$\begin{aligned} &D_{31,12,4} : D_{12,23,4} : D_{23,31,4} \\ &= \pm D_{124} D_{314} D_1 : D_{234} D_{124} D_2 : \pm D_{314} D_{234} D_3 . \end{aligned}$$

From the similar proportion with like signs in which ξ_5 replaces ξ_4 the equality of the double ratios is apparent.

Schottky ([63] p. 279; [58]) effects a similar translation for the sextic of genus three, J^6, the locus of nodes of cubics on P_7^2. For this the canonical contact adjoints ξ_i, ξ_{ij} are to be replaced by the 28 degenerate cubic adjoints of J^6. He sets

$$\begin{aligned} F_{12} &= e_{12} (v_{128})^{1/2}, \\ G_{12} &= (34567) (v_{12} v_{8;\,34567})^{1/2}, \qquad (4) \\ H_1 &= (1234567) (1; 234567)\, v_{18} (v_{8;\,1234567})^{1/2}. \end{aligned}$$

Then (C.1.1), (C.2.1), (C.3.1), (C.3.1) of **45** multiplied respectively by $(7; 123)\, (v_{78})^{1/2}$, $(v_{8;\,1234})^{1/2}$, $(12; 34)\, (1234; 7)\, (v_{8;\,1234})^{1/2}$, $(567)^2\, (1234; 567)\, (12; 34)\, (1234; 7)\, v_{8;\,567}\, (v_{8;\,1234})^{1/2}$ yield:

$$(5)\qquad \Sigma_1^3 \pm D_{237}\, F_{17} = 0,$$

$$(6)\qquad \Sigma_1^3 \pm F_{14}\ \ F_{23} = 0,$$

$$(7)\qquad G_{56} \pm F_{13}\, F_{24}\, D_{237}\, D_{147} \pm F_{23}\, F_{14}\, D_{137}\, D_{247} = 0,$$

$$(8)\qquad H_7 \cdot F_{56} \pm G_{13}\, G_{24}\, D_{147}\, D_{237} \pm G_{14}\, G_{23}\, D_{137}\, D_{247} = 0.$$

From (5) one may infer that the F_{ij} behaves like the line joining p_i, p_j and this is confirmed by (6). From (7) G_{56} appears as the conic on $p_1, \cdots, p_4, p_7$. In (8) $H_7 \cdot F_{56}$ appears as a quartic with nodes at p_5, p_6, p_7 and simple points at $p_1, \cdots, p_4$ which has the line on p_5, p_6 as a factor whence H_7, the residual factor, is the cubic with node at p_7 and on $p_1, \cdots, p_6$. The various relations, coupled with the definitions (4), give rise to a great variety of projective relations among these various curves. From (4) alone there is derived the identity in v:

$$(9)\qquad F_{ij}\, F_{kl}\, G_{ik}\, G_{jl} - F_{ik}\, F_{jl}\, G_{ij}\, G_{kl} = 0.$$

This is one of a number of analogous forms into which the equation of the sextic J^6 can be thrown. It has however the extraneous factor D_{mno}.

The projective relations embodied in (B.i.j) can be transferred at once from their expression above in ξ_i, ξ_{ij} into identities involving the degenerate cubics on P_7^2 by setting

$$(10)\qquad \varrho\, \xi_i = D_i \cdot H_i, \quad \varrho\, \xi_{ij} = F_{ij} \cdot G_{ij} \qquad [\varrho = (v_{8;\,1234567})^{1/2}].$$

This birational transformation from E^4 to J^6 is obtained from the definitions (2) and (4) of the loci involved.

Turning to the self-associated Q_8^3 in space, with the same conventions as in (1), we set

$$(11)\qquad \begin{aligned} D_{ijkl} &= (ijkl)\, e_{ijkl}; \quad D_{i,j} = (iklmnop)\ (i; jklmnop); \\ D &= (ijklmnop) \qquad\qquad (i, j, \cdots = 1, \cdots, 8). \end{aligned}$$

If the two types of modular relations **45** (A.1), (A.2) are multiplied, the first by $(1234; 56)\, e_{56}^2$ and then by $(1234; 78)\, e_{78}^2$, the second by $(12; 3456)\ (8; 123456)^2\ (34; 56)$ and then by $(12; 3456)\ (7; 123456)^2\ (34; 56)$, they are converted into

$$\text{(A.1)} \qquad D_{1456}\,D_{2356} \pm D_{2456}\,D_{1356} \pm D_{3456}\,D_{1256} = 0,$$

$$\text{(A.1)}' \qquad D_{1478}\,D_{2378} \pm D_{2478}\,D_{1378} \pm D_{3478}\,D_{1278} = 0,$$

$$\text{(A.2)} \; D_{8,7} \pm D_{2468}\,D_{2358}\,D_{1458}\,D_{1368} \pm D_{1468}\,D_{1358}\,D_{2458}\,D_{2368} = 0,$$

$$\text{(A.2)}' \; D_{7,8} \pm D_{1357}\,D_{1467}\,D_{2367}\,D_{2457} \pm D_{2357}\,D_{2467}\,D_{1367}\,D_{1457} = 0,$$

the duplication being due to $e_{ijkl} = e_{mnop}$.

The relation (A.1) shows that the D_{ijkl} behave like the determinants formed from the coördinates of eight points $q_1, \cdots, q_8$ in space. The set Q_8^3 is projectively defined by the double ratios formed from the ratio of two terms in a relation of this type. Since (A.1) and (A.1)$'$ arise from the same theta relation, the ratio of two terms in one is the same as that of the corresponding terms in the other whence

$$D_{1456}\,D_{2356} / D_{2456}\,D_{1356} = D_{1478}\,D_{2378} / D_{2478}\,D_{1378}.$$

This states that there is a quadric on $q_1, \cdots, q_4$ with generators $q_5\,q_6$ and $q_7\,q_8$. Similarly there is a quadric on Q_8^3 with generators $q_5\,q_6$ and $q_4\,q_8$ and therefore a pencil of quadrics on Q_8^3 with generator $q_5\,q_6$. Since there must also be a pencil on Q_8^3 with generator $q_5\,q_7$, there must be a net on Q_8^3, i. e., Q_8^3 is a self-associated set. From the form of $D_{8,7}$ in (A.2) we infer that $D_{8,7}$ is that invariant of degree 4 in q_8 and 2 in $q_1, \cdots, q_6$ whose vanishing expresses that there exists a quadric cone with vertex at q_8 and on $q_1, \cdots, q_6$.

Evidently the factors e_{ij}, e_{ijkl} are the discriminant factors of the set Q_8^3. Thus $D_{ijkl} = (i\,j\,k\,l)\,e_{ijkl}$ vanishes and $q_i, \cdots, q_l$ are coplanar if $e_{ijkl} = 0$; then $e_{mnop} = 0$ and the remaining four points are also coplanar. Furthermore D_{ijkl} vanishes if $e_{ij} = 0$, i. e., if the points q_i, q_j coincide in some direction. If the quadric cone implied by the vanishing of $D_{8,7}$ exists for a self-associated Q_8^3, q_7 must coincide with q_8 and $e_{78} = 0$. But the function $D_{7,8}$ of the coördinates of Q_8^3 will also vanish simply if any two of $q_1, \cdots, q_6$ coincide and doubly if any one coincides with q_8. Thus the occurence and multiplicity of the factors e in (11) is accounted for. A useful relation is

$$\text{(12)} \qquad D_{i,j}\,D_{j,i} = D^2.$$

If we set

$$(13)\qquad \begin{aligned} E_i &= (i;\, jklmnop), \quad \text{then} \\ D_{i,j}/D &= D/D_{j,i} = E_i/E_j. \end{aligned}$$

Thus the ratios of the eight E's can be expressed in terms of the $D_{i,j}$ and D and $D_{i,j}$, D^2 can in turn be expressed in terms of the D_{ijkl}. But also the 56 $D_{i,j}$ can be expressed rationally in terms of the $E_1, \cdots, E_8$ and D which are connected by the relation

$$(14)\qquad D^2 = E_1 \cdot E_2 \cdot \cdots \cdot E_8.$$

To one deduction we shall recur, namely

$$(15)\qquad D_{1,2}\, D_{2,3}\, D_{3,1} - D_{2,1}\, D_{3,2}\, D_{1,3} = 0.$$

This is in no sense an identity. For given $q_1, \cdots, q_7$ it is a sextic equation satisfied by the eighth base point q_8.

In order to obtain projective relations from the identities among the linear terms of the odd thetas, let

$$(16)\qquad \begin{aligned} P(123) &= (123)\,(v_{123})^{1/2}, \\ P(1^2 23456) &= (1;\, 23456)^2 (23456)\,(v_{78}\, v_{1;\,23456})^{1/2}, \\ P(1^2 \cdots 4^2 567) &= (1234)^4 (1234;\, 567)^2 (567)\,(v_{1234}\, v_{567;\,8})^{1/2}, \\ P(1^3 2^2 \cdots 7^2) &= (1;\, 234567)^6 (234567)^4\, v_{1;\,234567}. \end{aligned}$$

The theta relations identify these with the P-surfaces determined by $q_1, \cdots, q_7$. For if **45** (C.1), (C.2), (C.3) are multiplied respectively by $e_{78}^2\,(123;\, 78)\,(v_{78})^{1/2}$, $(1234;\, 5)\,(v_{1234;\,5})^{1/2}$, $(2356)\,(1456)\,(136)\,(246)\,(v_{6;\,1234})^{1/2}$, they become

$$(\text{C.1})\qquad D_{2378}\, P(178) \pm D_{1378}\, P(278) \pm D_{1278}\, P(378) = 0,$$

$$(\text{C.2})\qquad P(145)\, P(235) \pm P(245)\, P(135) \pm P(345)\, P(125) = 0,$$

$$(\text{C.3})\qquad \begin{aligned} &\pm D_{2356}\, D_{1456}\, P(136)\, P(246) \\ &\pm D_{1356}\, D_{2456}\, P(236)\, P(146) - P(6^2\, 12345) = 0. \end{aligned}$$

Then (C.1) shows that $P(ijk)$ behaves like the plane on q_i, q_j, q_k and this is confirmed by (C.2) while (C.3) shows that $P(i^2jklmn)$ behaves like the quadric cone with vertex at q_i and on $q_j, \cdots, q_n$. This is to be expected since the functions v_{ij} as defined in (2) are the canonical contact adjoints of the quartic envelope E^4. Then the combination $(v_{ijk})^{1/2}$ has simple zeros at the contacts of a contact cubic of E^4 of the system which maps E^4 into the nodal locus $C^6(y)$ (cf. **44**), and these are the zeros on $C^6(y)$ of the plane section containing the bisecants $q_i q_j$, $q_i q_k$, $q_j q_k$ of $C^6(y)$. Similarly the cone $P(i^2jklmn)$ contains the five bisecants $q_i q_j, \cdots, q_i q_n$ as well as the cubic curve on these six points which cuts $C^6(y)$ at the two points on the bisecant $q_o q_p$. If

$$e_{12}\, e_{1256}\,(v_{58}\, v_{68})^{1/2} \pm e_{28}\, e_{1347}\,(v_{51}\, v_{61})^{1/2} \pm e_{18}\, e_{2347}\,(v_{52}\, v_{62})^{1/2} = 0$$

is multiplied by $(1234)^3(1234567)(1234;567)(12;56)e_{56}(v_{1234}v_{78})^{1/2}$ it becomes

$$(\text{C.1})' \quad \begin{aligned} D_{1256}\, P(1^2 2^2 3^2 4^2 567) = &\pm D_{2,8}\, D_{1347}\, P(1^2 23456)\, P(234) \\ &\pm D_{1,8}\, D_{2347}\, P(2^2 13456)\, P(134). \end{aligned}$$

This expresses the cubic P-surface with nodes at $q_1, \cdots, q_4$ and on q_5, q_6, q_7 in terms of simpler P-surfaces. Transformation by A_{1234} shows that the relation is an identity throughout the entire space and not merely on $C^6(y)$ as its derivation might imply. The transform of this last relation by I_{1567} would furnish an expression for the remaining P-surface, $P(1^3\, 2^2 \cdots 7^2)$. By setting $y = q_8$ in (C.1)′ the discriminant condition which expresses that a cubic surface with nodes at $q_1, \cdots, q_4$ is on $q_5, \cdots, q_8$ turns out to be $E_1^2\, E_2^2\, E_3^2\, E_4^2\, D_{5678}$ and thus coincides with a coplanar condition.

If the relation **45** (B′) is multiplied by (12345) (5; 1234) $(v_{12345})^{1/2}$ it becomes

$$(\text{B.1}) \quad \begin{aligned} &D_{2345}\, P(125)\, P(135)\, P(145)\, P(234) \\ \pm\, &D_{1345}\, P(125)\, P(235)\, P(245)\, P(134) \\ \pm\, &D_{1245}\, P(135)\, P(235)\, P(345)\, P(124) \\ \pm\, &D_{1235}\, P(145)\, P(245)\, P(345)\, P(123) = 0. \end{aligned}$$

It may be proved that this projective relation is an identity throughout the space. Thus the various relations (B) yield a variety of projective relations which may be developed by similar methods.

We have seen that P_7^2 and Q_7^3 are associated and therefore satisfy a bilinear identity. We prove that

(17) *The identical bilinear relation between the associated P_7^2 and Q_7^3 is*

$$\Sigma_1^7\, \xi_i \cdot (\eta\, q_i) / (i; jklmno) \equiv 0.$$

Let η be the plane $q_5\, q_6\, y$ and ξ the line $p_4\, p_7$ in the assumed relation, $\sum \lambda_i\, \xi_i \cdot (\eta\, q_i) \equiv 0$. The relation then reduces to $\Sigma_1^3\, \lambda_1\, D_{147}\, P(156) = 0$. But from (C.1) $\Sigma_1^3\, D_{2356}\, P(156) = 0$. Hence $\lambda_1\, D_{147} : \lambda_2\, D_{247} = D_{2356} : D_{1356}$ which according to (2) and (11) is satisfied by the values λ given in (17).

It thus appears that the theta relations contain implicitly the geometric properties of the figures P_7^2 and Q_8^3 and their related loci of genus three. When properly manipulated they furnish valuable material supplementary to the usual analytic geometry.

47. Schottky's parametric equation of Cayley's dianode surface. For those values of the variables u_1, u_2, u_3 of the theta functions of genus three for which one of the odd or even functions of the first order, say the even function ϑ, vanishes, certain of the four term identities reduce to three term identities subject to the underlying relation $\vartheta(u) = 0$. These simpler identities are used by Schottky[62] to establish a parametric representation for Cayley's dianode sextic surface (cf. 44). He writes apparently in ignorance of the previous work of Cayley and Rohn. The major results are reproduced here.

According to the same argument by which 45 (A) was deduced from 30 VI there follows from 30 II the theorem that the products of the two odd thetas in each of four pairs of a Steiner complex are linearly related. If ϑ_{ij} is one of the eight odd thetas in this relation and u is increased by the half period P_{ij} then ϑ_{ij} is converted into ϑ and, for u such that $\vartheta(u) = 0$, this term disappears. In this way it appears that when $\vartheta = 0$ the following three triads are linearly related:

$$(1) \quad \begin{array}{ll} \text{(a)} & \vartheta_{38}\ \vartheta_{3678},\ \vartheta_{48}\ \vartheta_{4678},\ \vartheta_{58}\ \vartheta_{5678}; \\ \text{(b)} & \vartheta_{1258}\ \vartheta_{3458},\ \vartheta_{1358}\ \vartheta_{2458},\ \vartheta_{56}\ \vartheta_{78}; \\ \text{(c)} & \vartheta_{35}\ \vartheta_{1248},\ \vartheta_{45}\ \vartheta_{1238},\ \vartheta_{58}\ \vartheta_{5678}. \end{array}$$

The index 8 remains isolated so that there are six more typical triads. These give rise however to relations which later are obvious deductions from the three given.

Let now

$$(2) \quad \begin{array}{ll} P_{ijk} = \vartheta_{ijk8}\ \vartheta_{i8}\ \vartheta_{j8}\ \vartheta_{k8}; & P_{i,j} = \Pi\cdot\vartheta_{ij}\ \vartheta_{i8}/\vartheta_{j8}; \\ P_{ijkl} = \Pi^2\cdot\vartheta_{lmn8}/\vartheta_{l8}\ \vartheta_{m8}\ \vartheta_{n8}; & P_i = \Pi^2\cdot\vartheta_{i8}^2 \\ (\Pi = \vartheta_{18}\ \vartheta_{28}\cdots\vartheta_{78};\quad i, j, \cdots = 1, \cdots, 7). & \end{array}$$

Then according to (1a) the 35 functions P_{ijk} are such that any three of the type P_{imn}, P_{jmn}, P_{kmn} are linearly related. All 35 can therefore be linearly expressed in terms of a properly chosen set of four, e. g. P_{234}, P_{134}, P_{124}, P_{123}. If such a set of four, or any four linearly independent combinations of them, be taken as homogeneous coördinates y of a point y in S_3, then P_{ijk} represents a plane. Since all the P_{ijk} with fixed index i can be expressed linearly in terms of three of them, say P_{ijk}, P_{ijl}, P_{ikl}, they all pass through a point q_i in S_3, Hence $P_{ijk} = 0$ represents a plane on the points q_i, q_j, q_k of a Q_7^3. Since the coördinates y are functions of two parameters (u_1, u_2, u_3 subject to $\vartheta = 0$) the point y runs over a surface T.

If the relation (1b) is multiplied by $\vartheta_{18}\cdots\vartheta_{48}\,\vartheta_{58}^2$, it takes the form $a\,P_{5,6} = b\,P_{135}\,P_{245} + c\,P_{125}\,P_{345}$. Hence $P_{5,6}$ is a quadric with node at q_5 and on $q_1, \cdots, q_4$. Due to the similar relation in which the index 7 replaces 4, $P_{5,6}$ is also on q_7, i. e. $P_{5,6}$ is the quadric cone with node at q_5 and on $q_1, \cdots, q_4, q_7$. If the relation (1c) is multiplied by $\vartheta_{18}^2\cdots\vartheta_{48}^2\,\vartheta_{68}\,\vartheta_{78}$, it becomes $a\,P_{1234} = b\,P_{123}\,P_{4,5} + c\,P_{124}\,P_{3,5}$. Hence $P_{1234} = 0$ is a cubic surface with nodes at $q_1, \cdots, q_4$, on q_6 and q_7, and, by virtue of the symmetry in its definition (cf. (2)), on q_5 also. Finally from the four term relation connecting $\vartheta_{68}\vartheta_{78}$, $\vartheta_{61}\vartheta_{71}$, $\vartheta_{62}\vartheta_{72}$, $\vartheta_{63}\vartheta_{73}$ by multiplication with $\vartheta_{18}^2 \cdots \vartheta_{58}^2 \vartheta_{68} \vartheta_{78}^3$ there arises $a P_7 = b G_{1,6} G_{7,1} + c G_{2,6} G_{7,2} + d G_{3,6} G_{7,3}$. Hence P_7 is a quartic surface with triple point at q_7, double points at $q_1, \cdots, q_5$ and by symmetry at q_6 as well. Thus the functions defined in (2) represent the P-surfaces determined by Q_7^3 (cf. **44** (4)).

Certain identical relations of the sixth order follow immediately from (2). These are of four types:

$$
(3) \quad
\begin{aligned}
&(a)\ P_{1,2} P_{2,3} P_{3,1} - P_{2,1} P_{3,2} P_{1,3} = 0;\\
&(b)\ P_{3,4} P_{124} P_{4567} - P_{4,3} P_{123} P_{3567} = 0;\\
&(c)\ P_{123} P_{456} P_7 - P_{4567} P_{1237} = 0;\\
&(d)\ P_{1,2} P_2 - P_{2,1} P_1 = 0.
\end{aligned}
$$

These are not identities in y. They must therefore be equations of the surface T with triple points at Q_7^3. One may see from one or another of the equations (3) that T contains the 21 lines $q_i q_j$ and the seven cubic curves on six points of Q_7^3, and hence must coincide with Cayley's dianode surface (cf. **44**). It will be observed that (3a) is the same equation as **46** (15) if in the latter q_8 is replaced by y whereas T does not contain q_8. The individual term $P_{1,2} P_{2,3} P_{3,1}$ represents a sextic surface with triple points at Q_7^3 of the form

$$\alpha T + (\alpha_0 A_0 + \alpha_1 A_1 + \alpha_2 A_2) B + \sum \alpha_{ijk} A_i A_j A_k = 0.$$

Under I^{15}, $P_{1,2} P_{2,3} P_{3,1}$ interchanges with $P_{2,1} P_{3,2} P_{1,3}$ while T changes sign if B, A_0, A_1, A_2 are unaltered (after removal of the factor which contains the P-surfaces of I^{15}). Hence for proper choice of signs in the quadric cones $P_{i,j}$,

$$
\begin{aligned}
P_{1,2} P_{2,3} P_{3,1} - P_{2,1} P_{3,2} P_{1,3} &= 2\alpha T,\\
P_{1,2} P_{2,3} P_{3,1} + P_{2,1} P_{3,2} P_{1,3} &= 2B(\alpha_0 A_0 + \alpha_1 A_1 + \alpha_2 A_2)\\
&\quad + 2\sum \alpha_{ijk} A_i A_j A_k,
\end{aligned}
$$

and the latter surface is on q_8. Thus the ambiguity is due to the indetermination of sign in the quadric cone $D_{i,j}$.

Schottky defines an L-surface to be a quartic surface with nodes at Q_7^3. He observes the degenerate L-surfaces (cf. 44 (4))

$$(4)\quad L_{i8} = P_i = \Pi^2 \cdot \vartheta_{i8}^2; \quad L_{ij} = P_{i,j} \cdot P_{j,i} = \Pi^2 \cdot \vartheta_{ij}^2; \quad L_{ijk8} = P_{ijk} \cdot P_{mnop} = \Pi^2 \cdot \vartheta_{ijk8}^2,$$

and remarks that the linear system (∞^6) of L-surfaces cuts out on T the curves defined by the linear aggregate of theta squares of which seven are linearly independent when $\vartheta = 0$. He notes that the 21 products L_{ij} and the 35 products L_{ijk8} will have as double curves respectively the elliptic quartic curve E_{ij} of intersection of $P_{i,j}$ and $P_{j,i}$, and the elliptic cubic curve E_{ijk8} of intersection of P_{ijk} and P_{mnop}. To these we should add the 7 elliptic loci E_{i8} which consist of the directions at q_i on the surface P_i. The 63 curves then play the role of a conjugate set under regular Cremona transformation of Q_7^3. Schottky also remarks that the 28 F-curves of Q_7^3 (cf. 44 (5)), F_{ij} and F_{i8}, are exceptional curves on T since each is defined by $u = P_{ij}$, P_{i8}, a half period for which $\vartheta(u) = 0$. For, it is clear from (4) that if a theta square vanishes for one of these half periods, the corresponding degenerate L-surface contains the corresponding F-curve.

The memoir contains, along with the conventional way (cf. 44 (8)) of setting up the system of L-surfaces, a definition of the involution I^{15} from the proof that L-surfaces on y pass through y' and the following construction of I^{15}: planes on y, y' cut out on the elliptic quartic curve C^4 through them and Q_7^3 the same involution as is cut out on C^4 by planes on the tangent to C^4 at the eighth base point q_8. Though the fixed points, q_8 and T, of I^{15} are determined the transformation is not discussed in the light of the Cremona theory. It appears first as a Cremona transformation in an article of S. Kantor[38]. We give finally a brief proof of Schottky's theorem:

(5) *If B is a properly chosen L-surface and A_0, A_1, A_2 are quadrics of the net on Q_7^3 then $T^2 = 4B^3 - g_2 B - g_3$ where g_2, g_3 are polynomials of degrees 4, 6 in A_0, A_1, A_2.*

In $A_0 : A_1 : A_2 = \lambda : \mu : 1$ a particular elliptic curve C^4 on Q_8^3 is determined by λ, μ. On C^4 change the original parameter u to $v = u - u_8$. The parameter of u_8 is then $v = 0$ and the pairs of I^{15} on C^4 are $v + v' = 0$. Since these pairs are cut out on C^4 by a pencil $B - \varrho A_2^2 = 0$ then $\varrho = B/A_2^2$ is an elliptic function on C^4 with zeros $\pm v$ and double pole at $v = 0$, i. e., $\varrho = a\wp(v) + b$. If we set $B' = (B - bA_2^2)/a$ then $\wp(v) = B'/A_2^2$. Also T^2/A_2^6 has double zeros at the half period points and six-fold pole at the origin and is therefore $k(\wp'(v))^2$. Incorporating this constant with T^2 then $T^2 = 4B^3 - g_2 B - g_3$. This expression for T^2 is valid along the general curve C^4 and therefore throughout the space. The g_2, g_3 are the loci of equianharmonic and harmonic curves C^4 respectively. If the curves C^4 are projected from q_8 into a net of cubics on P_7^2 with parameters $\lambda : \mu : 1$ then the invariants S, T of the net, which are the g_2, g_3 of a particular curve, are of degrees 4, 6 in $\lambda : \mu : 1 = A_0 : A_1 : A_2$.

With reference to the mapping given in the next section we add some theorems:

(6) *Each of the* 63 *degenerate L-surfaces,* L_m *in* (4), *meets* T *in a curve of order* 24 *made up of the double elliptic curve* E_m *of* L_m *and of the* 12 *curves* F (cf. 44 (5)) *which meet* E_m. *The* 12 *points thus cut out on* E_m *are six pairs of a hessian correspondence* ($u' = u + \omega/2$) *on* E_m.

For, we observe first that two curves are understood to meet at q_i ($i = 1, \cdots, 7$) only when they have the same direction at q_i. Then E_{18}, the ∞^1 directions at q_1 on the triple point of L_{18} at q_1, has a direction in common with F_{j8} and F_{j1} ($j = 2, \cdots, 7$). The six pairs of directions on F_{j8}, F_{j1} are in a hessian correspondence ([26] pp. 170–2). The situation thus existing with respect to L_1 passes over under regular transformation to a congruent set $Q_7'^3$ into a similar situation with respect to L_m'. Hence L_m and Q_7^3 are similarly related.

(7) *The points of* T *are in* (2,1) *correspondence with the pairs of points on a planar quartic curve* f^4; *and the two pairs on* f^4 *which correspond to one point of* T *are complementary pairs on a line section of* f^4.

If x_1, x_2 are two arbitrary points of f^4 then (cf. [67] p. 231) $\vartheta(u_\alpha^{x_1} + u_\alpha^{x_2} + k) = 0$, and conversely, if $\vartheta(u) = 0$, the two points x_1, x_2 are determined for which $u \equiv u_\alpha^{x_1} + u_\alpha^{x_2} + k$. Since $\vartheta(-u)$ then vanishes also, there is a pair of points x_3, x_4 such that $-u \equiv u_\alpha^{x_3} + u_\alpha^{x_4} + k$. Then

$$u_\alpha^{x_1} + u_\alpha^{x_2} + u_\alpha^{x_3} + u_\alpha^{x_4} + 2k \equiv 0.$$

In the linear series defined by this congruence (cf. **34**) two points are arbitrary and the series must be the canonical g_2^4. Thus complementary pairs x_1, x_2 and x_3, x_4 on a line define values $\pm u$ for which $\vartheta(u) = 0$ and therefore a point on T and conversely.

The geometric construction of this correspondence is obvious on the nodal locus $C^6(y)$. A C^4 on a point P of T is contained on four nodal quadrics with nodes on $C^6(y)$ which correspond to the four points on a line section of f^4. The generators of these four cones with nodes N_{ij} cut out on C^4 involutions $u + u' \equiv \omega_{ij}$ [$\omega_{ij} = (i\,\omega_1 + j\,\omega_2)/2;\ i, j = 0, 1$]. The product of two of these is also the product of the other two and is a hessian correspondence $u' \equiv u + \omega/2$ which sends q_8 into P. This hessian correspondence is effected by the harmonic perspectivity with the opposite edges of the tetrahedron N_{ij} as lines of fixed points. Thus

(8) *On any C^4 on Q_7^3 the tetrahedron of nodes of nodal quadrics on C^4, and the tetrahedron of fixed points of I^{15} on C^4, are desmic.*

48. The generalized Kummer surface ($p = 3$). The theta manifold M_3^{24} in S_7 (cf. **32**), the map of pairs of points $\pm u$ in a parallelotop by the theta squares, as a generalized Kummer surface, has a set of 64 four-fold points, determined by the half periods, $u = 0$, $u = P_{ij}$, $u = P_{ijkl}$. There is also a set of 64 S_6's which touch M_3^{24} along an M_2^{12}, these being the S_6's obtained by equating to zero a particular one of the 64 $\vartheta_m(u)$'s. We denote them by $M_2^{12}(\vartheta_m)$. The 64 points and 64 M_2^{12}'s are transitively permuted by the half period group G_{64} of M_3^{24} so that each has the same projective relation to M_3^{24} as any other in its set.

Since the values $\pm u$ for which a particular ϑ_m, say $\vartheta(u)$, is zero can be spread over the points of Cayley's dianode sextic surface T with triple points at Q_7^3 and then the linear system defined by the remaining theta squares (cf. 47 (4)) is represented on T by the linear system of L-surfaces, there follows that

(1) *The $M_2^{12}(\vartheta)$ along which M_3^{24} is tangent to the S_6 defined by $\vartheta^2(u)=0$ is the map of Cayley's dianode sextic surface T with triple points at Q_7^3 by its linear system of adjoint quartic L-surfaces.*

Thus two L-surfaces meet in a 16-ic curve with four-fold points at Q_7^3, and such curves meet T in $16\cdot6-7\cdot4\cdot3$ or 12 variable points which correspond to the intersections of M_2^{12} by a variable S_5. The transcendental definition of $M_2^{12}(\vartheta)$ is replaced in this way by a purely algebraic definition.

To the 28 half periods for which $\vartheta(u)$ vanishes there correspond on T the 28 rational curves F_{ij} (44 (5)). An L-surface which contains a point of F_{ij} outside Q_7^3 contains the whole curve whence F_{ij} maps into one of the four-fold points, $u=P_{ij}$, of M_3^{24} on $M_2^{12}(\vartheta)$ and the individual points of F_{ij} map into directions about P_{ij}. We prove that

(2) *The 28 four-fold points of M_3^{24}, $u=P_{ij}$, on $M_2^{12}(\vartheta)$ are double points of M_2^{12}.*

The multiplicity of $u=P_{12}$ on M_2^{12} is $12-t$ if an S_4 on P determined by two S_5's in the S_6 of $M_2^{12}(\vartheta)$ meets M_2^{12} in t points outside P_{12}. Two L-surfaces on F_{12} meet again in a curve $\Gamma(1^3\,2^3\,3^4\cdots7^4)^{15}$. A plane π on F_{12} meets each L-surface again in a curve $\Delta\,(12)^3$ and the two curves Δ meet in 7 points outside F_{12}. Hence Γ meets π in 15 points of which 6 are at p_1 or p_2, 7 are not on F_{12}, and two are on F_{12}. Then T which contains F_{12} meets Γ in $6\cdot15-2\cdot3\cdot3-5\cdot3\cdot4-2=10=t$ points outside F_{12} and Q_7^3, and P_{12} is a double point of $M_2^{12}(\vartheta)$.

There is on T, in addition to the 28 rational curves F, a set of 63 elliptic curves E_{i8}, E_{ij}, E_{ijk8} which are simple curves on T and double on the respective degenerate L-surfaces. Thus E_{1238} is the intersection of the plane $P(123)^1$

and the cubic surface $P(1234^2 \cdots 7^2)^3$. It is mapped by L-surfaces into a normal elliptic sextic curve in the S_5 common to the S_6's determined by $\vartheta^2(u)$ and $\vartheta^2_{1238}(u)$. Hence (cf. **47** (6))

(3) *The S_6 containing $M_2^{12}(\vartheta)$ is cut by the S_6's determined by the remaining theta squares in a set of 63 S_5's each of which is tangent to $M_2^{12}(\vartheta)$ along a normal elliptic sextic curve E^6 which contains 12 of the 28 points P_{ij} on M_2^{12}, the 12 being composed of six pairs in a hessian correspondence on E^6.*

The Veronese surface V_2^4 is the map of the plane by the linear system (∞^5) of all conics [70, 71], the quadratic system built from the lines of a net. The existence among the L-surfaces of the quadratic system built from the quadrics A_0, A_1, A_2 of the net on Q_7^3 shows that

(4) *The map of the space S_3 of Q_7^3 by the system of L-surfaces is a quartic cone $M_3^4(O)$ in S_6 with vertex at O, the map of q_8, whose section by an S_5 not on O is a Veronese V_2^4. To the ∞^2 triads of points of T on elliptic quartics C^4 there correspond on $M_2^{12}(\vartheta)$ triads of points on generators of $M_3^4(O)$. The $M_2^{12}(\vartheta)$ is the complete intersection of $M_3^4(O)$ and the cubic spread in S_6,*

$$T^2 = 4B^3 - g_2 B - g_3 = 0$$

(*cf.* **47** (5)).

The $M_3^4(O)$ itself is, like V_2^4, the complete intersection of six quadrics.

The $M_2^{12}(\vartheta)$ contains a system of ∞^2 space sextics of genus four of the special type cut out on a quadric cone by a cubic surface. For any quadric, say A_2, of the net on Q_7^3 cuts T in a curve $(1^3, \cdots, 7^3)^{12}$ which is mapped by L-surfaces into a sextic curve on $M_2^{12}(\vartheta)$. The 12-ic is on the L-surfaces, $A_0 A_2$, $A_1 A_2$, A_2^2 whence its map in S_6 is a space curve which is cut out in its space by the quadric $A_0^2 \cdot A_1^2 - (A_0 A_1)^2 = 0$, a cone with vertex at O and the cubic surface $4B^3 - g_2 B - g_3 = 0$. Space sextics of this type are discussed in **51**.

49. Irrational and rational invariants of the planar quartic. It is well known that a plane section of the enveloping cone of a cubic surface, $(hx)^3 = 0$, from a point y on it is a general quartic curve. The equation of this cone is

$$(1) \quad (Ax)^4 \equiv 3[(hy)(hx)^2]^2 - 4[(hy)^2(hx)] \cdot [(hx)^3] = 0.$$

Let the quartic cone, $(Ax)^4 = 0$, be cut by the plane, $(\beta x) = 0$, in a quartic curve, $(qx)^4 = 0$. An invariant of degree k of $(qx)^4 = 0$ is a symbolic product $\prod(qq'q'')$ and the corresponding invariant of the section by (βx) of $(Ax)^4$ is a symbolic product $\prod(AA'A''\beta)$ of degree k in the coefficients A and of degree $4k/3$ in β and thus of degrees $2k$, $2k$, $4k/3$ in the coefficients h, in y, and in β respectively. Since the section by a particular plane β not on y is not material, the symbolic product must have an extraneous factor $(\beta y)^{4k/3}$ and an essential invariantive factor of degree $2k$ in the coefficients h and of order $2k/3$ in y, i. e.

(2) *The invariants of degree k of a ternary quartic are equal to covariants of a cubic surface of degree $2k$ and order $2k/3$ multiplied by λ^k where λ is an undetermined constant independent of k.*

Such a covariant of $(hx)^3$ has the symbolic form, $\prod(hh'h''h''') \cdots (h^{(r)}y) \cdots$, with $4k/3$ determinant factors. In Cremona's hexahedral form of the cubic surface (cf. 40 (4), (5)) with six variables $a, \cdots, f$ and surface $a^3 + \cdots + f^3 = 0$ the variables are subject to two linear relations, $a + \cdots + f = 0$ and $\bar{a}a + \cdots + \bar{f}f = 0$. According to Clebsch's principle of transference the quaternary symbolic determinants are then to be replaced by senary determinants $(hh'h''h'''1\bar{a})$. On expanding the symbolic form and noting that the coefficients h are now numerical there results:

(3) *An invariant of degree k of the ternary quartic is equal to λ^k times a covariant of the Cremona hexahedral cubic surface of degree $4k/3$ in $\bar{a}, \cdots, \bar{f}$ and of degree $2k/3$ in $a, \cdots, f$.*

Since the surface is mapped by cubic curves on P_6^2 for which the $\bar{a}, \cdots, \bar{f}$ are linear in $p_1, \cdots, p_6$, and the $a, \cdots, f$ are linear in $p_1, \cdots, p_6$ and of degree 3 in a variable point p_7, and since a section of the enveloping cone from y, the map of p_7, is projective to the envelope E^4 with nodes at P_7^2, there follows:

(4) *An invariant of degree k of the quartic envelope E^4 with nodes at P_7^2 is λ^k times an expression of degree $2k$ in each of the points of P_7^2 which vanishes $2k/3$ times for each coincidence of two points.*

The essential factors may be determined for certain invariants of the quartic which involve special positions of the double tangents. Thus the discriminant of degree 27 in the coefficients of E^4 should be of degree 54 in the coördinates of each point of P_7^2. The discriminant factor δ_{18} (cf. 43) is of degree two in each of $p_2, \cdots, p_7$ and δ_{1238} is linear in each of p_1, p_2, p_3. Hence $\Delta = \delta_{18}^2 \cdot\cdot\cdot\cdot\cdot \delta_{78}^2 \delta_{1238}^2 \cdot\cdot\cdot\cdot\cdot \delta_{5678}^2$ is of degree 54 in each point. Since $\delta_{38}, \cdots, \delta_{78}$ and $\delta_{1238}, \cdots, \delta_{1278}$ each vanish simply for the coincidence of p_1, p_2, Δ vanishes to the order 20 for this coincidence. Since the normal order (cf. (4)) is 18, the discriminant factor δ_{12}^2 is implied in Δ.

Again the quartic curve has an undulation when the contacts of a double tangent coincide, and E^4 has the dual singularity. This gives rise to two types of condition on P_7^2: (a) the line $p_1 p_2$ touches the conic on $p_3, \cdots, p_7$; (b) a cubic of the net has a cusp at p_7. In case (a) the conic is an $f(x^2, p_3^2, \cdots, p_7^2)$. Letting $x = p_1 + \lambda p_2$, the discriminant in λ is an $f(p_1^2, p_2^2, p_3^4, \cdots, p_7^4)$. The product of the 21 conditions (a) is an $f(p_1^{72}, \cdots, p_7^{72})$. In (b) the cusp locus p_7 of cubics on $p_1, \cdots, p_6$ is desired. This is, on the cubic surface, the parabolic curve cut out by the hessian. The hessian is an $f(\bar{a}^2, a^4)$ or, in terms of P_7^2, an $f(p_1^6, \cdots, p_6^6, p_7^{12})$. The product of the seven conditions (b) is an $f(p_1^{48}, \cdots, p_7^{48})$. Hence the undulation condition of the quartic is an invariant of degree 60. By a similar argument Morley[48] finds that an invariant of degree 54 vanishes if the quartic curve contains inscribed five-lines.

In order to locate the extraneous factor λ^k in (4) an equation of E^4 itself free of extraneous factors is needed. This is a form (cf.[48]) symmetric in P_7^2 of degrees 4 and 10 in η and p_i respectively, say

(5) $$E^4 = f(\eta^4, p^{10}) = 0.$$

It may be characterized more completely by giving its behavior for variable p_7.

(6) *The form* E^4 *in* (5), *for given* η *and* P_6^2 *and variable* p_7, *is a rational curve of order* 10 *with four-fold points at* P_6^2 *and five-fold tangent* η. *The line* η *and rational* 10-*ic are the transforms respectively of a rational quintic curve* Q_η^5 *with linear parameters* η *and nodes at* Q_6^2 *associated to* P_6^2, *and its perspective conic* K, *by the quintic Cremona transformation with double F-points at* Q_6^2, P_6^2.

For, $E^4 = 0$ in (5) is the locus of the ninth base point, p_7, of pencils of cubics on P_6^2 which touch η. On mapping the plane by cubic curves on P_6^2 into a cubic surface C^3, η becomes a twisted cubic N^3 and p_7 becomes the further intersection with C^3 of a tangent line of N^3. The locus of these tangent lines is a quartic surface which touches C^3 along N^3 and meets C^3 in a rational space sextic. The total intersection is the map from the plane of a 12-ic with four-fold points at P_6^2 from which η^2 must factor leaving the rational 10-ic. The transformation mentioned sends this 10-ic into a conic K and η into the rational quintic Q_η^5 with nodes at Q_6^2. The pencils of cubics on P_6^2 which touch η become pencils on Q_6^2 which touch Q_η^5. But W. Stahl[69] has proved that the pencil of adjoint cubics on two points of a rational quintic is on the point of intersection of the two corresponding tangents of the conic perspective to the quintic. Hence K is the perspective conic of Q_η^5 and touches Q_η^5 at five points.

The invariants of the quartic envelope E^4 in (5) are of degree $3l$ in the coefficients. Thus the one of lowest degree 3 is of degree 30 in $p_1, \cdots, p_7$. According to (4) its effective factor is of degree 6. If $p_2, \cdots, p_7$ are on a conic $(a\eta)^2$, then E^4 becomes $(a\eta)^2 \cdot (p_1\eta)^2$ (cf. [48]) and the first invariant has a zero of the second order. When formed for E^4 in (5) it must contain the factor δ_{18}^2 and similarly the factors $\delta_{28}^2, \cdots, \delta_{78}^2$. Hence

(7) *An invariant of degree* $3l$ *of the quartic envelope* E^4 *in* (5) *contains an extraneous factor* $\delta_{18}^{2l}\cdot\ldots\cdot\delta_{78}^{2l}$ *of degree* $24l$ *in the coördinates of* P_7^2 *and an essential factor of degree* $6l$ *in the coördinates of* P_7^2.

The invariants thus far considered have been rational and integral. They take the same values to within a factor of proportionality ϱ^l when formed for any one of the 288 sets congruent to P_7^2. Such invariants can be formed by symmetrizing simpler *irrational* invariants which are permuted under congruent transformation among the members of a conjugate set. An irrational invariant of P_7^2 under congruent transformation should be of the same degree in each point in order that the ratio of any two in a conjugate set may be independent of factors of proportionality in the coördinates of individual points. It should also be a projective invariant of the seven points and therefore be made up of determinants $|ijk|$. Its degree, and behavior with respect to P_7^2, should be such that it is transformed under congruent transformation, to within a factor depending only on the transformation, into an invariant of the same degree and behavior with respect to the congruent Q_7^2. The simplest polynomial in the coördinates of P_7^2 which satisfies these requirements is one of degree 3 in each point which, when any point is regarded as variable, becomes a cubic curve on the other six, i. e. which vanishes at least once for every coincidence δ_{ij}. The following are examples of such irrational covariants with their values in terms of $a, \cdots, f$; $\bar{a}, \cdots, \bar{f}$ and d_2 (cf. **40** (3), (8), (9), (11); also[17] II (60)):

$$(8)\quad \begin{aligned} 8\,|531|\,|461|\,|342|\,|562|\,|547|\,|217|\,|367| &= -(\overline{cf}+d_2)(c+f) \\ &\equiv [cf+]; \\ 8\,|523|\,|462|\,|341|\,|567|\,|541|\,|217|\,|367| &= (\overline{cf}-d_2)(c+f) \\ &\equiv [cf-]; \\ 8\,\Delta_{78}\,|547|\,|217|\,|367| &= 2\,d_2\,(c+f) \\ &\equiv [cf]. \end{aligned}$$

Each of these has the proper degree in p_i and vanishes at least once for any coincidence. Recalling that the 63 discriminant conditions on P_7^2 are permuted like the points in the finite geometry, we observe that, apart from the 21 coincidences, each of these contains the seven discriminant conditions which correspond to the points on a Göpel plane (cf. 28), and that the three Göpel planes so determined have in common a null line whereas the three irrational invariants are linearly related. Hence

(9) *There are* 135 *irrational Göpel invariants which satisfy a set of* 315 *three term relations which correspond in the finite geometry to the sets of three Göpel planes on each of the* 315 *null lines. By virtue of these relations the Göpel invariants can be expressed in terms of* 15 *which are linearly independent. The* 15 *are subject to a set of* 63 *cubic relations which correspond to the points in the finite geometry or to the discriminant conditions of* P_7^2.

By permutation of the points alone the first two types (8) contribute 30 of the Göpel invariants and the last contributes 15 more. It may be proved (cf. [17]II pp. 382–3) that

$$(10)\quad \begin{aligned} &8\Delta_{28}\,|127|\,|245|\,|236| \\ &= -(\overline{ad}-d_2)(a+d)+(\overline{be}+d_2)(b+e) = [ad,\, be]. \end{aligned}$$

From this by the parallel substitutions 35 (4) (with a change of sign in $a, \cdots, f, d_2$ under odd permutation) the remaining 90 Göpel invariants are obtained. The whole set is obviously made up of the 6 terms $d_2\, a, \cdots, d_2 f$ subject to one relation, and the terms $\overline{cf}(c+f)$ subject to five relations (cf. loc. cit.). If a discriminant condition such as $d_2 = \Delta_{78}$ be isolated, fifteen of the Göpel invariants such as $[cf] = 2\,d_2\,(c+f)$ contain it as a factor. These are expressible in terms of the six, $d_2\, a, \cdots, d_2 f$, themselves subject to the one linear relation, $d_2\,(a+\cdots+f) = 0$, but also to the one cubic relation $d_2^3\,(a^3+\cdots+f^3) = 0$. There must be therefore one such cubic relation for each discriminant condition. The relations may then be denominated

$$R_{ij} = 0,\quad R_{ijkl} = R_{mnop} = 0 \quad (i, j, \cdots = 1, \cdots, 8).$$

In the notation introduced in (8) and (9) for the Göpel invariants the three term relations read as follows:

$$
(11)\quad
\begin{aligned}
&(a)\quad [cf] + [cf+] + [cf-] = 0,\\
&(b)\quad [cf] + [be] + [ad] = 0,\\
&(c)\quad [ad, be] + [ad-] + [be+] = 0,\\
&(d)\quad [ad, be] + [be, ad] + [cf] = 0,\\
&(e)\quad [ad, be] + [be, cf] + [cf, ad] = 0,\\
&(f)\quad [ab, de] + [bc, ef] + [ca, fd] = 0.
\end{aligned}
$$

In particular the 15 Göpel invariants which contain a given discriminant condition satisfy 15 of these relations; e.g. for Δ_{78} the fifteen relations are those in (11b). Furthermore the cubic relation $d_2^3(a^3 + \cdots + f^3)$ may be rewritten as

$$
(12)\quad d_2^3[(a+b)(a+c)(b+c) + (d+e)(d+f)(e+f)] = 0
$$

since $a + \cdots + f = 0$. The three Göpel invariants in each product may be characterized by the fact that two in the same product do not, while two from different products do, occur in a relation (11).

The bearing of the cubic relations is given by the theorem:

(13) *If a set of* 135 *constants g can be expressed by means of the relations* (11) *in terms of* 15 *constants h that are linearly independent, and if the constants h satisfy the* 63 *cubic relations* (9) *then the constants g are the Göpel invariants defined by a set of points P_7^2 or by any set congruent to P_7^2.*

If for example the 15 constants g_{78} of the form $d_2(a+b)$ associated with the discriminant condition Δ_{78} and subject to the 15 relations (11b) are known, and if furthermore they satisfy the cubic relation $R_{78} \equiv d_2^3(a^3 + \cdots + f^3) = 0$ then they are the linear invariants of the sextic line pencil from a point p_7 to points $p_1, \cdots, p_6$, and the double ratios in this pencil can be expressed by properly chosen ratios of two of these fifteen constants, as in $[ab]/[cd]$. If the four points

$p_1, \cdots, p_4$ are selected at a base, and if the 15 constants g_{18} subject to the cubic relation $R_{18} = 0$ are given then the position of the sextic line pencil $p_1 p_2, \cdots, p_1 p_7$ is determined. If also the 15 constants g_{28} subject to $R_{28} = 0$ are given, the position of the sextic line pencil $p_2 p_1, p_2 p_3, \cdots, p_2 p_7$ is determined and the position of the points p_5, p_6, p_7 is also determined. If again the 15 constants g_{38} subject to $R_{38} = 0$ are given, the position of the sextic line pencil $p_3 p_1, p_3 p_2, p_3 p_4, \cdots, p_3 p_7$ is determined and three conditions must be satisfied in order that $p_3 p_5, p_3 p_6, p_3 p_7$ may be on the points p_5, p_6, p_7 previously determined. We prove that these conditions are respectively $R_{67} = 0$, $R_{57} = 0$, $R_{56} = 0$. Thus the cubic relations play the double role of first ensuring the existence of the sextic line pencils and second ensuring their united position in P_7^2. It is sufficient to prove the case for p_5. If p_1, p_2, p_3 is the reference triangle and p_4 the unit point, while p_5 is $x:y:u$, the double ratios on vertices p_1, p_2, p_3 respectively are:

$$(14)\quad \begin{aligned} \frac{|124|\,|135|}{|125|\,|134|} &= \frac{y}{u} = \frac{\Delta_{18}|124|\,|135|\,|167|}{\Delta_{18}|125|\,|134|\,|167|} = \frac{-[ae,\, bc]}{[fd,\, ae]};\\ \frac{|125|\,|234|}{|124|\,|235|} &= \frac{u}{x} = \frac{\Delta_{28}|125|\,|234|\,|267|}{\Delta_{28}|124|\,|235|\,|267|} = \frac{[bd,\, ac]}{-[ef,\, bd]};\\ \frac{|134|\,|235|}{|135|\,|234|} &= \frac{x}{y} = \frac{\Delta_{38}|134|\,|235|\,|367|}{\Delta_{38}|135|\,|234|\,|367|} = \frac{-[cf,\, ab]}{[de,\, cf]}. \end{aligned}$$

The final expressions of these double ratios are obtained from a list of formulae conjugate to (10). From the identity $(y/u)(u/x)(x/y) = 1$ there follows:

$$(15)\quad [ae,\, bc]\,[bd,\, ac]\,[cf,\, ab] + [df,\, ae]\,[ef,\, bd]\,[de,\, cf] = 0,$$

a cubic relation on the constants g which is obviously satisfied if these are the Göpel invariants of a P_7^2. The six factors in (15) all contain the discriminant condition S_{67} (doubly, i. e., once more than it normally occurs); any one of the first three is coupled with any one of the second three in a relation (11e) or (11f); and no two in the same product are coupled in a relation (11). Hence (15) must be $R_{67}=0$ in a form like that of $R_{78}=0$ in (12).

The 15 coefficients α of the quartic spread L^4 in S_7 on which the theta manifold M_3^{24} is a locus of double points satisfy a system of cubic relations of the type (9) (cf. **33**(5), (6)) whence

(16) *The* 15 *coefficients* α *of the quartic spread* L^4 *in* **33** (5) *can be expressed linearly with numerical coefficients in terms of the Göpel invariants of* P_7^2 *and conversely. The leading coefficient* α *is itself the Göpel invariant* $[ad]$ *whose value in terms of the zero values of the even thetas is given in* **28** (9).

For α is invariant under a $G_{168\cdot 2^6}$ containing an invariant G_{2^6} generated by the involutions attached to the points $\{000;\, ijk\}$ of a Göpel plane (**33** (10)). The transition from the characteristic to the basis notation is effected (**33** (8)) by setting

$$(17)\qquad \{111;\, 111\} = \{P_{18}\, P_{4567}\, P_{67};\, P_{12}\, P_{34}\, P_{56}\}.$$

The points of the Göpel plane in the basis notation are then $P_{12}, P_{34}, P_{56}, P_{78}, P_{1278}, P_{3478}, P_{5678}$ and these correspond to the seven discriminant factors of $[ad]$. With $\alpha=[ad]$ the correspondence between Göpel invariants and linear forms in the coefficients α is set up by effecting on the linear forms the operations of the modular collineation group **33** (9) and on the Göpel invariants the corresponding operations of $G_{7,2}$. It is perhaps unnecessary to note that the $1, \cdots, 7$ notation in **33** is not related to that in P_7^2.

If P_7^2 is taken in the canonical form as above with $p_5, p_6, p_7 = x:y:u,\ z:t:u,\ r:s:u$ the Göpel invariants all contain the factor u^2 and are of effective degree 7 in the coördinates x, y, z, t, r, s, u of the point in Σ_6 which is the map of P_7^2. Hence

(18) *The Cremona group* $G_{7,2}$ *in* Σ_6 *is mapped by the linear system of order seven determined by the Göpel invariants upon a collineation group* G *in* S_{14} *which is identical with*

the modular group 33(9). *The space* Σ_6 *is mapped upon a modular manifold* M_6 *defined by a set of* 63 *cubic relations.*

Every invariant of the collineation group G is a rational invariant of the quartic envelope E^4 unless it vanishes on M_6. For example the sum of the squares of the Göpel invariants yields the first invariant of the quartic which in terms of the fifteen α's is, to within a numerical factor,

$$6\,\alpha^2 + 2\sum \alpha_1^2 + \sum \alpha_{234}^2. \tag{19}$$

CHAPTER V

GEOMETRIC ASPECTS OF THE ABELIAN MODULAR FUNCTIONS OF GENUS FOUR.

In the present chapter a variety of geometric situations are presented which have as a common foundation an algebraic curve G_4 of genus four. As a rule the absolute projective invariants of the figures discussed can be expressed either rationally or irrationally in terms of the birational moduli of G_4 and therefore are described as modular functions. They also are related in a transcendental way, which in certain cases is described, to the moduli of the theta functions defined by G_4. Most of the topics are discussed with reference to sets of points and their behavior under regular Cremona transformation. A natural departure would therefore be a study of the set P_8^2 in the plane. This brings to light a certain *special* sextic of genus four whose properties are more easily apprehended by comparison with those of the general G_4. For this reason we begin with a discussion of a birationally general plane sextic curve of genus four, first observed by Caporali ([8] pp. 358–62), whose significance was pointed out by Wirtinger ([75]; [76] pp. 115–7).

50. Wirtinger's plane sextic curve of genus four.

As a starting point consider the general algebraic form

$$(\alpha x)^2 (b\tau)(\beta t) = 0 \tag{1}$$

in which x is a ternary variable and t, τ are digredient binary variables. The form depends upon $6\cdot 2\cdot 2 - 1 = 23$ constants or upon $23 - 8 - 3 - 3 = 9$ absolute projective constants. It is convenient to interpret t, τ as a point on a quadric A in space with generators t, τ, and at times to replace the bilinear combinations of t_0, t_1; τ_0, τ_1 by the coördinates y of a point in S_3 (cf. 17 (6)); in which case (1) becomes

$$(\alpha x)^2 (\gamma y) = 0. \tag{1.1}$$

The latter form depends upon $23-8-15 = 0$ absolute constants so that the transition from (1.1) to (1) amounts to the choice in S_3 of a quadric A with isolated generators.

When τ in (1) is fixed and t varies the conic (1) in S_2 runs through a pencil with four base points x. For one of these points x the equation (1) is satisfied for given τ and any t whence the form factors into forms linear respectively in t and τ and

$$(2) \qquad f^4 = (\alpha x)^2 (\alpha' x)^2 (b b') (\beta \beta') = 0.$$

Thus for fixed τ and variable t the pencil of conics (1) has base points $p_1, \cdots, p_4$ on the quartic curve f^4. If for fixed t the conic meets f^4 in $q_1, \cdots, q_4$ then for this t and variable τ the pencil (1) is on $q_1, \cdots, q_4$. Hence

(3) *The form* (1) *determines on the quartic curve* f^4 *a linear series* g_1^4 *of quadrupels* p_τ *(for fixed* τ*), and a linear series* $g_1'^4$ *of quadrupels* q_t *(for fixed* t*), which are residual in the* g_5^8 *cut out on* f^4 *by conics. Conversely two such residual* g_1^4*'s on* f^4 *determine a form* (1) (cf. 14 (8)).

Let t, τ be a point on A for which the conic (1) is a pair of lines $(\xi x)\cdot(\xi' x)$ on x. Then

$$(4) \qquad (\alpha \alpha' \alpha'')^2 (b \tau) (b' \tau) (b'' \tau) (\beta t) \beta' t) (\beta'' t) = 0$$

and t, τ is on the section of the quadric A by a cubic surface. The point x is then a diagonal point, both of a quadrupel p_τ and of a quadrupel q_t, and its locus W, the sextic of Caporali and Wirtinger, is birationally equivalent to the space sextic curve (4). In order to prove that the space sextic is general consider the line equation of the conic (1)

$$(5) \qquad (\alpha \alpha' \xi)^2 (b \tau) (b' \tau) (\beta t) (\beta' t) = 0,$$

or the more general form

$$(5.1) \qquad (\alpha \alpha' \xi) (\alpha \alpha' \xi') (b \tau) (b' \tau) (\beta t) (\beta' t) = 0.$$

For given ξ, ξ', (5.1) is the equation on A of a quadric section which meets (4) in 12 points t, τ. For each of these points the envelope (5) is $\varrho (\xi x)^2 = 0$ where x is the point of W which corresponds to t, τ on (4). Hence (5.1) vanishes at the 12 points x where ξ, ξ' cut W and the points on A correspond to these 12 points x. When $\xi = \xi'$ these 12 become six coincident pairs and the section (5) of A is by a contact quadric of the space sextic (4). Conversely given a general space sextic of genus four, G_4, on a quadric A, one of its 255 contact systems of quadrics cuts A in a system of curves (5) which contains a ternary parameter ξ quadratically (cf. 14). The reciprocal quadratic ternary form breaks up into the product of (4) and (1); and (1) in turn determines G_4. Hence

(6) *The Wirtinger sextic W, the locus of the vertices of the diagonal triangles of the quadrupels of a g_1^4 (or its residual $g_1'^4$) on a general ternary quartic curve f^4, is a birationally general G_4. It has 9 absolute projective constants and is differentiated from the projectively general plane sextic of genus four (13 absolute constants) by the fact that its line sections are the contacts of its canonical space sextic with one of the 255 systems of contact quadrics* [12].

The two g_1^3's, coresidual to each other in the canonical series of W, which on (4) are cut out by the two systems of generators of A, are, on W, the vertices of the diagonal triangles of g_1^4 and $g_1'^4$ respectively.

The pencil of conics (1) for fixed τ and variable t determines a planar quadratic involution of pairs x, x' such that $(\alpha x)(\alpha x')(b\tau)(\beta t) = 0$ for every t whence

$$(7) \qquad (\alpha x)(\alpha' x)(\alpha x')(\alpha' x')(b b')(\beta \beta') = 0.$$

If x is at a diagonal point of p_τ, an F-point of the involution, x' is any point of the opposite diagonal line and the conic (7) in variables x' is a pair of lines whence

(8) *The equation of W, of degree 6 in the coefficients of* (1), *is the discriminant of the conic* (7) *in variables x'.*

This furnishes an equation of W in the form of a symmetric three-row determinant whose elements are conics. Such a determinant isolates the system of contact quartic adjoints of W associated with the half period determined by the contact system (6) of G_4. The planar quartic f^4 is also a two-row determinant which isolates the residual g_1^4's. But also the space sextic G_4 in (4) is expressed as a symmetric three-row determinant whose elements are bilinear forms in t, τ and thereby the contact system (5) appears. If in this latter determinant the bilinear forms are replaced by linear forms in y as in (1.1) the symmetric three-row determinant is the equation of a four-nodal cubic surface due to Cayley.[10] Evidently there is one such Cayley surface on G_4 for each contact system of quadrics. These were noticed by P. Roth[57], the gist of whose article follows in quite different form.

The form (1.1) determines for every x in S_2 a plane in S_3 which envelopes a surface S, and for every y in S_3 a conic in S_2 which belongs to a web R. The class of S is four since the conics of R determined by y, y' meet in four points. The locus of points y for which the conic of R has a node at x is

$$(9) \qquad (\alpha\alpha'\alpha'')^2 (\gamma y)(\gamma' y)(\gamma'' y) = 0.$$

For any node x there is in general one such nodal conic since $(\alpha x)(\alpha x')(\gamma y)$ vanishes identically in x'. If η is a plane on y which determines this conic with node at x then

$$(10) \qquad (\alpha\alpha'\alpha'')(\gamma\gamma'\gamma''\eta)(\alpha x)(\alpha' x)(\alpha'' x) = 0.$$

The web R contains 4 line squares, $(\xi^i x)^2$, the four common tangents of the pencil of line conics apolar to R. At each of their six intersections ξ_{ij} there is a pencil in R with node at ξ_{ij} whence the point y in (10) is indeterminate and the web of cubic curves (10) for variable η is on the six points ξ_{ij}. Hence the cubic surface (9) is a Cayley cubic surface, C^3, the map of the plane by the system (10), with four nodes at which the directions correspond to points on ξ^i, and containing the edges of the nodal tetrahedron, the points of an edge corresponding to directions about ξ_{ij}. The three lines of the diagonal triangle of the four lines ξ^i make up a cubic of the web (10) corresponding to the tritangent plane η of C^3.

A point x' in (1.1) determines a plane η in S_3 which cuts C^3 in the map from the plane of the cubic curve (10),

$$(\alpha\alpha'\alpha'')(\gamma\gamma'\gamma''\gamma''')(\alpha x)(\alpha' x)(\alpha'' x)(\alpha''' x')^2 = 0,$$

which has a node at x'. For, the polar line of x' is

$$(\alpha\alpha'\alpha'')(\gamma\gamma'\gamma''\gamma''')(\alpha x)(\alpha' x')(\alpha'' x')(\alpha''' x')^2$$
$$\equiv (\gamma\gamma'\gamma''\gamma''')(\alpha x')(\alpha' x')(\alpha'' x')(\alpha''' x')\cdot(\alpha'\alpha''\alpha''')(\alpha x)/3 \equiv 0.$$

Hence the envelope S, the map of the plane by the web R, is the quartic surface of Steiner (Roman surface) reciprocal to the Cayley cubic surface, C^3.

If y is on C^3, the conic $(\alpha x)^2(\gamma y) = 0$ is a line pair ξ', ξ'' on x and

(11) $$(\alpha\alpha'\xi)^2(\gamma y)(\gamma' y) = 0$$

is a conic proportional to $(x\xi)^2$. The pair ξ', ξ'' are partners in the involutorial quadratic correlation determined by the apolar pencil of R. Hence $(\alpha\alpha'\beta)^2(\gamma y)(\gamma' y) = 0$ is a quadric which cuts C^3 in a rational sextic curve, the map of the conic $(\beta x)^2 = 0$ in S_2. If $(\beta x)^2$ is in R and is, say $(\alpha'' x)^2(\gamma'' y') = 0$, this quadric is the polar quadric of y' as to C^3. If $(\beta x)^2$ is a line pair, the quadric cuts C^3 in the two cubic curves on C^3 which map the lines of the pair. If the lines of the pair coincide at ξ as in (11), the quadric (11) touches C^3 along the cubic curve on C^3 which is the map of ξ. Moreover (11) is a quadric cone whose planes correspond in (1.1) to points x on ξ. For, if ξ is $x + \lambda x'$ the quadratic locus of planes (1.1) is

$$[(\alpha x)^2 + 2\lambda(\alpha x)(\alpha x') + \lambda^2(\alpha x')^2](\gamma y) = 0$$

and the locus of points y for which the two planes of this cone coincide is

$$[(\alpha x)^2(\alpha' x')^2 - (\alpha x)(\alpha' x)(\alpha x')(\alpha' x')](\gamma y)(\gamma' y)$$
$$= (\alpha x)(\alpha' x')(\alpha\alpha'\overline{xx'})(\gamma y)(\gamma' y) = (\alpha\alpha'\overline{xx'})^2(\gamma y)(\gamma' y)/2$$
$$= (\alpha\alpha'\xi)^2(\gamma y)(\gamma' y)/2.$$

Hence

(12) *If y is a point of C^3 the conic $(\alpha x)^2(\gamma y)$ is a line pair ξ', ξ''. The points of either line are mapped by (1.1) upon the planes of a quadric cone with vertex at y which touches C^3*

along the cubic curve on C^3 *which is the map by* (10) *of the line. The planes of these two cones are the planes of* C^3 *on* y. *The equations of the cones are given by* (11) *for* $\xi = \xi'$, $\xi = \xi''$.

Into this mapping in S_3 of point x in S_2 upon the point y of C^3 by (10) and upon the tangent plane at y by (1. 1)let the quadric A with generators t, τ be inserted so that the plane $(\gamma\, y)$ becomes the plane section $(b\,\tau)(\beta\, t)$ of A and the form (1. 1) reverts to (1). Then A cuts C^3 in the space sextic G_4 whose equation on A is (4) which is the map from S_2 of the locus of nodes of conics of R determined by points on G_4, i. e. of W. Moreover A as an envelope has in common with C^3 (or its reciprocal, S) the planes of an octavic curve O^8 of class 8 and genus 3 which is the map from S_2 of points of the quartic curve f^4. The pencils of planes on generators t or τ have four planes in common with C^3 or O^8 and thus the two residual g_1^4's are marked on O^8. As x runs over a line ξ in S_2 its corresponding point y runs over a cubic curve $K^3(\xi)$ on C^3 which cuts A and therefore G_4 in the points which correspond to the points of ξ on W; also the corresponding plane (tangent to C^3 at y) runs over the planes of a quadric cone $Q(\xi)$ with vertex y' on C^3 and tangent to C^3 along K^3 (i. e. a contact cone of G_4) with four planes in common with A and O^8 which correspond to the four points where ξ meets f^4. The cone $Q(\xi)$ with vertex at y', and the enveloping cone of A from y', will touch along two common generators if ξ is a double tangent of f^4. Then $Q(\xi)$ will have two points of contact with A and the pencil determined by $Q(\xi)$ and A will contain a pair of planes η, η'. Since $Q(\xi)$ touches G_4 on A at 6 points the planes η, η' also touch G_4 at 6 points and are a pair of tritangent planes of G_4. Since also the six edges of the nodal tetrahedron of C^3 cut A each in two points which are the map of the same point ξ_{ij} on the plane, there follows:

(13) *The birationally general plane sextic* W *of genus four has the projective peculiarity (four conditions) that its six nodes are the six vertices of a four line. The map*

of the plane by the canonical adjoints of W is one of the 255 four-nodal Cayley cubic surfaces C^3 on G_4, the map of W. The planes common to C^3 and the quadric A on G_4 determine in the plane the quartic curve f^4 and its residual g_1^{4}'s. The enveloping cone of C^3 from a point on it breaks up into two quadric cones which are contact cones of G_4 determined by lines ξ of S_2 in (11). *The quadric cone thus determined by a double tangent ξ of f^4, and the quadric A, determine a pencil which contains one of the 28 pairs of tritangent planes in the contact system of quadrics.*

W. P. Milne[45] obtains these and other results by synthetic methods. We quote one theorem:

(14) *The ∞^2 cubic curves determined by the six contacts of the quadrics of a contact system of G_4 meet in four points, the nodes of a Cayley cubic surface on G_4.*

For, these curves are the maps $K^3(\xi)$ of lines ξ in S_2 each of which meets the four lines ξ^i. We observe also that the cones $Q(\xi)$ are on the nodes of C^3 whence the linear system

$$(\alpha\alpha'\beta)^2(\gamma y)(\gamma' y) = 0 \tag{15}$$

is the linear system (∞^5) of quadrics on the four nodes each of which meets C^3 in the map of a conic $(\beta x)^2$. Supplementing further results contained in the memoirs cited, the following theorems may be mentioned:

(16) *The nodal tetrahedra of the 255 Cayley cubic surfaces on G_4 are the only tetrahedra whose edges are bisecants of G_4.*

For, if T is such a tetrahedron with planes y_0, y_1, y_2, y_3 the linear system (∞^3) of Cayley surfaces with nodes at the vertices of T cuts G_4 in a g_3^6, or a g_2^6 if one Cayley surface is on G_4. Since G_4 is on five linearly independent cubic surfaces, all cubic surfaces cut it in a complete g_{14}^{18} and the complete involution obtained by fixing the 12 points on the edges of T is a g_r^6. If r were 3 this involution would be the canonical involution of plane sections whereas a trihedral of T cuts G_4 in the 12 points and the 6 points on the edges of the trihedral which is not a plane section. Hence $r=2$ and there is a Cayley cubic on G_4 with nodal tetrahedron T.

(17) *A regular cubic Cremona transformation with F-points at the nodes of C^3 transforms G_4 into the W-sextic which appears as the general plane section of a sextic surface with four four-fold points.*

For, if T is chosen as above and $A=\sum a_i y_i^2+\sum a_{ij}y_i y_j=0$, the transformation has the form $y_i y_i'=1$. The transform of C^3 is a plane and the transform of A is

$$\sum a_i y_j'^2 y_k'^2 y_l'^2 + y_0' y_1' y_2' y_3' \left(\sum a_{ij} y_k' y_l'\right) = 0.$$

The latter is the general sextic surface of its type with six absolute constants to which the plane contributes three more. The section of the surface by the plane has nodes at the vertices of a four line and is the W-sextic birationally equivalent to G_4.

(18) *The 28 pairs of tritangent planes of G_4 are pairs of an involutorial cubic Cremona transformation for which the planes of T are F-planes. These pairs cut the edges of T in 28 pairs of an involution defined by the pair of nodes on the edge, and the pair of points of G_4 on the edge. The 28 lines on the respective pairs are in a cubic complex.*

With T and A as above, the system of contact quadrics is in a linear system of the form $\sum b_{ij} y_i y_j=0$, and with A determines a system (∞^6) whose apolar net has the form

$$\alpha_0 \eta_0^2+\cdots+\alpha_3 \eta_3^2=0 \qquad (\alpha_0 a_0+\cdots+\alpha_3 a_3)=0.$$

The pairs of planes in the system (∞^6) are the pairs apolar to this net and are pairs of the Cremona involution $\eta_i \eta_i'=a_i$ for which the lines joining pairs are known to lie in a cubic complex. The 28 pairs of tritangent planes are in the system (∞^6) (cf. (13)). Let π be any plane which cuts G_4 in six points on the conic πA. It corresponds in the plane to an adjoint cubic c of W which cuts W in the corresponding six points. On c the inscribed four-line with vertices at the nodes of W determines six points for which $\sum u_{ij} = \omega/2$. Hence the six points of W on c are on a conic $(\beta x)^2 = 0$ which maps into the intersection of C^3 by $B = (\alpha\alpha'\beta)^2(\gamma y)(\gamma' y) = 0$. Then B and A meet π in the same conic πA and there is in the pencil of B and A a pair of planes π, π'. Since $A, \pi\pi', B$ are in a pencil, there follows that, given any adjoint cubic c of W, there exists another c' such that

$$W = cc' + (\xi^0 x) \cdots (\xi^3 x) \cdot (\beta x)^2 \tag{19}$$

where $(\xi^i x)$ is the four line. Then the tangents to $c \cdot c'$ at ξ_{ij} are in the involution containing the pair of tangents of W and the pair of lines ξ^i, ξ^j.

If C_1^3 and C_2^3 are two Cayley cubics on G_4, a member of their pencil on a point of A contains A as a factor, i. e. $C_1^3 + k_2 C_2^3 = \pi A$. Then C_1^3 and C_2^3 cut the plane π in the same cubic curve c_{12} and the nodal tetrahedra of C_1^3, C_2^3 cut π in two inscribed four-lines of c_{12}. Two cases are possible according as these two four-lines in c_{12} correspond to the same or to different half periods on c_{12}. Either of these cases can occur. For given π, c_{12}, and two inscribed four-lines of c_{12} belonging to the same or to different systems on c_{12}, two tetrahedra T_1, T_2 can be found (each in ∞^4 ways) which cut π in the inscribed four lines. For each T_i a C_i^3 is uniquely determined with nodal tetrahedron T_i and plane section c_{12}. The C_1^3 and C_2^3 with common curve c_{12} meet in a further curve G_4.

Two half periods of G_4 associated with surfaces C_1^3, C_2^3 determine a third, associated with C_3^3, in either the syzygetic or the azygetic way. Two cases are possible:

$$\begin{aligned} C_2^3 + k_{23} C_3^3 &= \pi_{23} A, & k_2 C_2^3 + k_3 C_3^3 &= \pi A, \\ C_3^3 + k_{31} C_1^3 &= \pi_{31} A, & l_3 C_3^3 + l_1 C_1^3 &= \pi A, \\ C_1^3 + k_{12} C_2^3 &= \pi_{12} A; & m_1 C_1^3 + m_2 C_2^3 &= \pi A. \end{aligned}$$

In the first case the three surfaces are not in a pencil, whereas in the second case they are in a pencil with πA. If the first case occurs the two inscribed four lines must belong to the same half period of c_{23} on π_{23}. Otherwise on c_{23} two half periods would be isolated but not the third—a lack of symmetry not to be expected. Presumably then the second case occurs when the nodal tetrahedra of C_1^3, C_2^3, C_3^3 meet π in inscribed four lines of c_{123}, one from each of the three systems. With respect to the first case we prove

(20) *Given two tetrahedra T_1, T_2 in general position there are four conics which touch their eight faces. The plane π of any one of these conics is cut by T_1, T_2 in two four-lines of the same system inscribed in a cubic curve c_{12}. The two Cayley cubics, C_1^3, C_2^3, with nodal tetrahedra T_1, T_2 respectively and plane section c_{12} meet again in a G_4.*

For, two four-lines of the same system inscribed in c_{12} touch a conic and if two four-lines touch a conic, the two are inscribed in a cubic c_{12} and belong to the same system. Thus it is necessary only to find a conic which touches the faces of T_1, T_2, the dual of the problem of finding a quadric cone on eight points.

Certain particular cases of the W-sextic are important. As we have seen the W-sextic in the plane is determined by the space figure of a quadric A and a Cayley cubic surface C^3, and on W there is an isolated half period or contact system. The half periods are associated with the discriminant factors of the curve. When one is isolated the others are either syzygetic or azygetic with respect to it. In the present case G_4 can acquire a node in one of two ways: either A touches C^3 or passes through a node of C^3. If A touches C^3, G_4 has a node and the curve of class 8 common to A and C^3 has a double plane whence in S_2 both W and f^4 have nodes. If A passes through a node of C^3 the corresponding line ξ^i factors out of W which then is a quintic with nodes at the points ξ_{jk}. The curve f^4 however has no singularity.

The particular case which is treated more completely in the next section is that for which A is a quadric cone with vertex y^0 and coincident sets t, τ of generators. It is characterized by the vanishing of one of the 136 even thetas for the zero argument. Retaining C^3 as the map of the plane the octavic locus of planes common to A and y^0 is the map from the plane of a conic $k = (\alpha x)^2 (\gamma y^0)$ in the system R. It t is a parameter on this conic and τ a parameter for the generators of A then, for $t = x$ on k, the plane $(\alpha x)^2 (\gamma y) = 0$ on y^0 contains two generators τ whereas each generator τ is on four planes of C^3 determined by four points x on k. Thus there is a relation

$$(21) \qquad (\alpha t)^4 (a \tau)^2 = 0$$

which expresses that generator τ of A is on the plane of C^3 determined by $t = x$ on k. The quadric A no longer divides the system R bilinearly but rather into an isolated conic k and a system quadratic in τ which cuts k in the quadratic system (21) of four-points. Originally a generator of A was on four planes and three points of C^3, the maps respectively of a quadrupel on f^4 and its diagonal triangle on W. The W-sextic still remains with its six nodes at the vertices of a four-line but f^4 is now the doubly covered conic k, i. e., a hyperelliptic curve of genus three with 8 branch points, $(\alpha t)^4 (\alpha' t)^4 (a a')^2 = 0$. These determine on C^3 the eight common tangent planes of C^3 and A. Hence

(22) *The locus of the diagonal triangles of a quadratic system of four-points on a conic k is a W-sextic whose canonical space curve G_4 is on a quadric cone.*

As x in S_2 runs over the line ξ two planes of the quadric cone $Q(\xi)$ with vertex at y' on C^3 correspond to the two points where ξ meets k. If these two points are branch points of (21), the two planes are tangent planes of the cone A, and as before there is a pair of tritangent planes in the pencil of $Q(\xi)$ and A, i. e.

(23) *The lines ξ which determine as in* (13) *the* 28 *tritangent plane pairs in the contact system of G_4 on a quadric cone are the lines joining pairs of the* 8 *branch points on k of the correspondence* (21).

A general form of type (21) in digredient variables t, τ with 14 constants and 8 absolute constants is just sufficient to determine the entire geometric apparatus. For if the normconic k in S_2 is selected and a parameter t is installed upon it, the form determines three linearly independent four points on k and thereby, to within multiples of k, three conics which with k define the system R. On mapping S_2 by the system R upon the planes of C^3, W maps upon G_4 and the quadric cone A on G_4 is determined.

In the next section we find a birationally equivalent form of this G_4—namely, the locus of the ninth node of a plane sextic with eight nodes given at P_8^2, a 9-ic curve with triple points at P_8^2. When the individual points of P_8^2 are given, this 9-ic curve can be mapped upon G_4 in such wise that all of the 120 tritangent planes are rationally known.

51. The planar set, P_8^2, and the space sextic of genus four on a quadric cone. If the F-points of the 35 types of Cremona transformation listed in 6 (10), including the projectivity, are selected in all possible ways from a set P_8^2, the transformations give rise to $2 \cdot 8640$ sets Q_8^2 congruent to P_8^2. But, as with P_7^2, the symmetric type E_{17} of order 17 with eight 6-fold F-points produces a set Q_8^2 which is projective to P_8^2. For, the same method as was used with P_7^2 (cf. 43) shows that if Q_8^2 is congruent to P_8^2 on a cubic K with canonical elliptic parameter u for which $P_8^2 = u_1, \cdots, u_8$ then Q_8^2 is projective to the set $-u_1, \cdots, -u_8$ on K and therefore to P_8^2 itself. If the projective sets P_8^2 and Q_8^2 are superposed, E_{17} becomes an involution I_{17} discovered by Bertini.[4] Hence there are only 8640 projectively distinct sets Q_8^2 congruent in some order to P_8^2, which are permuted by Cremona transformation according to a group of order 8! 8640.

The group $g_{8,2}$ has the order $8!\,8640 \cdot 2$ and in it the projectivity and I^{17} constitute an invariant g_2. The P-curves of the set P_8^2 are paired under I^{17} and this pairing is invariant under the larger group. The 120 pairs are

$$(1)\quad \begin{aligned} O_{190} &= P(1)^0 \cdot P(1^3 2^2 \cdots 8^2)^6; \\ O_{129} &= P(12)^1 \cdot P(1\,2\,3^2 \cdots 8^2)^5; \\ O_{678} &= P(1 \cdots 5)^2 \cdot P(1 \cdots 5\,6^2 7^2 8^2)^4; \\ O_{120} &= P(1^2 3 \cdots 8)^3 \cdot P(2^2 3 \cdots 8)^3; \end{aligned}$$

these having respectively 8, 28, 56, 28 conjugates under $G_{8!}$. They constitute the 120 degenerate sextics with nodes at P_8^2. In the notation $1, 2, \cdots, 8, 9, 0$ the indices 9, 0 are isolated. The 120 discriminant conditions on P_8^2 are also of four types under $G_{8!}$, namely:

$$\begin{gathered}\delta_{12}, \quad \delta_{123}, \quad \delta_{123456}, \quad \delta_{1^2 2345678}; \quad \text{or} \\ \Delta_{12}, \quad \Delta_{1230}, \quad \Delta_{7890}, \quad \Delta_{19},\end{gathered} \tag{2}$$

these representing respectively 28, 56, 28, 8 conjugates whose vanishing implies that two points coincide, three are on a line, six on a conic, and seven on the cubic with node at the eighth. With each of these (cf. **6** (19)) there is associated one of a conjugate set of generators of $g_{8,2}$, say I_{12}, I_{1230}, I_{7890}, I_{19}.

In the basis notation for $p = 4$ with subscripts $1, 2, \cdots, 9, 0$ there are 136 even theta functions, 10 of type ϑ_i and 126 of type $\vartheta_{ijklm} = \vartheta_{nopq}$; 120 odd functions of type ϑ_{ijk}; and 255 half periods of types P_{ij}, P_{ijkl}. If the even function ϑ_0 be isolated, the half periods divide into 120 which satisfy $\vartheta_0(u) = 0$ and 135 which do not. In the finite geometry an E-quadric is not on 120 points and the sub-group of the modular group which leaves it unaltered is generated by the involutions attached as in **22** (10) to these 120 points which are of types P_{12}, P_{1230}, P_{7890}, P_{19} with reference to $G_{8!}$ on the indices $1, \cdots, 8$. If then we identify the discriminant conditions (2) with the points of the finite space not on Q_0, the pairs of F-curves (1) with the odd quadrics Q_{ijk} and the Cremona transformations I above with the involutions attached to the corresponding points P, it is a simple matter to verify that the discriminant conditions and P-curves permute under Cremona transformation just as the points and O-quadrics permute in the finite geometry. It is indeed sufficient to check this for the single transformation I_{1230} since the statement is obvious for $G_{8!}$. The modular group has the order $2^{16} \cdot 255 \cdot 63 \cdot 15 \cdot 3 = 8!\, 8640 \cdot 136$ (cf. **22** (7)) and the subgroup which leaves Q_0 unaltered has the order $8!\, 8640$ whence the Cremona group in Σ_8 defined by P_8^2 is isomorphic with the subgroup of the group of period transformations mod. 2 for $p = 4$, which leaves an even theta function unaltered.

The curve of genus 4 which defines this group is the locus of fixed points of I^{17}. The involution is determined by the web of sextic curves with nodes at P_8^2. If K_0, K_1 are two cubic curves on P_8^2, $\alpha_{00} K_0^2 + 2\alpha_{01} K_0 K_1 + \alpha_{11} K_1^2 = 0$ is a net of sextic curves with nodes at P_8^2, each sextic consisting of two irreducible cubics. Then $P(12)^1 \cdot P(123^2 \cdots 8^2)^5$ is a reducible sextic, not included in the net, which with the net determines a web w on P_8^2. The web w contains all the sextic curves with nodes at P_8^2. For, a system (∞^4) would contain at least a pencil with a fixed part $P(123^2 \cdots 8^2)^5$ and a residual pencil $P(12)^1$. If S is a general curve of w, and K_0 the cubic of the pencil on a point x, then, for proper λ, $S + \lambda K_1^2$ as well as the entire net $\beta(S + \lambda K_1^2) + \beta_0 K_0^2 + \beta_1 K_0 K_1 = 0$ is on x. The base points of this net outside P_8^2 are at the two intersections x, x' of $S + \lambda K_1^2$ and K_0. Hence the net in w which is on x is also on x' and x, x' is a pair of a Cremona involution I. Each sextic S of w is invariant under I. In particular each $(\alpha_0 K_0 + \alpha_1 K_1)^2$ and therefore each cubic, $\alpha_0 K_0 + \alpha_1 K_1$, is invariant. If x is on the sextic $P(1^3 2^2 \cdots 8^2)^6$, x' is at p_1 whence I has the same F-points and P-curves as I^{17} and coincides with it.

If u is the canonical elliptic parameter on K with $u_1, \cdots, u_8$ as the parameters of P_8^2 and u_9 that of p_9, the ninth base point of the pencil K, then $\sigma = u_1 + \cdots + u_8 + u_9 \equiv 0$. The involution cut out on K by sextics with nodes at P_8^2 is $2(\sigma - u_9) + u + u' \equiv 0$ or $u + u' \equiv 2u_9$. Thus K is projected into itself from $-2u_9$, the tangential point of u_9, and p_9 is a fixed point on each K, i. e., a fixed point such that all the directions on it are fixed. The three remaining fixed points on K are $u_9 + \omega_1/2$, $u_9 + \omega_2/2$, $u_9 + (\omega_1 + \omega_2)/2$. The locus of these fixed points has a triple point at p_1 with the same directions at p_1 as $P(1^3 2^2 \cdots 8^2)$. Hence G_4 must be a 9-ic with triple points at P_8^2, the locus of points p_9' which with P_8^2 make up a "half period set of nine points".

Such a half period set is a set of 9 nodes of a sextic curve whence (cf. [20] pp. 251–54)

(3) *The Bertini involution* I^{17} *is the projection of each member of a pencil of cubics on* P_8^2 *and* p_9 *into itself from the tangential point of* p_9. *In addition to the isolated fixed point* p_9, I^{17} *has a locus of fixed points* G_4^9 *with triple points at* P_8^2, *the locus of the ninth node of sextics with given nodes at* P_8^2.

If the pencil K is $\tau_0 (\alpha x)^3 + \tau_1 (\beta x)^3 = 0$, the locus of the tangential point of p_9 is

(4) $$R = (\alpha x)^3 \cdot (\beta x)(\beta p_9)^2 - (\beta x)^3 \cdot (\alpha x)(\alpha p_9)^2 = 0.$$

This quartic curve, necessarily rational, is on P_8^2 and has a triple point at p_9. For, the hessian curve of the pencil is cubic in $\tau_0 : \tau_1$ and three cubics of the pencil have a flex at p_9. The set P_8^2, p_9 is respectively the simple and triple F-points of a Jonquières involution J^5 whose locus, H_3^5, of fixed points has R as the polar of p_9, and has the equation

(5) $$H_3^5 = (\alpha x)^3 \cdot (\beta x)^2 (\beta p_9) - (\beta x)^3 \cdot (\alpha x)^2 (\alpha p_9) = 0.$$

Evidently H_3^5 is the locus of the contacts of tangents from p_9 to cubics of the pencil K. The four contacts on a particular cubic are $-u_9/2 + P\, (P = 0,\ \omega_1/2,\ \omega_2/2,\ (\omega_1+\omega_2)/2)$ and their diagonal triangle is the three-point of G_4^9 on this cubic. The four contacts are also on the polar conic of p_9 as to the cubic. The pencil of polar conics with the pencil of cubics generates H_3^5, and the base points of the pencil of conics are at p_9 and at the three further intersections with H_3^5 of the tangents at its triple point. Hence

(6) *The locus of the tangential point of* p_9 *is the rational quartic* R. *The* G_4^9 *of fixed points of* I^{17} *is the locus of diagonal triangles of the four-points cut out on* H_3^5 *by the pencil* K *or by the pencil of polar conics of* p_9 *as to* K.

The web w is the system of canonical adjoints of G_4^9 which maps G_4^9 upon its canonical space sextic G_4^6. The plane of w is at the same time mapped in $(2, 1)$ fashion upon a quadric cone A with a pair of I^{17} corresponding to a point of A (cf. [73]). For, if

$$(7) \qquad y_0 = S,\ y_1 = K_0^2,\ y_2 = K_0 K_1,\ y_3 = K_1^2,$$

the cone A is $y_1 y_3 - y_2^2 = 0$ and the cubics K map into generators of the cone. The ∞^1 tangent planes of A are the tritangent planes of G_4^6 which arise from the coincidence of the two g_1^3's in the canonical involution. The 120 proper tritangent planes of G_4^6 are those sections of A which correspond to the 120 degenerate sextics (1) in w. Thus the tritangent planes of the particular canonical G_4^6 on a quadric cone can be rationally isolated in terms of the eight isolated points P_8^2. The Cremona group in Σ_8 determined by P_8^2 is the Galois group of the tritangent planes of this particular space sextic. Weber[72] has proved that the transcendental condition satisfied by G_4^6 in this case is $\vartheta_0(0) = 0$.

Any cubic surface on G_4^6 furnishes a cubic polynomial in the sextics of w which is the equation of G_4^9 taken twice. The general curve of order $3k$ with k-fold points at P_8^2 has the form

$$(8) \qquad \begin{gathered} G_4^9 \left[\sum S^a K_0^b K_1^c\right] + \sum S^d K_0^e K_1^f = 0 \\ (2a+b+c = k-3;\quad 2d+e+f = k). \end{gathered}$$

Its transform under I^{17} is obtained by changing the sign of G_4^9.

With the isolation of the individual tritangent planes of G_4^6 all contact systems are rationally known. For all such systems contain members which break up into groups of tritangent planes. Since on the special G_4^6 the even theta function $\vartheta_0(u)$ is of special character, the half periods divide into two classes according as they do or do not satisfy $\vartheta_0(u) = 0$. These classes are:

$$(9) \qquad \begin{gathered} P_{i0},\quad P_{ijkl},\quad P_{ijk9},\quad P_{09}; \\ P_{ij},\quad P_{i9},\quad P_{ijk0},\quad P_{ij90} \quad (i, j, \cdots = 1, \cdots, 8). \end{gathered}$$

They contain respectively 135 and 120 half periods. The pencil of lines on p_1, say $F(1;s)^1$ where s is the linear parameter of the pencil, is converted by I^{17} into an $F(1^3 2^4 \cdots 8^4; s)^{11}$. The line on p_1 meets G_4^9 in 6 variable points which also are on its transform. The product of the line and its transform is a 12-ic with four-fold points at P_8^2 which meets G_4^9 doubly at 6 variable points and thus is the map on the plane of a contact quadric section of G_4^6. The product contains s quadratically and furnishes a quadratic system (∞^1) of contact quadrics. For the seven particular values of s for which $F(1;s)^1$ passes through $p_2, \cdots, p_8$ the product degenerates further into $O_{129}\, O_{290}, \cdots, O_{189}\, O_{890}$, i. e., the system (∞^1) of contact quadrics contains 7 of the 28 pairs of tritangent planes in the system P_{10}. The members of this system P_{10} can be expressed as a quadratic system (∞^2) in terms of the parameters ξ_0, ξ_1, ξ_2 of a line in a plane π; and the quadratic systems (∞^1) contained within it arise from lines ξ on a point of π. It is easily verified that Cremona transformation with F-points in P_8^2, applied to the (∞^1) system $F(1;s)^1 \cdot F(1^3 2^4 \cdots 8^4; s)^{11}$, yields $8 \cdot 135$ systems (∞^1) which lie 8 at a time in the 135 systems (∞^2) determined by the first class (9) of half periods.

Of the 135 systems $(\infty)^2$ the one associated with P_{90} is symmetrical with respect to P_8^2. Its eight systems (∞^1) are

$$(10)\quad S_i = F(i^3 jklmnop; s)^4 \cdot F(ij^3 k^3 l^3 m^3 n^3 o^3 p^3; s)^8 \qquad (i = 1, \cdots, 8).$$

The system S_i determines a point r_i in the plane π of ξ; and the line $\xi = r_i r_j$ determines that pair of tritangent planes in the two systems S_i, S_j. These pairs arise in S_i from the seven values of s for which a member of the pencil $F(i^3 j \cdots p; s)^4$ contains the factors $P(ij)^1, \cdots, P(ip)^1$. This pencil is projective to the complementary pencil in (10) which, under the Geiser involution I^7 with F-points at $p_j, \cdots, p_p$ and corresponding pair p_i, p_9, becomes the line pencil on p_9 while the seven members mentioned become the lines from p_9 to $p_j, \cdots, p_p$. Similarly in the system S_j the pairs of tritangent planes have parameters projective to those of the lines from p_9 to $p_i, p_k, \cdots, p_p$. Hence in π the pencil of lines ξ from r_i to $r_k, \cdots, r_p$ is projective to that from r_j to the same points and the eight points r_i in π are on a conic, the conic k with parameter t in 50 (21) (cf. also 50 (23)).

Following out the geometric developments of the preceding section this quadratic system P_{09} is supposed to be written as a form $f(\xi^2, x^{12})$. For given point r on π, the lines ξ on r determine a system (∞^1) whose envelope is a curve of order 24 which consists of G_4^9 taken twice and a sextic of the web w. This envelope is the point equation in variables r of the conic, $f(\xi^2, x^{12}) = 0$, of lines ξ. Eliminating $(G_4^9)^2$ the equation of the envelope is $\varphi(r^2, x^6) = 0$. For variation of r in $\varphi = 0$ the sextics of w map upon the plane sections of A which envelope the Cayley cubic surface C^3 associated with the quadratic system P_{09}. For r a point with parameter t on k the sextic breaks up into two cubics τ of the pencil on P_8^2. In particular for $r = r_i$ and $t = t_i$, the parameter of the line $p_9 p_i$ in the pencil on p_9, the sextic is the square of the cubic τ which touches the line $p_9 p_i$ at p_i. For, the pair of projective pencils in (10) generates the curve G_4^9 and a residual cubic K_i while the envelope of (10) as a system quadratic in s is the square of the locus thus generated. The particular member

$$P(ij)^1 \cdot P(i^2klmnop)^3 \cdot P(ijk^2 \cdots p^2)^5 \cdot P(j^2k \cdots p)^3$$

contains that point u of K_i in which $P(ij)^1$ meets $P(j^2k \cdots p)^3$. Then $u + u_i + u_j \equiv 0$ and $u + 2u_j + u_k + \cdots + u_p \equiv 0$ while $u_i + u_j + u_k + \cdots + u_p + u_9 \equiv 0$, i. e., $2u_i + u_9 \equiv 0$ and K_i touches $p_9 p_i$ at p_i. From this there follows:

(11) *The form* $(\alpha t)^4 (a\tau)^2 = 0$ (cf. 50 (21)) *which defines the Cayley cubic surface associated with the system* P_{90} *of contact quadrics of* G_4^6 *on the quadric cone* A *is that which expresses that the cubic* τ *of the pencil on* P_8^2 *touches the line* t *of the pencil on* p_9.

We observe that $(\alpha t)^4 (a\tau)^2 = 0$ is the equation of the curve H_3^5 in (5) if the coördinate system is chosen so that

t as before is a line on p_9 while τ is the polar conic of p_9 as to the cubic τ.

It was mentioned above that the set $p_1, p_2, \cdots, p_8$ was congruent to $p_9, p_2, \cdots, p_8$ under a Geiser I^7. Thus the 8640 sets P_8^2 congruent to a given set are distributed 9 at a time in 960 sets P_9^2 and (cf. also [17] II (47))

(12) *The infinite number of projectively distinct sets P_9^2 congruent to a given set reduces to* 960 *when the given set is the base of a pencil of cubics.*

Each of the 8640 sets P_8^2 determines in the above fashion (11) one of the 135 contact systems and each contact system is determined in $8640/135 = 64$ ways. The 64 sets P_8^2 which define the contact system P_{09} are obtained from P_8^2 by the Jonquières group G_{64} of **38** (15) attached to H_3^5.

The 120 systems of contact quadrics of G_4^6 associated with the second type of half-period (9) are entirely different in character from the 135 systems just discussed. The contacts are cut out on G_4^9 in the plane of P_8^2 by nets of elliptic curves. The nets of elliptic curves determined by P_8^2 which cut G_4^9 in six variable points are paired under I^{17} into four pairs of types which are distinct under permutation of the points of P_8^2, namely:

$$(13)\qquad \begin{aligned} &F(1 \cdots\cdots\cdots 7; \xi)^3 \cdot F(1^3 \cdots 7^3\, 8^4; \xi)^9;\\ &F(1 \cdots\cdot 6\, 7^2\, 8^2; \xi)^4 \cdot F(1^3 \cdots 6^3\, 7^2\, 8^2; \xi)^8;\\ &F(1\,2\,3\,4^2 \cdots 8^2; \xi)^5 \cdot F(1^3\, 2^3\, 3^3\, 4^2 \cdots 8^2; \xi)^7;\\ &F(1^3\, 2\, 3^2 \cdots 8^2; \xi)^6 \cdot F(1\, 2^3\, 3^2 \cdots 8^2; \xi)^6. \end{aligned}$$

These are typical of 8, 28, 56 and 28 respectively; and the four given in (13) are associated with P_{89}, P_{7890}, P_{1230}, P_{12} respectively. For example in the first type ξ is the ternary parameter of the cubic of the net $F(1 \cdots 7; \xi)^3$ of cubics on $p_1, \cdots, p_7$ which meets G_4^9 in six variable points. The transform of this net by I^{17} is $F(1^3 \cdots 7^3; \xi)^9$ which for the same ξ cuts G_4^9 in the same six points. The product is a 12-ic, quadratic in ξ, with four-fold points at P_8^2 and therefore quadratic in sextics w, which touches G_4^9 at six points. Its map in S_3 is a quadric which touches G_4^6 on the cone A in six points. In $F(1\cdots 7;\xi)^3$ there are 28 degenerate curves one of which is $P(12)^1 \cdot P(34567)^2$ whose transform by I^{17} is $P(12\,3^2\cdots 8^2)^5 \cdot P(1^2\,2^2\,3\cdots 7\,8^2)^4$. Hence the contact system contains the pair of tritangent planes (cf. (1)) $O_{129}\,O_{128}$ and is associated with P_{89}.

We discuss this type P_{89} further as a sample of the 120 contact systems. Taking again the parameter ξ as a line ξ in a plane Π, the G_4^9 is transformed into the W-sextic whose line sections are the contacts of the system of quadrics, P_{89}. As ξ rotates about the various points r in Π, the various quadratic systems (∞^1) in the system (∞^2) are obtained. When r is fixed and ξ on r has the parameter s, the quadratic system (∞^1), $F(1\cdots 7;s)^3 \cdot F(1^3\cdots 7^3\,8^4;s)^9$, has for envelope the square of the 12-ic curve generated by its two component pencils. This 12-ic is composed of G_4^9 and the cubic K of the pencil $F(1\cdots 7;s)^3$ which lies in the pencil $F(1\cdots 789)^3$. The residual envelope is therefore K^2, the section of the cone A by one of its own tangent planes rather than as before a tangent plane of the Cayley cubic surface C^3. Hence

(14) *If a space sextic G_4^6 is on a quadric cone A and, say $\vartheta_0(0)=0$, the 120 Cayley cubic surfaces on G_4^6, associated with the 120 half periods P for which $\vartheta_0(P)=0$, collapse as quartic envelopes into the tangent planes of A.*

In spite of this collapse of the Cayley surface many of its properties may be observed on the cone A. It is apparent first of all that, with reference to the system P_{89} of contact quadrics of G_4^6, there must be an isolated tritangent plane, O_{890}. For, the condition $\vartheta_0(0)=0$ separarates one system (∞^5) of contact cubics from the 135 others. A partial system (∞^2) contained in this system (∞^5), and therefore determining it, is that cut out on G_4^9 by the lines of its plane. For, such a line and its transform by I^{17} make up a curve of order 18 which is mapped by the web w upon the section of A by a contact cubic surface. The system (∞^2) to which the system (∞^5) reduces when the three contacts of O_{890} are fixed is the contact system P_{89}.

The properties of the system P_{89} are brought out by a study of three mappings. The first of these, M, is that already employed in which the pairs of I^{17} in the plane P_8^2 are mapped on the points of the cone A. The second, N, is that in which the pairs of the Geiser involution, I^7, determined by $p_1, \cdots, p_7$ are mapped on the points r of a plane π in such wise that a line ξ of π becomes a cubic curve of the net, $F(1 \cdots 7; \xi)^3$. The third, N', is a mapping of the points r of π upon A which is defined later. We observe first that the 12-ic, $F(1 \cdots 7; \xi)^3 \cdot F(1^3 \cdots 7^3 8^4; \xi)^9$ is mapped by M upon a contact quadric section of A. If ξ is such that $F(1 \cdots 7; \xi)^3$ is a cubic K of the pencil on p_8, p_9 then $F(1^3 \cdots 7^3 8^4; \xi)^9 = K \cdot P(1^2 \cdots 7^2 8^3)^6$; and this quadric section of A is by the tritangent plane O_{890}, and the tangent plane of A along the generator which is the map of K. Thus the system (∞^3) obtained by adjoining A to the system P_{89} contains, in addition to the 28 pairs of tritangent planes mentioned under (13), the ∞^1 pairs of tritangent planes composed of O_{890} and a variable tangent plane of A. The space cubic curve on the six contacts of such a contact quadric is made up of the conic section of A by the plane O_{890}, and the contact generator. The ∞^1 cubic curves of this sort are all on the vertex of A and the three contacts of O_{890}; and it will appear later that

(15) *The ∞^2 space cubic curves on the six contacts of quadrics of the system P_{89} all pass through the four vertices of a tetrahedron T formed by the node of A and the three contacts of the tritangent plane O_{890}* (cf. 50 (14)).

In the mapping, N, the curve G_4^9 becomes a W-sextic which has a triple point at r^0, the map of the pair, p_8, p_9, of I^7. The three points on W at r^0 correspond to the three points on G_4^9 at p_8 and therefore to the three contacts of O_{890} with G_4^6. The pencil of cubics K becomes the pencil of lines on r^0. The transform of G_4^9 by I^7 is an $F(1^3 \cdots 7^3 9^4)^9$ which meets G_4^9 in 18 points outside P_8^2 of which 12 are the fixed points of I^7 where $J(1^2 \cdots 7^2)^6$ meets G_4^9. The remaining 6 points are made up of 3 pairs of I^7 which map into the remaining three nodes r^1, r^2, r^3 of W. The web of canonical adjoints of W consists of cubics with a node at r^0 and simple points at r^1, r^2, r^3. The two g_1^3's in such a web are cut out by the pencil of lines on r^0 and the pencil of conics on $r^0, \cdots, r^3$. If they coincide, as in this case they must, the nodes r^1, r^2, r^3 of W must lie on a line ξ^0. Let N' be the mapping of the plane by these canonical adjoints of W. The lines ξ on r^0 map into the generators of A and W maps into G_4^6. The directions at r^0 map into the points of the conic section of A by the tritangent plane O_{890}, and the points of the lines r^0r^1, r^0r^2, r^0r^3 map into directions on A about the three points of contact. The points of ξ^0 map into directions at the node of A. Since a line ξ in general position cuts $\xi^0, r^0r^1, r^0r^2, r^0r^3$ and cuts W in 6 points, it maps into a cubic curve on A on the vertices of T (cf. (15)) and the 6 contacts of a quadric of a system P_{89}. If ξ is on r^0 this cubic curve breaks up into a generator and the conic on O_{890}, a result already obtained by the mapping M. The directions about the points r^1, r^2, r^3 map respectively into the generators of A on the contacts of O_{890}. This same figure in π is obtained from the space curve by Cremona transformation (cf. 50 (17)) whence

(16) *The regular cubic Cremona transformation with F-points at the node of A and the contacts of* O_{890} *transforms A into a plane* π, *and* G_4^6 *into the W-sextic with a triple point at* r^0 *and three double points* r^1, r^2, r^3 *on a line. The cubic curves* (15) *pass into the linear sections of W.*

The proof that this inversion of the mapping N' can be made is easily supplied.

It is clear that the G_4^9 in the plane of P_8^2 is determined in π by the choice of a planar quartic curve f^4 with an isolated Aronhold set of double tangents, and a point r^0. For then the mapping N can be inverted. The connection between f^4, r^0, and the W-sextic, which we do not pursue, involves the twelve tangents from r^0 to f^4, a topic recently discussed by Zariski ([77] pp. 317–8).

The G_4^9 in the plane of P_8^2 is also determined from the space 12-ic referred to at the close of 48, the intersection of the dianode sextic surface by the quadric A_i on its triple points, Q_7^3. If this 12-ic curve is projected from its triple point at q_7, the triple points at $q_1, \cdots, q_6$ and the two generators A_i on q_7 yield the eight triple points of G_4^9.

The individual contact quadric systems of the particular G_4^6 on the cone A have each been identified and related to the curve G_4^6 in much the same way as was done in **50** for the general curve G_4^6. A closer study of this particular case would doubtless bring to light relations among these contact systems which would furnish indications of the relations sought in connection with **50** (20).

Schottky[60] has obtained the coördinates of P_8^2 and the equation of G_4^9 directly from the modular functions for $p = 4$ on the assumption that $\vartheta_0(0) = 0$ (cf. **58**).

52. Special planar sets. The ten nodes of a rational sextic. In the plane the number of types of Cremona transformations with $\varrho \geqq 9$ F-points is infinite and a general set of points P_ϱ^2 is therefore congruent to an infinite number of projectively distinct sets ([17] II (7)). Nevertheless certain special sets exist which are self-congruent under infinitely many types and as a consequence are congruent to only a finite number of projectively distinct sets which are permuted among themselves under regular transformation. Such a special set defines an infinite Cremona group in the plane.

In this section three special planar sets of this character are discussed: (a) P_9^2, the nine base points of a pencil of cubics; (b) P_9^2, the nine nodes of an elliptic sextic; and (c) P_{10}^2, the ten nodes of a rational sextic. This account is for the most part descriptive. Details of proof may be found in memoirs of the author ([17] II; [20]; [19]).

We have seen (cf. **51** (12)) that a set, P_9^2, of base points of a pencil of cubics is congruent to 960 sets, or, in some order, to 9! 960 ordered sets. It is in fact self-congruent under the Jonquières involution J^5 with 4-fold F-point at p_9 and simple F-points at $p_1, \cdots, p_8$; and also under the Bertini involution I^{17} with fixed point at p_9 and 6-fold F-points at $p_1, \cdots, p_8$. Let $J^5 = F_9$ and $I^{17} = E_9$ to indicate the symmetry in $p_1, \cdots, p_8$. The product $D_9 = F_9 E_9$ is of infinite period and F_9, E_9 generate an infinite dihedral group within which they are involutions belonging to different conjugate sets. This dihedral group contains an exemplar of every Cremona transformation with 9 or less F-points of which 8 are of like order ([17] II (40)). These dihedral groups, formed for each of the points in turn, generate the entire ternary Cremona group under which P_9^2 is self-congruent, i. e. the Cremona group $i_{9,2}$ which transforms a pencil of cubics into itself. In fact if we set $C_{2,1} = E_2 E_1$ then the transformations comprised under

$$(1)\qquad \begin{array}{c} D_2^{\nu_2} D_3^{\nu_3} \cdots D_9^{\nu_9} C_{2,1}^{\varrho_2} C_{3,1}^{\varrho_3} \cdots C_{9,1}^{\varrho_9} \\ (\varrho_2, \cdots, \varrho_9 = 0, 1, 2; \quad \varrho_2 + \cdots + \varrho_9 \equiv 0 \bmod. 3) \end{array}$$

for all integer values of $\nu_2, \cdots, \nu_9$ constitute an invariant abelian subgroup of $i_{9,2}$ of index two and the remaining elements are products of (1) and a fixed involution such as E_9 (cf. [17] II (42) in which the limitation $\varrho_2 + \cdots + \varrho_9 \equiv 0$ mod. 3 is overlooked). If the ordered planar sets P_9^2 are mapped as in 7 upon points P of a space Σ_{10} the aggregate of sets congruent to P_9^2 is mapped upon an infinite aggregate of points in Σ_{10} conjugate under the Cremona group $G_{9.2}$ in Σ_{10}. The special sets P_9^2 which are base points of a pencil of cubics are mapped upon points of a manifold M_8 in Σ_{10}. The transformations of $G_{9,2}$ which arise from the ternary Cremona group $i_{9,2}$ above constitute the subgroup $I_{9,2}$ of $G_{9,2}$ for which M_8 is a locus of fixed points. Thus $I_{9,2}$ is an invariant subgroup of $G_{9,2}$ whose factor group $F_{9,2}$ of order $9!\,960$ is the finite group of permutations of the points on M_8 effected by $G_{9,2}$. This factor group is represented in the plane by the 960 ordered sets congruent to the set of base points P_9^2 and is the group of the tritangent planes of the space sextic G_4^6 dicussed in 51.

In the case (b) when P_9^2 is the set of nine nodes of an elliptic sextic it is self-congruent under the nine Bertini involutions $E_1, \cdots, E_9$ and therefore also under the transformations $C_{i,1}$ $(i = 2, \cdots, 9)$ in (1). It is not however self-congruent under $F_1, \cdots, F_9$ and the number of projectively distinct sets congruent to P_9^2 is enlarged to $2^8 . 960$ ([17] II (47)). The Bertini transformation is that of lowest order for which the linear transformation S of 4 (1) reduces to the identity mod. 2 when the two sets of F-points are properly ordered. Hence as elements of the arithmetic group $g_{9,2}$ of 6 the $E_1, \cdots, E_9$ as well as their transforms by any element of $g_{9.2}$ reduce to the identity mod. 2. Moreover the set P_9^2 must be self-congruent under the transform of a Bertini involution. The square of the element $D_2 = F_2 \cdot E_2$ in (1) is $F_2 E_2 F_2 \cdot E_2$, the product of E_2 and its transform by the involution F_2. Thus P_9^2 is self-congruent under all the transformations (1) for which $\nu_2, \cdots, \nu_9$ are even. Hence

(2) *An elliptic sextic with nodes at P_9^2 is invariant under a ternary Cremona group $i_{9,2}^{(2)}$, generated by Bertini involutions and their conjugates, which has an invariant subgroup of index two comprised of the elements* (1) *for which $\nu_2, \cdots, \nu_9$ are even. The remaining elements are obtained by adding a factor E_1. This group is isomorphic with the invariant subgroup $g_{9,2}$* (2) *of $g_{9,2}$ whose elements are reducible to the identity mod.* 2. *The factor group $g_{9,2}^{(2)}$ of $g_{9,2}$* (2) *under $g_{9,2}$ is isomorphic with that subgroup of order* $9!\, 2^8 . 960$ *of the modular group* ($p = 5$) *which leaves two even theta functions unaltered.*

The order and character of the factor group $g_{9,2}^{(2)}$ is derived in [19]. With $2p+2 = 12$ subscripts $1, 2, \cdots, 9, 0, \alpha, \beta$, and invariant even theta characteristics $Q_{0\alpha}, Q_{0\beta}$, and therefore invariant period characteristic $P_{\alpha\beta}$, the group is generated by the involutions attached to the points in the finite geometry not on either quadric $Q_{0\alpha}, Q_{0\beta}$. It is sufficient to identify I_{ij} with the transposition (p_i, p_j) $(i, j = 1, \cdots, 9)$ and I_{0123} with the quadratic transformation A_{123}. The factor group has an invariant subgroup of order $9!\, 960$ which arises from the same transformations as in case (a). The additional $2^8 \cdot 960$ congruent sets can be obtained from the 960 of case (a) by applying the eight involutions $F_2, \cdots, F_9$ in any combination. The effect of the Cremona transformation on P_9^2 is determined most easily by taking the set on a cubic curve with canonical elliptic parameters $u_1, \cdots, u_9$ and $\sigma \equiv u_1 + \cdots + u_9$. Then, in case (a), $\sigma \equiv 0$; and, in case (b), $2\sigma \equiv 0$. Clearly a similar situation is present when P_9^2 is the set of r-fold points of an elliptic curve of order $3r$. This method is not applicable in the next case.

In case (c) let P_{10}^2 with points $p_1, p_2, \cdots, p_9, p_0$ be the ten nodes of a rational sextic $S(t)$ with parameter t. This set is subject to three projective conditions and has 9 absolute constants. For if P_8^2 is chosen, p_9 must lie on the curve G_4^9 (**51** (3)) and, for fixed p_9 on G_4^9, p_0 must be one of 12 points on G_4^9 ([20] pp. 251–54). There is but one rational sextic with nodes at P_{10}^2 and $S(t)$ is therefore invariant under any Cremona transformation for which P_{10}^2 is self-congruent. If E_{90} is the Bertini involution determined by P_8^2 for which p_9, p_0 are fixed points, then P_{10}^2 is self-congruent, and $S(t)$ is invariant, under the Cremona group h generated by the 45 Bertini involutions E_{ij} $(i, j = 1, \cdots, 9, 0)$ and by their conjugates under Cremona transformation with F-points within P_{10}^2. Since these generators correspond to elements of $g_{10,2}$ which are congruent to the identity mod. 2, the group h corresponds to a subgroup $g'_{10,2}(2)$ of that subgroup $g_{10,2}(2)$ of $g_{10,2}$ which reduces to the identity mod. 2. In order to prove that $g'_{10,2}(2)$ coincides with $g_{10,2}(2)$ we introduce a particular set of 527 P-curves of P_{10}^2 which in the customary notation, called hereafter the *signature*, are of the following types:

$$(3) \quad P(1)^0, \quad P(12)^1, \quad P(12345)^2, \quad P(1^2 2 \cdots 7)^3, \quad P(1^3 2 \cdots 9)^4.$$

Two P-curves, $P(1^{r_1} 2^{r_2} \cdots 0^{r_0})^r$ and $P(1^{s_1} 2^{s_2} \cdots 0^{s_0})^s$, for which $r_i \equiv s_i$ and $r \equiv s$ mod. 2 are said to be congruent mod. 2. Thus the aggregate (3) comprises

$$\binom{10}{1} + \binom{10}{2} + \binom{10}{5} + \binom{10}{6} + \binom{10}{9} = 527$$

incongruent types since in the fourth type $i_1^2 \equiv i_8^2$, and in the fifth $i_1^3 i_2 \equiv i_1 i_2^3$.

Any two P-curves of P_{10}^2 whose signatures are congruent mod. 2 are conjugate or *equivalent* under an element of the group h ([20] pp. 246–47). For example the P-curves, $P(8)^0$ and $P(1^2 \cdots 7^2 8^3)^6$, are conjugate under E_{90}. This pair is transformed by the quadratic transformation A_{678} into the pair $P(67)^1$, $P(1^2 \cdots 5^2 678^2)^5$, and the latter pair is equivalent under that element of h which is the transform of E_{90} by A_{678}. Proceeding in this way all the equivalences mentioned can be proved. Let C be any Cremona transformation whose corresponding element in $g_{10,2}$ is congruent to the identity mod. 2. Then C transforms $P(0)^0$ into a P-curve, P_1, congruent mod. 2 to $P(0)^0$. An element h_1 exists in h which transforms $P(0)^0$ into P_1. Hence Ch_1^{-1} leaves $P(0)^0$, the directions at p_0, unaltered and therefore is found in the group $i_{9,2}^{(2)}$ of (2) and is a product of Bertini involutions and their conjugates. Then C itself is a product of the same character. Hence

(4) *The infinite group $i_{10,2}^{(2)}$ of ternary Cremona transformations under which a rational sextic $S(t)$ with nodes at P_{10}^2 is invariant is generated by the Bertini involutions with F-points in P_{10}^2 and their conjugates. It is simply isomorphic with that invariant subgroup $g_{10,2}(2)$ of $g_{10,2}$ which is congruent to the identity mod. 2.*

Since P_{10}^2 is self-congruent under $i_{10,2}^{(2)}$, the projectively distinct sets congruent to it are obtained from transformations which correspond to elements in the factor group $g_{10,2}^{(2)}$ of $g_{10,2}(2)$ in $g_{10,2}$. This factor group (cf. **27** (8)) has the order $10!\,2^{13} \cdot 31 \cdot 51$ and is isomorphic with the subgroup of the modular group ($p = 5$) which leaves one even theta characteristic unaltered and permutes the remaining 527 as the 527 classes of congruent P-curves (whose exemplars are given in (3)) are permuted under Cremona transformation. The subgroup which leaves one of these P-curves, say $P(0)^0$, unaltered has the order $10!\,2^{13} \cdot 3 = 9!\,2^8 \cdot 960$ which is the number of projectively distinct ordered sets P_9^2 congruent to the nodal set P_9^2 under transformation for which p_0 is an ordinary point. Hence the factor group $g_{10,2}^{(2)}$ gives rise to

$10!\,2^{13}\cdot 31\cdot 51$ projectively distinct ordered sets P_{10}^2 congruent to the nodal set or to $2^{13}\cdot 31\cdot 51$ unordered sets. Hence

(5) *A general rational plane sextic with ten nodes can be transformed by Cremona transformation into precisely* $2^{13}\cdot 31\cdot 51$ *projectively distinct sextics. Under such transformation these distinct projective types (with ordered nodes) are permuted according to that group of the odd and even theta characteristics* ($p=5$) *which has an invariant even characteristic.*

The number of discriminant conditions on the nodal set P_{10}^2 is finite. For example there is a transformation in $i_{10,2}^{(2)}$ which leaves $S(t)$ unaltered but converts $P(1\cdots 56^2 7^2 8^2)^4$ into the congruent $P(12345)^2$. If then the discriminant condition, $\delta(1\cdots 59 6^2 7^2 8^2)^4=0$, were satisfied by P_{10}^2, the same set P_{10}^2 would satisfy the congruent condition, $\delta(123459)^2=0$. Hence (cf. [20] pp. 249–50; [34] § 4)

(6) *The number of discriminant conditions—infinite in the case of a general set* P_{10}^2*—is finite for the nodal* P_{10}^2 *of* $S(t)$*, a set subject to three conditions. Any two discriminant conditions whose signatures are congruent mod.* 2 *impose the same fourth condition on the ten nodes. The members of this finite aggregate of* 496 *conditions are permuted under Cremona transformation as the odd theta characteristics are permuted under the group of* (4).

This equivalence of discriminant conditions yields an infinite variety of theorems relating to the nodes of $S(t)$ of the following type ([20] p. 250):

(7) (a) *If the jacobian of the net of cubics on seven nodes of* $S(t)$ *passes through one, it passes through all three of the remaining nodes.* (b) *If an adjoint quartic of* $S(t)$ *has a triple point at one node of* $S(t)$*, there will exist an adjoint quartic with a triple point at any one node of* $S(t)$.

For later specific reference let the basis notation for $p=5$ be taken with subscripts 1, 2, ···, 9, 0, α, β and let the invariant even characteristic in (4) be that of $E_{\alpha\beta}$. The notation for the group is then symmetric in 1, 2, ···, 9, 0.

The 527 classes of congruent P-curves are then, named as their exemplars occur in (3),

$$(3.1) \qquad E_{1\alpha}, \quad E_{12}, \quad E_{12345\alpha}, \quad E_{2\cdots 7}, \quad E_{0\beta}.$$

The 496 discriminant conditions on P_{10}^2 are of the following types:

$$(8) \qquad \begin{gathered} \delta(12) = 0, \quad \delta(123)^1 = 0, \quad \delta(1, \cdots, 6)^2 = 0, \\ \delta(1^2 2 \cdots 8)^3 = 0, \quad \delta(1^3 2 \cdots 90)^4 = 0. \end{gathered}$$

The first three comprise respectively $\binom{10}{2}$, $\binom{10}{3}$, and $\binom{10}{6}$, conditions. The last two by virtue of (7a) and (7b) comprise only $\binom{10}{7}$ and $\binom{10}{10}$ conditions. They correspond respectively to the odd theta characteristics:

$$(8.1) \qquad O_{12\alpha\beta}, O_{123\beta}, O_{7890}, O_{190\alpha}, O.$$

Since the even theta characteristic $E_{\alpha\beta}$ is invariant these odd characteristics are permuted just as the 496 points, $E_{\alpha\beta} + O_m = P_{\alpha\beta m}$, in the finite geometry ($p = 5$) which are not on the quadric $Q_{\alpha\beta}$. Thus the discriminant conditions are permuted respectively as the half periods

$$(8.2) \qquad P_{12}, P_{123\alpha}, P_{7890\alpha\beta}, P_{190\beta}, P_{\alpha\beta}.$$

The generating involutions of the factor group are, in the finite geometry, the involutions attached to these points (cf. **22** (10)). In the plane of $S(t)$ they are respectively the transposition (p_1, p_2), the quadratic transformation

$$A_{123} = T(123)^2, \quad T(1^2 \cdots 6^2)^5, \quad T(1^6 2^3 \cdots 8^3)^{10},$$

and $T(1^{12} 2^4 \cdots 9^4 0^4)^{17}$ (cf. **6** (4), (8), (9)).

A transformation τ of the group $i_{10,2}^{(2)}$ in (4) which leaves the rational sextic $S(t)$ unaltered must be represented on the sextic by the binary transformation of the parameter

$$(9) \qquad \tau: \quad t' = (at + b)/(ct + d).$$

The transformations thus induced on $S(t)$ by $i_{10,2}^{(2)}$ constitute an infinite discontinuous group $\gamma_{10,2}$. It may be that Cremona transformations exist for which $S(t)$ is a locus of fixed points. This is not very likely since such transformations would no doubt have properties so striking that they would not thus far have escaped notice. If however they exist they form an invariant subgroup of $i_{10,2}^{(2)}$ whose factor group is the group $\gamma_{10,2}$. The Bertini involution E_{90} with F-points at P_8^2 and fixed points at p_9, p_0 is in $i_{10,2}^{(2)}$. Since E_{90} leaves $S(t)$ unaltered, its fixed points on $S(t)$ are the intersections of its fixed locus G_4^9 outside P_8^2. But, of these six intersections, four are found at the nodes p_9, p_0 at which the nodal directions and the nodal parameters t of $S(t)$ interchange. The remaining two points, at which $S(t)$ is tangent to a cubic curve on P_8^2, are the proper fixed points of the involution τ determined on $S(t)$ by E_{90}. These constitute the jacobian pair of the nodal parameters at p_9, p_0.

The P-curve of the set P_{10}^2, $P(1)^0$, has two directions at p_1 in common with $S(t)$. Hence every P-curve has just two points in common with $S(t)$ outside P_{10}^2. The two P-curves, $P(9)^0$ and $P(0)^0$, have no points in common with each other. Moreover any two P-curves, P', P'', which have no common points outside P_{10}^2 can be transformed simultaneously into $P(9)^0$ and $P(0)^0$. For if P'' is transformed into $P(0)^0$, P' is transformed into a P-curve, P''' which has no direction at p_0. Hence P''' can be transformed by a transformation with an ordinary point at p_0, and therefore with invariant $P(0)$, into $P(9)^0$. Hence ([20] § 3)

(10) *Any two P-curves of* P_{10}^2 *which have no intersections outside* P_{10}^2 *determine a generating involution (a Bertini involution or its conjugate) of* $i_{10,2}^{(2)}$ *and also a generating involution* τ *of* $\gamma_{10,2}$. *The two P-curves meet* $S(t)$ *outside* P_{10}^2 *in two pairs of points in the involution* τ, *whose jacobian pair is the pair of fixed points of* τ.

The binary collineation group $\gamma_{10,2}$ is thus intimately related to the aggregate of pairs of points cut out on $S(t)$ by the aggregate of P-curves of P_{10}^2. A P-curve with signature

$P(1^{r_1}\cdots 9^{r_9}0^{r_0})^r$ cuts $S(t)$ in a pair of points whose parameters t are determined by a binary quadratic, say $q(1^{r_1}\cdots 9^{r_9}0^{r_0})^r$. With reference to a norm-conic $N(t)$ in a plane this quadratic q is represented by a point. Hence the P-curves determine an infinite aggregate of points in the plane of $N(t)$ which divide into 527 sets of conjugate points under the discontinuous collineation group $\Gamma_{10,2}$ induced in the plane by the group $\gamma_{10,2}$ on the conic $N(t)$. We shall find later (cf. 55, 56) notable properties of this aggregate.

53. Special sets in space. The ten nodes of a symmetroid. The special set, Q_8^3, of eight base points of a net of quadrics has been discussed in 44 with particular reference to the 36 sets congruent to it. At present the group of regular Cremona transformations with invariant net under which Q_8^3 is self-congruent is more pertinent. This group, $i_{8,3}$, is discussed under case (a) as a basis for the further discussion of (b), the half-period set Q_8^3, (c), the nine nodes, and (d), the ten nodes, of a Cayley symmetroid.

Unless otherwise specified all of the Cremona transformations employed are *regular*, i. e. products of collineations and $y_i y_i' = 1$ $(i = 0, \cdots, 3)$. Three types are particularly useful: the Kantor involution I^{15} (type T^{15} in 44 (1)); and the *dilated* Geiser and Bertini planar involutions. If a planar transformation is expressed as a product $\prod A_{ijk}\cdot\pi$ of quadratic transformations A_{ijk} with F-points in P_r^2, and the permutation π of the points of P_r^2, then the spatial transformation, whose expression is the product $\prod A_{0ijk}\cdot\pi$ of cubic transformations A_{0ijk} with F-points in the set $Q_{r+1}^3 = q_0 + Q_r^3$ and the same permutation π of the points of Q_r^3, is called the *dilation* into S_3 of the planar transformation (cf. [20] § 4). The dilations into S_3 of the Geiser and Bertini transformations are (in the notation of 44 (1)) respectively:

$$(1)\qquad \begin{array}{c|ccc} & & 1 & 7 \\ \hline & 15 & -7 & -3 \\ 1 & 14 & -6 & -3 \\ 7 & 6 & -3 & -2,\,-1 \end{array}\;; \qquad \begin{array}{c|ccc} & & 1 & 8 \\ \hline & 33 & -16 & -6 \\ 1 & 32 & -15 & -6 \\ 8 & 12 & -6 & -3,\,-2 \end{array}\,.$$

If the eight F-points of the dilated Geiser transformation are a half period set on their elliptic quartic curve the direct and inverse F-points are projective. When they coincide in the identical order the space transformation is an involution ([20] (28)). The similar fact is true of the dilated Bertini transformation when its nine F-points are a half-period set, i. e. any eight are a half period set on their elliptic quartic ([20] (29)). The elements of $g_{m,8}$ determined by the Kantor involution and the dilated Bertini involution reduce mod. 2 to the identity. This evidently is not true of the dilated Geiser involution.

In the case (a) when Q_8^3 is the set of base points of a net of quadrics, let C_8 be the Kantor involution with ordinary point at q_8, and G_8 the dilated Geiser involution with F-point of order 14 at q_8. Then $D_8 = C_8\, G_8$ is of infinite period and C_8, G_8 generate an infinite dihedral group which contains all transformations for which $q_1, \cdots, q_7$ are a symmetrical set of F-points. Under these and like transformations at the other points q, the set Q_8^3 is self-congruent. Since also Q_8^3 is congruent to projectively equivalent sets under A_{1234} and A_{5678} (cf. **44** (9)), it is self-congruent under $A_{1234}\, A_{5678}$ and therefore under

$$(2) \qquad E_{1,8} = A_{8234}\, A_{1567} \cdot A_{8567}\, A_{1234}.$$

This product $E_{1,8}$ is also defined by the fact that $E_{1,8}\,(18)$ is T'^{9} in **44** (1). In terms of the products

$$(3) \qquad D_8 = C_8\, G_8, \qquad C_{2,1} = C_2\, C_1$$

there follows ([17] II pp. 376–7):

(4) *The group $i_{8,3}$ of regular Cremona transformations with an invariant net of quadrics and self-congruent base-points Q_8^3 has an invariant abelian subgroup of index two whose elements can be expressed in a single way in the form*

$$D_2^{\nu_2}\, D_3^{\nu_3} \cdots D_8^{\nu_8}\, C_{3,1}^{\varrho_3}\, C_{4,1}^{\varrho_4} \cdots C_{8,1}^{\varrho_8}\, E_{3,1}^{\sigma_3}\, E_{4,1}^{\sigma_4} \cdots E_{8,1}^{\sigma_8}$$
$$(\varrho_3, \cdots, \varrho_8, \sigma_3, \cdots, \sigma_8 = 0, 1).$$

These multiplied by C_1 complete the group.

This group $i_{8,3}$ gives rise to an invariant subgroup of $G_{8,3}$ in Σ_9 whose factor group is the modular group ($p = 3$) discussed in **44**.

In the case (b) when Q_8^3 is a half period set, or the 8 nodes of an azygetic 8-nodal quartic surface, each point is on the Cayley dianode sextic surface determined by the other seven. The theorem corresponding to (4) now reads ([17] II (46)):

(5) *The group* $i_{8,3}^{(2)}$ *of regular Cremona transformations with an invariant azygetic 8-nodal quartic surface and self-congruent nodal* Q_8^3 *has an invariant abelian subgroup of index two whose elements are*

$$D_2^{\nu_2}\, D_3^{\nu_3} \cdots D_8^{\nu_8}\, C_{3,1}^{\varrho_3}\, C_{4,1}^{\varrho_4} \cdots C_{8,1}^{\varrho_8} \quad (\varrho_3, \cdots, \varrho_8 = 0, 1).$$

These multiplied by C_1 *complete the group. The set* Q_8^3 *is congruent to* $36 \cdot 2^6$ *projectively distinct sets which are permuted under regular Cremona transformation according to a group* $f_{8,3}^{(2)}$ *of order* $8!\, 2^6 \cdot 36$.

This group $f_{8,3}^{(2)}$ is a factor group in the modular group ($p = 4$) of order $10!\, 2^8 \cdot 51$. The subgroup which leaves one of the 255 half periods, or points in the finite geometry S_7, unaltered, has the order $8!\, 2^9 \cdot 9$ and is generated by the involutions attached to all the points of S_7 which are syzygetic with the invariant point, say P_{90} in the basis notation ([19] (16)). Among these is the involution I_{90} attached to the invariant point P_{90} itself which evidently must be invariant in the subgroup. The factor group of the subgroup with respect to this invariant g_2 is $f_{8,3}^{(2)}$.

A more explicit identification of $f_{8,3}^{(2)}$ may be made by means of the finite number, $2 \cdot 63$, of discriminant conditions of the set Q_8^3. If this half period set is taken with canonical parameters $u_1, \cdots, u_8$ on its elliptic quartic curve so that $u_1 + \cdots + u_8 \equiv \omega/2$, the three types of discriminant conditions and their identification with points in S_7 syzygetic with P_{90} are:

$$(6) \qquad P_{12}\colon\ u_1 - u_2 \equiv 0; \qquad P_{1234}\colon\ u_1 + \cdots + u_4 \equiv 0;$$
$$P_{1290}\colon\ u_1 - u_2 + \omega/2 \equiv 0.$$

These represent respectively the coincidence of q_1, q_2; the coplanar condition of $q_1, \cdots, q_4$; and the condition that a quadric with node at q_1 (or at q_2) is on $q_3, \cdots, q_8$. That all discriminant conditions reduce to these is proved as follows. Under the Kantor involution C_8 for which Q_8^3 is self-congruent the directions at q_1 become the points of the P-surface $P(1^3 2^2 \cdots 7^2)^4$, or $P(1)^0 \equiv P(1^3 2^2 \cdots 7^2)^4$. On transforming this equivalence by A_{1234} and A_{1256} the further equivalences, $P(234)^1 \equiv P(1^2 2345^2 6^2 7^2)^3$ and $P(12^2 3456)^2 \equiv P(134567^2)^2$, under transforms of C_8 for which also Q_8^3 is self-congruent, appear. From the last two there follows that $\delta(2348)^1 = 0$ implies $\delta(23481^2 5^2 6^2 7^2)^3 = 0$ and that $\delta(12^2 34568)^2 = 0$ implies $\delta(134567^2 8) = 0$. This coincidence of later discriminant conditions with the initial ones given in (6) is sufficient to show that all are reducible to those in (6). If furthermore the involutions attached to P_{ij} in the finite geometry are identified with the transpositions $(q_i\, q_j)$ in Q_8^3, and those attached to P_{ijkl} $(i, j, k, l = 1, \cdots, 8)$ with the cubic transformation A_{ijkl} then the discriminant conditions (6) are permuted under Cremona transformation as their corresponding points in the finite geometry are permuted under $f_{8,3}^{(2)}$.

The transformations of $i_{8,3}^{(2)}$ which, as elements of $g_{8,3}$, reduce to the identity mod. 2 constitute an invariant subgroup i' of $i_{8,3}^{(2)}$ of index two which is generated by the Kantor involutions $C_1, \cdots, C_8$ and their conjugates. The factor group f' of i' in $g_{8,3}$ has the order $8!\, 2^6 \cdot 36 \cdot 2$ ([19] p. 337). The additional element in $i_{8,3}^{(2)}$ which reduces the order of this factor group to that of $f_{8,3}^{(2)}$ arises from the dilated Geiser involutions $G_1, \cdots, G_8$ which figure in the products $D_1, \cdots, D_8$. These involutions all reduce mod. 2 to the same element in f' which corresponds in the finite geometry to I_{90}.

The cases (c) and (d) of respectively nine nodes, Q_9^3, and ten nodes, Q_{10}^3, of a Cayley symmetroid are related in much the same way as the set P_8^2 and the set P_9^2 of base points of a pencilof cubics on P_8^2. The principal facts with reference to the determination of these nodal sets have been given by

Cayley[11] and Rohn ([55] §§ 9, 10, 11; [54]; [56]). The symmetroid is the quartic surface whose equation can be put in the form of a symmetric four-row determinant with linear forms as elements. The ten nodes are the points for which the first minors vanish. The enveloping cone from any node breaks up into two cubic cones on the other nodes, i. e., nine nodes are projected from any one into the base points of a pencil of cubics. If this happens at one node of a 10-nodal surface it happens at every node and the surface is a symmetroid. Any eight of the nodes are a half period set. Only seven of the nodes can be chosen at random. If $q_1, \cdots, q_7$ are given nodes the remaining three lie on the Cayley dianode surface (cf. **44**) determined by Q_7^3. If q_8 is fixed on this surface the dianode of $q_1, \cdots, q_6, q_7$ and of $q_1, \cdots, q_6, q_8$ meet in a curve of order 36 from which the 15 lines $q_i q_j$ $(i, j = 1, \cdots, 6)$ factor, as well as the cubic curve on $q_1, \cdots, q_6$, leaving as the significant factor the "dianodal curve" of Cayley. The dianodal curve thus determined by the half period set Q_8^3 has the order 18, has triple points with coplanar tangents at each point of Q_8^3, and meets each of the infinite number of F-curves of the second kind determined by Q_8^3 in two points outside Q_8^3. If q_9 is chosen on this dianodal curve, there are 13 ten-nodal quartic surfaces with nodes at Q_9^3, *one* of which is a symmetroid. Thus the pairs of remaining nodes of symmetroids with given nodes at Q_8^3 are pairs of an involution on the dianodal curve of Q_8^3.

There is but one symmetroid with given nodes and if nine of these are given the tenth is uniquely determined. Also ([20] (26))

(7) *A symmetroid Σ is transformed by regular Cremona transformation with $\varrho \leqq 10$ F-points at its nodal Q_{10}^3 into a symmetroid Σ' with nodal $Q_{10}'^3$ congruent to Q_{10}^3.*

For, A_{1234} with F-points at $q_1, \cdots, q_4$ transforms Σ into a four-nodal quartic surface Σ' with additional nodes q_i' corresponding to the nodes q_i of Σ $(i = 5, \cdots, 9, 0)$. Also A_{1234} is a quadratic transformation on the net of lines on q_1 and, if the lines from q_1 to the other nodes of Σ are base lines of a pencil of cubic cones, the net of lines on q_1' has the same property and Σ' is also a symmetroid. Furthermore ([20] § 5)

(8) *A symmetroid is invariant under a Kantor or a dilated Bertini involution whose F-points are in its nodal* Q_{10}^3. *The dilated Geiser involution whose eight F-points are in the nodal* Q_{10}^3 *leaves the symmetroid unaltered but interchanges the remaining two nodes.*

From this there follows

(9) *The dianodal curve of* Q_8^3 *in* (5) *is a locus of fixed points for those elements of the group* $i_{8,3}^{(2)}$ *which correspond to elements of* $g_{8,3}$ *congruent to the identity mod.* 2. *The other elements of* $i_{8,3}^{(2)}$ *effect on the dianodal curve the involution of pairs which make up with* Q_8^3 *the nodes of a symmetroid.*

Let now Q_9^3 be the set of nine nodes of a symmetroid Σ. The set is self-congruent and Σ invariant under the Kantor and dilated Bertini involutions defined by the set (cf. (8)), as well as the conjugates of the involutions obtained by regular transformation with F-points within the set. These conjugates generate the group $i_{9,3}^{(2)}$ which corresponds to the subgroup $g_{9,3}(2)$ of $g_{9,3}$ which is congruent mod. 2 to the identity ([20] pp. 261–2). The factor group $g_{9,3}^{(2)}$ of $g_{9,3}(2)$ with respect to $g_{9,3}$ is isomorphic with the modular group for $p=4$ ([19] p. 337). Hence Q_9^3 is congruent to only $2^8 \cdot 51 \cdot 10$ projectively distinct unordered sets. Any two P-surfaces whose signatures are congruent mod. 2 are equivalent under some operation of $i_{9,3}^{(2)}$ for which Q_9^3 is self-congruent. All P-surfaces defined by Q_9^3 are then equivalent to one of 255 types ([20] p. 262) which can be put into one-to-one correspondence with the 255 half periods ($p=4$) as follows:

$$(10)\qquad \begin{array}{llll} P_{10} & : P(1)^0; & P_{1230} & : P(123)^1; \\ P_{1789} & : P(1^2 23456)^2; & P_{29} & : P(1^3 2^2 3 \cdots 8)^3. \end{array}$$

From the equivalence of P-surfaces there follows as before the identity of discriminant conditions. These also reduce to a set of 255 distinct conditions in correspondence with half periods as follows:

$$(11)\quad \begin{array}{ll} P_{12} \ : \delta(1\,2)^0 = 0; & P_{1234} : \delta(1\,2\,3\,4)^1 = 0; \\ P_{1890} : \delta(1^2 2 \cdots 7)^2 = 0; & P_{20} \ : \delta(1^3 2^2 3 \cdots 9)^3 = 0. \end{array}$$

If now the transposition $(q_i\, q_j)$ is identified with the involution in the finite geometry attached to P_{ij}, and the Cremona transformation A_{ijkl} $(i, \cdots, l = 1, \cdots, 9)$ with the involution attached to P_{ijkl} then the points P in the finite geometry are permuted under the modular group precisely as the corresponding classes of P-surfaces in (10), or the corresponding discriminant conditions in (11), are permuted under regular Cremona transformation. It is sufficient to verify this (obviously true of the transpositions) for the generator A_{1234}.

When for case (d) the tenth node of Σ is added to $\overset{3}{Q}_9$ to make up the nodal set $\overset{3}{Q}_{10}$, the Kantor involution which has F-points at $q_1, \cdots, q_8$ and interchanges q_9, q_0 and which did not appear in the group $\overset{(2)}{i}_{9,3}$ of $\overset{3}{Q}_9$, comes into play. Under it the set $\overset{3}{Q}_9 = q_1, \cdots, q_8, q_9$ is congruent to the set $\overset{3}{Q}_9 = q_1, \cdots, q_8, q_0$. Hence the $2^8 \cdot 51 \cdot 10$ projectively distinct sets congruent to $\overset{3}{Q}_9$ are distributed 10 at a time in $2^8 \cdot 51$ sets $\overset{3}{Q}_{10}$, or

(12) *A symmetroid can be transformed by regular Cremona transformation into only* $2^8 \cdot 51$ *projectively distinct symmetroids. These projectively distinct types permute under such transformation according to the modular group* $(p = 4)$. *The* 255 *discriminant conditions on the nodal* $\overset{3}{Q}_{10}$ *are permuted like the half periods.*

It should be noted that there are 2. 255 discriminant conditions of the types in (11) that may be formed for $\overset{3}{Q}_{10}$. But the Kantor involutions show that a condition of type $\delta(1^3 2^2 3 \cdots 9)^3 = 0$ implies $\delta(20)^0 = 0$ and that one of type $\delta(1^2 2 \cdots 7)^2 = 0$ implies $\delta(1\,8\,9\,0)^1 = 0$. Thus there are now under permutations of the ten indices only two types of discriminant conditions

$$(13)\qquad P_{12} : \delta(1\,2)^0 = 0; \qquad P_{1234} : \delta(1\,2\,3\,4)^1 = 0.$$

These identities among discriminant conditions lead to theorems such as the following:

(14) *If two nodes of a symmetroid coincide the cubic cone with vertex at any one of the remaining nodes and on the ten nodes has a double generator on the coalescent nodes. If four nodes of a symmetroid are coplanar there is a quartic cone with vertex at any one of the four nodes and on the remaining six nodes.*

An expression of the figure Q_{10}^3 of ten nodes of a symmetroid in terms of modular functions ($p = 4$) given by Schottky will be found in 59. The remaining sections of this chapter are devoted to the geometric relations which connect the rational sextic and symmetroid with allied figures.

54. Geometric relations connecting the rational sextic and the symmetroid with the jacobian of a web of quadrics. This section contains a study of the projective relations which exist among a variety of figures in space defined by a planar rational sextic. In addition to the birationally related jacobian (of a web of quadrics) and symmetroid these figures include the rational sextic curve in space and the pair of cubic space curves. A certain pairing of planar rational sextics appears which is important for the purposes of 56. If the underlying rational sextic is transformed by Cremona transformation into a projectively different type, the related figures also are subject to transformations which are discussed in the next section.

The simplest point of departure is the figure of two cubic curves in space with digredient parameters τ and t respectively ([24] § 4). These as point loci will be denoted by $C_1(\tau)$, $C_2(t)$; as loci of planes by $\overline{C}_1(\tau)$, $\overline{C}_2(t)$. The two curves each depend on 12 constants and the pair depends upon $24 - 15 = 9$ absolute constants. On each point τ of $C_1(\tau)$ are three planes of $\overline{C}_2(t)$ and vice versa; dually on each plane τ of $\overline{C}_1(\tau)$ are three points of $C_2(t)$ and vice versa. The incidence conditions are therefore

$$(1)\quad F = (a\tau)^3 (\alpha t)^3 = 0; \qquad \overline{F} = (\overline{a}\tau)^3 (\overline{\alpha} t)^3 = 0.$$

By reason of the duality the forms F, $\overline{F}$ are mutually related. Each is the same covariant of the other ([24] § 3). The general form F in digredient variables has 9 absolute constants and determines in turn the two curves, $C_1(\tau)$, $\bar{C}_2(t)$. For if the planes of $\bar{C}_2(t)$ are taken in the canonical form, 17 (1), the points of $C_1(\tau)$ which satisfy the incidence condition (1) are perfectly general.

Let Q_1, Q_2 be the two nets of point quadrics on C_1, C_2 respectively; $\bar{Q}_1, \bar{Q}_2$ the two nets of quadric envelopes on $\bar{C}_1, \bar{C}_2$ respectively. The pencils of the net Q_i are the pencils on C_i and a bisecant of C_i; the pencils of the net $\bar{Q}_i$ are the pencils on $\bar{C}_i$ and an axis of $\bar{C}_i$. The net Q_i will cut $C_j\,(i, j = 1, 2;\, i \neq j)$ in an involution, I_2^6, of hexads of points. An I_2^6 on a binary domain may be visualized as the line sections of a projectively definite rational plane sextic. The nets $Q, \bar{Q}$ thus determine a tetrad of rational sextics, namely: $S_1(\tau)$, whose line sections are cut out on $C_1(\tau)$ by the net Q_2; $S_2(t)$, cut out on $C_2(t)$ by the net Q_1; $\bar{S}_1(\tau)$, cut out on $\bar{C}_1(\tau)$ by the net $\bar{Q}_2$; and $\bar{S}_2(t)$, cut out $\bar{C}_2(t)$ by the net $\bar{Q}_1$. The parameters, t_1, t_2, of a node of $S_2(t)$ are a neutral pair of I_2^6 whence the points t_1, t_2 of $C_2(t)$ lie on a pencil of the net Q_1 and therefore lie on a bisecant of C_1, i. e., a common bisecant of C_1, C_2. Conversely a common bisecant determines a neutral pair of I_2^6 whence (cf.[44] p. 312 for references to original sources):

(2) *The curves C_1, C_2 have ten common bisecants which determine on C_1, C_2 the pairs of nodal parameters of $S_1(\tau)$, $S_2(t)$ respectively; dually $\bar{C}_1, \bar{C}_2$ have ten common axes which determine on $\bar{C}_1, \bar{C}_2$ the pairs of nodal parameters of $\bar{S}_1(\tau)$, $\bar{S}_2(t)$ respectively.*

If one planar sextic, $S_2(t)$, is given its line sections determine on $C_2(t)$ ∞^2 hexads of points cut out by a net Q_1 each member of which is determined to within members of the net Q_2. Thus a system ∞^5 is defined which is apolar to a web $\bar{Q}$ of quadric envelopes. But Reye[53] has proved that in the system apolar to a web $\bar{Q}$ there are precisely two nets Q_1, Q_2 which are on cubic curves C_1, C_2. Hence

(3) *The rational sextics of the plane can be arranged in tetrads $S_1(\tau)$, $S_2(t)$; $\bar{S}_1(\tau)$, $\bar{S}_2(t)$ in such wise that given any one the other three are projectively determined.*

In such a tetrad the two in either pair will be called *paired sextics*; these have digredient parameters. Two with like parameters, t or τ, will be called *counter sextics*; these in the space figure are of dual character. Any other pair will be called a *diagonal* pair.

The line sections of $S_2(t)$ in a plane π are in projective correspondence with the quadrics Q_1. In Q_1 the cones on $C_1(\tau)$, in one-to-one correspondence with their vertices τ, form a quadratic system. There is therefore on π a conic $K(\tau)$ whose lines correspond to these cones. To a pencil of lines in π on a point p there corresponds a pencil in Q_1, which contains two cones corresponding to the two tangents of $K(\tau)$ on p. If in particular p is a node of $S_2(t)$, the pencil in Q_1 is the pencil on a common bisecant of C_1, C_2 and the two cones have nodes at τ_1, τ_2, the parameters of a node of the paired sextic, $S_1(\tau)$. Hence

(4) *Given a rational plane sextic, $S_2(t)$, there exists in its plane a covariant conic, $K(\tau)$, such that the ten pairs of parameters on $K(\tau)$ of the ten nodes of $S_2(t)$ furnish the nodal parameters of the sextic $S_1(\tau)$. Conversely if with reference to a norm-conic, $K(\tau)$, we mark the ten points determined by the ten pairs of nodal parameters of a rational sextic $S_1(\tau)$ then these ten points are the nodes of another rational sextic $S_2(t)$ which is the sextic paired with $S_1(\tau)$.*

It may be remarked that only six pairs of nodal parameters of $S_1(\tau)$ (with 9 absolute constants) can be chosen at random and that then five sextics $S_1(\tau)$ are determined ([44] p. 316 cf. also 56). In the plane however eight nodes may be chosen at random and one degree of freedom remains for the choice of the ninth. In the first case a choice of $K(\tau)$ is also implied. The relations between the binary conditions on the nodal parameters and the ternary conditions on the nodes, which for a single sextic would be quite complicated, are quite simple for the paired sextics.

The sextic $S_2(t)$ and its covariant conic $K(\tau)$ are so related that a tangent τ of $K(\tau)$ cuts out 6 points t of $S_2(t)$ and a point t of $S_2(t)$ is on two tangents τ of $K(\tau)$. This

(6, 2) relation in t, τ is obtained from $C_1(\tau)$, $C_2(t)$ as follows. The condition that plane τ of C_1 is on point t of C_2 is $(\bar{a}\tau)^3(\bar{\alpha}t)^3 = 0$. The pencil in Q_1 on t of C_2 contains the two nodal quadrics with vertices at τ_1, τ_2, the meets of C_1 and the bisecant from t to C_1. Then τ_1, τ_2 are the hessian pair of the three planes of C_1 on t, i. e. the hessian of the cubic $(\bar{a}\tau)^3(\bar{\alpha}t)^3$ in τ. Hence

(5) *The parametric equations of the rational sextics of the tetrad* (3) *referred to their covariant conics, $K(t)$, $K(\tau)$, as norm-conics respectively are*

$$\begin{aligned}
(\alpha\alpha')^2(\alpha t)(\alpha' t)(a\tau)^3(a'\tau)^3 &= 0,\\
(\bar{a}\bar{a}')^2(\bar{a}\tau)(\bar{a}'\tau)(\bar{\alpha}t)^3(\bar{\alpha}'t)^3 &= 0,\\
(\bar{\alpha}\bar{\alpha}')^2(\bar{\alpha}t)(\bar{\alpha}'t)(\bar{a}\tau)^3(\bar{a}'\tau)^3 &= 0,\\
(a a')^2(a\tau)(a'\tau)(\alpha t)^3(\alpha' t)^3 &= 0.
\end{aligned}$$

It may be proved easily (cf. [24] pp. 158–59) that:

(6) *The forms F, $\bar{F}$ in* (1) *represent the direct and inverse linear transformations of the parametric coördinate system in space determined by $C_1(\tau)$ into that determined by $C_2(t)$. Specifically, the point and plane determined by $(c\tau)^3$ with reference to $C_1(\tau)$ are the point $(ac)^3(\alpha t)^3$ and the plane $(\bar{a}c)^3(\bar{\alpha}t)^3$ with reference to $C_2(t)$; the point and plane determined by $(\gamma t)^3$ with reference to $C_2(t)$ are the point $(\bar{a}\tau)^3(\bar{\alpha}\gamma)^3$ and the plane $(a\tau)^3(\alpha\gamma)^3$ with reference to $C_1(\tau)$.*

Let t_1, t_2, t_3 be three points of the sextic, $S_2(t)$, on a line. Then there is a quadric q_1 in Q_1 on the points t_1, t_2, t_3 of $C_2(t)$ which with the net Q_2 on $C_2(t)$ determines a web which meets the plane t_1, t_2, t_3 in the net on these three points. Hence one quadric $q_1 + q_2$ of the web contains the plane t_1, t_2, t_3. This plane meets $C_1(\tau)$ in three points τ_1, τ_2, τ_3 which also are points in which the quadric q_2 of Q_2 meets $C_1(\tau)$ whence τ_1, τ_2, τ_3 are three points of the paired sextic $S_1(\tau)$ on a line. This mutual relation between the two linear triads must, according to (6), be as follows:

(7) *There is a one-to-one correspondence between the linear triads on two paired rational sextics; if $(\gamma t)^3$ is the linear triad on $S_2(t)$ $\{\bar{S}_2(t)\}$ then $(a\tau)^3(\alpha\gamma)^3\{(\bar{a}\tau)^3(\bar{\alpha}\gamma)^3\}$ is the*

linear triad on the paired sextic $S_1(\tau)\{\bar{S}_1(\tau)\}$; *if* $(c\tau)^3$ *is the linear triad on* $S_1(\tau)\{\bar{S}_1(\tau)\}$ *then* $(c\bar{\alpha})^3(\bar{\alpha}t)^3\{(ca)^3(\alpha t)^3\}$ *is the linear triad on the paired sextic* $S_2(t)\{\bar{S}_2(t)\}$.

A particular plane τ_0 of C_1 is cut by the ∞^1 planes of C_1 in a line conic which is to be identified with $K(\tau)$. A point x in τ_0 is determined by the two tangents of $K(\tau)$ which arise from the two further planes τ_1, τ_2 of C_1 on x. These two planes meet in an axis l_x $(x = \tau_1, \tau_2)$ which cuts τ_0 in x. Thus the congruence of axes l_x of C_1 determines a projective correspondence among all the planes of C_1. According to (6) the plane t of $C_2(t)$ is the plane $(a\tau)^3(\alpha t)^3$ referred to $C_1(\tau)$ and the point $x = \tau_1, \tau_2, \tau_0$ is on this plane if $(a\tau_1)(a\tau_2)(a\tau_0)(\alpha t)^3 = 0$. Hence the planes t of C_2 cut each plane τ of C_1 in a rational cubic of lines

$$(8)\qquad (\pi x)(a\tau)(\alpha t)^3 = (a\tau_1)(a\tau_2)(a\tau)(\alpha t)^3 \qquad (x = \tau_1, \tau_2).$$

Selecting a fixed plane τ_0 of C_1 as a base, the ∞^1 (for variable τ) rational cubic envelopes (8) on the ∞^1 planes τ are projected by means of the axes l_x to form in τ_0 a family (with parameter τ) of ∞^1 rational line cubics (with parameter t). Since for fixed t there is one axis l_x in the plane t of C_2 which for variable τ is met by all the lines (8), it follows that the cubic curves (2) in τ_0 are all perspective to the rational curve (i. e., line t of the envelope is on point t of the rational curve) cut out on τ_0 by the ∞^1 axes of C_1 which lie in planes t of C_2. Since for given t these axes are determined by the quadratics, $(aa')^2(a\tau)(a'\tau)(\alpha t)^3(\alpha' t)^3 = 0$, this rational curve is the sextic $\bar{S}_2(t)$ referred to its covariant conic $K(\tau)$. In particular the ten common axes of C_1, C_2 which are on two planes t cut τ_0 in the nodes of $\bar{S}_2(t)$. Hence

(9) *Upon a plane* τ_0 *of* C_1, *the axes of* C_1 *which lie in the planes of* $C_2(t)$ *cut out the rational sextic* $\bar{S}_2(t)$, *the ten common axes of* C_1, C_2 *cut out the nodes of* $\bar{S}_2(t)$, *the planes of* C_1 *cut out the covariant conic* $K(\tau)$ *of* $\bar{S}_2(t)$, *and the planes of* C_2 *cut the* ∞^1 *planes of* C_1 *in* ∞^1 *rational cubic envelopes which are projected upon* τ_0 *by the congruence of axes of* C_1 *into the* ∞^1 *perspective cubics of* $\bar{S}_2(t)$.

The general form (8) in digredient variables x (ternary), t, τ (binary), has 23 constants and $23 - 8 - 3 - 3 = 9$ absolute constants. In terms of its coefficients the parametric equation of $\bar{S}_2(t)$ is

$$(10) \qquad (\pi \pi' \xi)(a a')(\alpha t)^3 (\alpha' t)^3 = 0.$$

One may prove further ([24] § 9 (1)) that

(11) *The triads τ of cusps of the ∞^1 perspective cubic envelopes* (8) *of $\bar{S}_2(t)$ run over a sextic curve of genus four birationally equivalent to $\bar{F} = 0$ in* (1), *and constitute a g_1^3 in the canonical g_3^6. The residual g_1^3 is a system of triangles on the sextic whose sides envelop $K(\tau)$.*

The form (8) for given x and τ furnishes the three tangents t of the perspective cubic τ of $\bar{S}_2(t)$ which pass through x. If however x is at a point t_1 of $\bar{S}_2(t)$ then one of these tangents is $t = t_1$ for every τ and $(t t_1)$ is a factor of (8). If x is at x_i, one of the ten nodes of $\bar{S}_2(t)$ with nodal parameters $(q_i t)^2$, then

$$(12) \quad (\pi x_i)(a \tau)(\alpha t)^3 = (l_i \tau)(\lambda_i t) \cdot (q_i t)^2 \quad (i = 1, \cdots, 9, 0).$$

On each point of a common axis of C_1, C_2 there is a third plane τ of C_1 and t of C_2. As the point runs over the axis the (1, 1) relation between these planes τ and t is $(l_i \tau)(\lambda_i t) = 0$.

Let as before Q be the web of point quadrics apolar to the two nets $\bar{Q}_1$, $\bar{Q}_2$ of quadric envelopes on $\bar{C}_1$, $\bar{C}_2$; $\bar{Q}$ the web of quadric envelopes apolar to the nets Q_1, Q_2. According to the usual theory the jacobian J of the web Q is the locus of nodes of quadrics of the web or the locus of pairs y, y' of points apolar to the web. These pairs lie in an involutorial correspondence I on J. Let the net $\bar{Q}_1$ on $\bar{C}_1$ meet $\bar{C}_2(t)$ in hexads $t_1, \cdots, t_6$ of planes of the involution $\bar{I}_2^6$ cut out on $\bar{S}_2(t)$ by lines of the plane. Then the entire web $Q(y)$ will meet C_2 in sets of 6 points in the involution $\bar{I}_3^6$ which is the conjugate of (or apolar to) the $\bar{I}_2^6$ and which may be visualized as the plane sections of a rational space sextic curve $\bar{R}(t)$. For, $Q(y)$ is apolar to $\bar{Q}_2$ and therefore the polarized quadrics of $Q(y)$ will be represented on C_2 by polarized sextics; and $Q(y)$ is apolar to $\bar{Q}_1$ and therefore these sextics are apolar to the sextics cut out on $\bar{C}_2$ by $\bar{Q}_1$, i. e., to $\bar{I}_2^6$. Now the linear triad t_1, t_2, t_3 of $\bar{S}_2(t)$ can be supplemented by a triad t_4, t_5, t_6 to form a hexad apolar to $\bar{I}_3^6$ whence the polars of the triad t_1, t_2, t_3 as to four independent members of $\bar{I}_3^6$ will be linearly related, or the triad will be apolar to a unique hexad of $\bar{I}_3^6$ cut out say by $Q'(y)$. Then $Q'(y)$ can be represented as a sum of squares of the three planes t_1, t_2, t_3 of C_2 and has a node at the point $y(t_1, t_2, t_3)$ on these three planes. Conversely if $Q'(y)$ has a node y it cuts C_2 in a hexad whose catalecticant (the discriminant of $Q'(y)$) vanishes whence $Q'(y)$ can be expressed as a sum of three squares of planes t. Moreover the quadric of the net $\bar{Q}_1$ which touches two of these planes must touch the third since $\bar{Q}_1$ is apolar to $Q'(y)$ whence the planes t_1, t_2, t_3 are a triad of a hexad of $\bar{I}_2^6$. Thus J is the locus of points y on planes $t_1, t_2, t_3 \{\tau_1, \tau_2, \tau_3\}$ of $C_2(t) \{C_1(\tau)\}$ where t_1, t_2, t_3 and τ_1, τ_2, τ_3 satisfy the symmetric conditions,

$$(13) \qquad \bar{f}_2(t_1^4, t_2^4, t_3^4) = 0, \qquad \bar{f}_1(\tau_1^4, \tau_2^4, \tau_3^4) = 0,$$

which express that $t_1, t_2, t_3 \{\tau_1, \tau_2, \tau_3\}$ are a linear triad on the rational sextic $\bar{S}_2(t) \{\bar{S}_1(\tau)\}$. Moreover the two trihedrals t_1, t_2, t_3 and τ_1, τ_2, τ_3 on y, being self polar as to $Q'(y)$, are six planes of a quadric cone $q'(y)$ with vertex at y. Conversely if y is a point for which the two trihedrals have this property then, of the ∞^2 quadrics $\bar{Q}_1$ on τ_1, τ_2, τ_3, one is on t_1, t_2 and therefore contains all the planes of $q'(y)$ including t_3. Hence

(14) *The jacobian quartic surface J is the locus of points y for which the two sets of three planes of C_1, C_2 on y are planes of a quadric cone $q'(y)$. The two trihedrals on C_1, C_2 correspond to linear triads, paired as in* (7), *on $\bar{S}_1(\tau), \bar{S}_2(t)$. Referred to $C_2(t)$, or to $C_1(\tau)$, J is an involution surface with equations* (13).

By the term involution surface is meant a surface whose parametric equation when referred to a norm-curve is the equation of a binary involution.

As noted above there is one quadric $\bar{q}_1$ of the net $\bar{Q}_1$ on the planes τ_1, τ_2, τ_3; t_1, t_2, t_3 which contains the planes of the cone $q'(y)$. Similarly there is a quadric $\bar{q}_2$ of the net $\bar{Q}_2$ on the same 6 planes and cone. Each of the quadrics $\bar{q}_1, \bar{q}_2$ also contains the planes τ_4, τ_5, τ_6; t_4, t_5, t_6 and the cone $q'(y')$ on this second set of 6 planes. But if two quadric envelopes have two quadric cones in common a member of their pencil factors into the product y, y' of the vertices of the cones. Hence y, y' are corresponding points of I on J and the corresponding linear triads on either rational sextic make up a linear hexad on that sextic. Thus if, on $\bar{S}_2(t)$, t_1, t_2 are nodal parameters on a line with t_3, t_4, t_5, t_6 then as the line revolves about the node the point $y(t_1, t_2, t_3)$ runs along a common axis A_i of $\bar{C}_2(t)$, $\bar{C}_1(\tau)$ on J and the point $y'(t_4, t_5, t_6)$ runs over a cubic curve N_i on J. The curve N_i is the involution curve determined by triads referred to C_2 which are contained in the I_1^4 cut out on $\bar{S}_2(t)$ by lines on the node. If t_3, t_4 are the parameters of another node, N_i meets the axis $A_j(t_3, t_4)$ of $\bar{C}_2$ in the points $y'(t_3, t_4, t_5)$, $y'(t_3, t_4, t_6)$ and A_j is a bisecant of N_i. Hence

(15) *There are two systems of* ∞^2 *6-planes, one on each of the curves* $\bar{C}_1, \bar{C}_2$, *inscribed in* J. *Each point* y *of* J *determines one 6-plane of either system and these two 6-planes also have in common the opposite point* y' *of the 6-planes which corresponds to* y *under* I. *The ten common axes* A_i *of* $\bar{C}_1\, \bar{C}_2$ *on* J *correspond under* I *to ten cubic curves* N_i *such that* A_i *is a bisecant of* N_j $(i, j = 1, \cdots, 9, 0;\ i \neq j)$.

Let the web of quadrics Q in the $S_3(y)$ of J with dual coördinates y, η be given by the double quaternary form

$$(16) \qquad (\alpha y)^2 (\beta z) = 0$$

with z, ζ dual coördinates in another $S_3(z)$. The form has 40 coefficients and depends upon 39 constants or $39 - 15 - 15 = 9$ absolute projective constants. For every point z in $S(z)$, (16) determines a quadric in $S(y)$ and the locus of points z for which this quadric has a node y is the symmetroid

$$(17) \qquad \Sigma = (\beta z)(\beta' z)(\beta'' z)(\beta''' z)(\alpha\alpha'\alpha''\alpha''')^2 = 0,$$

a symmetric four-row determinant whose elements are linear forms in z. Any such determinant may be regarded as the discriminant of a quadric (16). For point z of Σ and node y of quadric (16) the equations,

$$(18) \qquad \alpha_i(\alpha y)(\beta z) = 0 \qquad (i = 0, 1, 2, 3),$$

are simultaneously satisfied. On eliminating z we have

$$(19) \quad J = (\beta\beta'\beta''\beta''')(\alpha\alpha'\alpha''\alpha''')(\alpha y)(\alpha' y)(\alpha'' y)(\alpha''' y) = 0,$$

the jacobian of the web (16). For corresponding points y, y' of J under I the equations,

$$(20) \qquad \beta_i(\alpha y)(\alpha y') = 0 \qquad (i = 0, 1, 2, 3),$$

hold simultaneously. For z, y corresponding points on Σ, J the equation of the point y is

$$(21) \qquad (\beta z)(\beta' z)(\beta'' z)(\alpha\alpha'\alpha''\eta)^2 = (y\eta)^2 = 0.$$

A point y determines in (16) a plane ζ but ζ on points z^1, z^2, z^3 is determined by any one of the 8 base points y of the net of quadrics $(\alpha y)^2(\beta z^i) = 0$ $(i = 1, 2, 3)$ on y. Thus (16) defines an (8,1) correspondence between points y and planes ζ of the spaces of J, Σ. If y is on J one of the three quadrics of the web on y has a node at y and two of the 8 points corresponding to ζ coincide at y. If z^0 is the point of Σ which furnishes this nodal quadric then the square of y is furnished by (21) and this substituted in (16) yields the plane ζ which corresponds to y in the form $(\beta z^0)(\beta' z^0)(\beta'' z^0)(\beta''' z)(\alpha\alpha'\alpha''\alpha''')^2 = 0$ which is the tangent plane of Σ at the point z^0 which corresponds to y. Hence J is mapped by (16) upon the planes ζ of Σ. If ζ is on z, z', y is on a quartic curve which meets J in 16 points, i.e., Σ is of class 16. Ten quadrics in the web (16) are pairs of planes whose nodal lines are the axes A_i on J.

There are therefore 10 points z^0 for which (21) vanishes identically and at which the tangent plane of Σ is indeterminate. Hence

(22) *The points of the jacobian surface J are mapped by* (16) *upon the planes of the symmetroid of class* 16 *and order* 4 *is such wise that point y and contact z of tangent plane correspond in* (18). *The ten lines A_i of J map into the planes of the ten tangent quadric cones of Σ at its ten nodes, i. e., the lines of J correspond to the directions on Σ at its nodes.*

Under the birational transformation B in (18) between J and Σ a quadric section of J by $(\delta y)^2$ corresponds to the section of Σ by its cubic adjoint surface,

$$(\beta z)(\beta' z)(\beta'' z)(\alpha \alpha' \alpha'' \delta)^2 = 0.$$

The sections of J by planes η correspond on Σ to a linear system Γ_3^6 of order 6 and genus 3 on the nodes of Σ which are contact curves of the cubic adjoints

$$(\beta z)(\beta' z)(\beta'' z)(\alpha \alpha' \alpha'' \eta)^2 = 0$$

whence on every plane section π of Σ there is isolated a system of contact cubics. The linear system Γ_3^6 is cut out on Σ by the linear system of adjoint cubics

$$(\beta z)(\beta' z)(\beta'' z)(\alpha \alpha' \alpha'' \eta)(\alpha \alpha' \alpha'' \eta') = 0$$

for variable η and fixed η'.

If ζ is a plane section of Σ determined by points z, z', z'' then the three quadrics $(\alpha y)^2 (\beta z^{(i)})$ determine a net and the involutorial Cremona transformation of pairs apolar to this net (i. e. I on J) is

$$(\alpha \alpha' \alpha'' \eta)(\beta \beta' \beta'' \zeta)(\alpha y)(\alpha' y)(\alpha'' y) = 0. \tag{23}$$

The fundamental sextic curve of this involution, the locus of nodes of the net, is the member of the linear system C_3^6 on J which corresponds under B to plane sections of Σ. Cubic surfaces on C_3^6 cut J in a residual system $C_3'^6$ of sextics of genus 3, the transform of plane sections of J

under I. For given sections ζ of Σ and η of J the cubic surface (23) cuts J in the curves C_3^6, C'^6_3 which correspond to ζ, η under B, I respectively. Hence if ζ corresponds to C_3^6 on J under B and η corresponds to Γ_3^6 on Σ under B but to C'^6_3 under I then for given ζ (23) is a cubic Cremona involution which on J is I with fundamental curve C_3^6; whereas for given η (23) is a cubic Cremona transformation of J into Σ which has C'^6_3 and Γ_3^6 for inverse fundamental curves. The inverse transformation is obtained from (21) in the form $(\beta z)(\beta' z)(\beta'' z)(\alpha\alpha'\alpha''\eta)(\alpha\alpha'\alpha''\eta') = 0$ for fixed η'. Thus on any plane section of J there are isolated two corresidual linear series g_3^6 cut out respectively by the linear systems C_3^6 and C'^6_3.

The points y on the two cubic curves $C_1(\tau)$, $C_2(t)$ correspond in (16) to the planes ζ of two rational sextic envelopes $\bar{R}_1(\tau)$, $\bar{R}_2(t)$. The planes ζ of these curves on a point z are given by the parameters of the points in which the quadric (16) cuts C_1, C_2 whence the curves $\bar{R}_1(\tau)$, $\bar{R}_2(t)$ are the space curves conjugate or apolar to $\bar{S}_1(\tau)$, $\bar{S}_2(t)$ respectively. The symmetroid Σ is the locus of points z for which the quadric (16) can be expressed as a sum of squares of three planes of C_1, or three planes of C_2 (these planes themselves being represented by perfect cubes), and therefore is the locus of points z whose point sections of $\bar{R}_1(\tau)$, or of $\bar{R}_2(t)$, are expressible as a sum of three sixth powers, or are *catalectic sextics*. If however z is a node of Σ the quadric (16) is a pair of planes η, η' on a common axis A_i of $\bar{C}_1$, $\bar{C}_2$. Since η, η' is apolar to the net $\bar{Q}_1$, and the net $\bar{Q}_2$, the planes η, η' are harmonic to the pairs of planes of $\bar{C}_1$ and $\bar{C}_2$ on A_i and the quadric can be expressed as a sum of squares of either pair of planes. Hence the sextic point-section of $\bar{R}_1(\tau)$, or of $\bar{R}_2(t)$, from a node of Σ is a *cyclic sextic*, i. e., a sextic reducible to a sum of two sixth powers of linear forms. The linear forms themselves determine a pair of nodal parameters on the conjugate plane sextics $\bar{S}_1(\tau)$, $\bar{S}_2(t)$. The catalecticant of a binary sextic is the particular symmetric four-row determinant for which the elements in a line perpendicular to the principal diagonal are all alike. Conner[25] shows that Σ can be transformed into such a symmetric determinant in just two ways corresponding to $\bar{R}_1(\tau)$ and $\bar{R}_2(t)$. These two rational space sextics are conjugate to the *paired* planar sextics $\bar{S}_1(\tau)$, $\bar{S}_2(t)$. Unlike the planar curves the one space sextic is localized when the other is given since each determines Σ.

Let $(q\,t)^5$ be a linear pentad on $\bar{S}_2(t)$ whose carrier meets $\bar{S}_2(t)$ again at t_1. The polar of $(q\,t)^5$ as to four independent point sections of $\bar{R}_2(t)$ is $\varrho_i\,(t\,t_1)$ $(i = 0, \cdots, 3)$ and $(q\,t)^5$ is therefore apolar to three independent point sections of $\bar{R}_2(t)$ and thus to all the point sections of $\bar{R}_2(t)$ by points of a plane ζ. The locus of planes ζ thus determined by the ∞^2 linear pentads on $\bar{S}_2(t)$ is the Stahl ([68] p. 56) *quadric* K of $\bar{R}_2(t)$. For, if $\bar{R}_2(t)$ for the moment is taken as $(\beta z)\,(b\,t)^6 = 0$, the condition that the three sextics $(\beta\,z^{(i)})\,(b\,t)^6 = 0$ $(i = 1, 2, 3)$ have a common apolar quintic is of degree two in each $z^{(i)}$ and these can occur only as $\zeta = \|z^{(1)}\,z^{(2)}\,z^{(3)}\|$. Two linear pentads $(q\,t)^5$, $(q'\,t)^5$ of $\bar{S}_2(t)$ with the same sixth point t_1, which determine planes ζ, ζ' of K, are apolar to point sections of $\bar{R}_2(t)$ by points on the line $\zeta\,\zeta'$. The equations of ζ, ζ' are $(\beta z)\,(b q)^5\,(b\,t) = \varrho\,(z\,\zeta)\cdot(t\,t_1)$ and $(\beta z)\,(b\,q')^5\,(b\,t) = \sigma(z\,\zeta')\cdot(t\,t)_1$. Hence the variable linear pentads $(q\,t)^5 + \lambda\,(q'\,t)^5$ on lines through a fixed point t_1 of $\bar{S}_2(t)$ determine a pencil of planes $\varrho\,(z\,\zeta) + \lambda\,\sigma(z\,\zeta')$ on K, necessarily a pencil on a generator t_1 of K. Again for given τ in (8) the line t_1 of the perspective cubic of $\bar{S}_2(t)$ cuts $\bar{S}_2(t)$ in t_1 and a linear pentad of the form (cf. [24] § 5(5)) $f(\overset{5}{t}, \overset{1}{\tau}, \overset{2}{t_1}) = (q\,t)^5$. Then $(\beta z)\,(b\,q)^5\,(b\,t) = f'(\overset{1}{z}, \overset{1}{\tau}, \overset{1}{t_1})\cdot(t_1\,t)$ and the plane $(z\,\zeta)$ of K is linear in t_1, and for variable t_1 turns about the generator τ of K. Hence

(24) *The planes ζ of the Stahl quadric K are in one-to-one correspondence with the linear pentads of $\bar{S}_2(t)$. As ζ turns about a t generator of K, the variable pentad of $\bar{S}_2(t)$ turns about the point t of $\bar{S}_2(t)$. As ζ turns about a τ generator of K, the pentad of $\bar{S}_2(t)$ runs over the lines of the perspective cubic envelope τ of $\bar{S}_2(t)$ in* (8). *K has a similar relation to $\bar{S}_1(\tau)$ with the roles of the generators reversed.*

For given τ the linear pentads, $f(t^5, \tau^1, t_1^2) = 0$ on the lines t_1 of the perspective cubic τ are apolar to the pencil of sextic point sections of points z on the generator τ of K. This quadratic system (for variable t_1) is in a net of binary quintics whence the pencil of sextics consists of the polars of a binary septimic. As τ varies this septimic must lie in a pencil

$$(25) \qquad (c\tau)(\gamma t)^7 = 0$$

since the polars

$$(26) \qquad (c\tau)(\gamma t_1)(\gamma t)^6 = 0$$

must be included in the linear system (∞^3) of point sections of $\bar{R}_2(t)$ (cf. (a) below).

If $t_1, \cdots, t_6$ is a linear hexad ξ of $\bar{S}_2(t)$ the three pentads obtained by dropping t_6, t_5, t_4 successively determine three planes ζ of K which meet in a point $z(t_1, t_2, t_3)$. The point section of $\bar{R}_2(t)$ from z is apolar to the three quintics and therefore is apolar to their common triad t_1, t_2, t_3 and is a catalectic sextic, or z is the point of the symmetroid Σ which corresponds to $y(t_1, t_2, t_3)$ on J. From the existence of the ∞^2 line sections of $\bar{S}_2(t)$ each containing six pentads and the similar behavior of Σ with respect to the paired sextics, $\bar{S}_1(\tau)$ and $\bar{S}_2(t)$, there follows (cf. (15)):

(27) *There are two systems* (∞^2) *of* 6-*planes circumscribed to the Stahl quadric K and inscribed in Σ. Then K is a rational covariant of Σ being part of the envelope of planes which cut Σ in Lüroth quartics* (*with inscribed five lines*). *A point* $z(t_1, t_2, t_3) = z(\tau_1, \tau_2, \tau_3)$ (*cf.* (7)) *of Σ belongs to a single* 6-*plane of each system and these two* 6-*planes have also their opposite points* $z'(t_4, t_5, t_6) = z'(\tau_4, \tau_5, \tau_6)$ *in common where z, z' are images in the involution I on Σ which corresponds to I on J.*

In this and the following sections certain theorems, indicated by letters, which refer to the configurations introduced above, are given without proof.* The first group of theorems relates to the Stahl quadric and the perspective curves of $\bar{S}_2(t)$.

* Proofs of these theorems will be found in a second article by the author under the title: *Geometric aspects of the abelian modular functions of genus four*; Amer. Jour. (1929).

(a) *The form* (26), *the polar of* (25), *is the equation*, $(bz)(\beta t)^6 = 0$, *of the rational space sextic envelope* $\overline{R}_2(t)$ *when the point* z *is referred to the Stahl quadric* K *of* $\overline{R}_2(t)$ *with generators* τ *and* $t_1 = t$.

Thus the general (1, 7) form (25) with 16 coefficients and 15 constants or $15-3-3=9$ absolute constants is sufficient to determine projectively $\overline{R}_2(t)$ and therefore the entire configuration of J, Σ and $\overline{S}_2(t)$.

(b) *The catalecticant of the sextic* (26) *in* t *is, for* $t_1 = t$ *a form* $(e\tau)^4(\varepsilon t)^4$ *which furnishes the equation on* K *of the octavic curve of intersection of* K *with* Σ. *For given* τ *in* $(e\tau)^4(\varepsilon t)^4 = 0$ *the quartic in* t *with roots* $t_3, \cdots, t_6$ *is the linear tetrad on* $\overline{S}_2(t)$ *cut out by the double tangent (with parameters* t_1, t_2*) of the perspective cubic* (8) *of* $\overline{S}_2(t)$. *For variable* τ *this double tangent of the perspective cubic envelops a rational curve* $R^{10}(\tau)$ *of class* 10. *Through a point* t *of* $\overline{S}_2(t)$ *the ten tangents of* $R^{10}(\tau)$ *contain a set of* 6 *for which* t *is in the pair* t_1, t_2 *and a set of* 4 *for which* t *is in the tetrad* $t_3, \cdots, t_6$. *The corresponding* 6 *and* 4 *values of* τ *are determined by* $(\overline{\alpha}\,\overline{\alpha}')^2(\overline{\alpha}t)(\overline{\alpha}'t)(\overline{a}\tau)^3(\overline{a}'\tau)^3 = 0$ (*cf.* (5)) *and* $(\varepsilon t)^4(e\tau)^4 = 0$.

(c) *The only linear tetrads of* $\overline{S}_2(t)$ *for which the four points on their line are projective to their four parameters on* $\overline{S}_2(t)$ *are the four residual points* $t_3, \cdots, t_6$ *on the double tangents* t_1, t_2 *of the perspective cubics of* $\overline{S}_2(t)$. *These tetrads are furnished by* $(\varepsilon t)^4(e\tau)^4 = 0$.

(d) *The locus on* J *which corresponds to the intersection of* K *with* Σ *is the transform under* I *of the locus of points* y *for which* $q'(y)$ *in* (14) *breaks up into two pencils of planes. These are points* y *at which the axis of* $\overline{C}_2$ *in the plane* τ *of* $\overline{C}_1$ *meets the axis of* $\overline{C}_1$ *in the plane* t *of* $\overline{C}_2$ *when* τ, t *satisfy* $(e\tau)^4(\varepsilon t)^4 = 0$.

(e) *The system of perspective envelopes of* $\overline{S}_2(t)$ *of class* $m+3$ *corresponds in* (1,1) *fashion to the system of curves on* K *of order* $m+1$ *of the form* $(k\tau)(\varkappa t)^m = 0$. *The equation of the system of perspective curves is* $(\pi x)(ak)(\alpha t)^3(\varkappa t)^m = 0$ (cf. (8)).

The next group of theorems centers about (15) and (22). We recall that, at the end of **52**, the binary quadratic $q(1^{r_1}\ldots 9^{r_9}\,0^{r_0})^r$ was defined to be the pair of parameters t of the two points in which $\overline{S}_2(t)$ is cut by the P-curve, $P(1^{r_1}\ldots 9^{r_9}\,0^{r_0})^r$, of its nodal set P_{10}^2 outside of the intersections at the points of P_{10}^2. Thus the nodal parameters of $\overline{S}_2(t)$ are $q(1)^0, \cdots, q(9)^0, q(0)^0$; the parameters of the pair of points cut out on $\overline{S}_2(t)$ by the line $p_1\,p_2$ are $q(12)^1$; etc. These pairs will be distinguished from the like-named pairs of points on the paired sextic, $\overline{S}_1(\tau)$, by the addition of the parameter t or τ as the case may be. Thus the parameters of the ten common axes A_i of $\overline{C}_1(\tau)$, $\overline{C}_2(t)$ on J are $q(i;\tau)^0$ and $q(i;t)^0$. The parameters τ and t cut out on A_i by the third plane of $\overline{C}_1(\tau)$ or $\overline{C}_2(t)$ on a point of A_i are related by the projectivity, $(\lambda_i\,t)(l_i\,\tau)=0$, in (12).

(f) *The cubic curve N_1 on J is cut by the axes A_j $(j=2,\cdots,9,0)$ in pairs of points whose parameters t are $q(1j;\,t)^1$ or whose parameters τ are $q(1j;\,\tau)^1$, the parameters t, τ being connected by the projectivity $(\lambda_i\,t)(l_i\,\tau)=0$ in* (12).

(g) *The* 45 *quadratics $q(ij;\,t)^1$ and the* 45 *quadratics $q(ij;\tau)^1$ arise from the two common solutions of the* 45 *pairs of projectivities, $(\lambda_i\,t)(l_i\,\tau)=0$ and $(\lambda_j\,t)(l_j\,\tau)=0$; i. e., $q(ij;\,t)^1=(\lambda_i\,t)(\lambda_j\,t)(l_i\,l_j)$ and $q(ij;\tau)^1=(\lambda_i\,\lambda_j)(l_i\,\tau)(l_j\,\tau)$ $(i,j=1,\cdots,9,0)$.*

(h) *The points of an axis A_i on J correspond to directions at the node D_i of Σ; the curve N_i on J corresponds to the rational octavic m_i^8 on Σ cut out by the quadric M_i on the nine nodes other than D_i. The polar of D_i as to the Stahl quadric K cuts K in a conic section whose equation on K is $(\lambda_i\,t)(l_i\tau)=0$. The octavic m_i^8 with parameter t (or τ) has for nodal parameters at D_j the quadratic $q(ij;\,t)^1$ (or $q(ij;\,\tau)^1$).*

These theorems are used in the next section to determine the behavior of the paired sextic $\overline{S}_1(\tau)$, the jacobian J, and the symmetroid Σ, when the sextic $\overline{S}_2(t)$ is transformed by Cremona transformation into a congruent sextic. The account given in Meyer's Apolarität ([44] §§ 30, 31, 32) has contacts with the exposition just given.

55. Associated Cremona transforms of the rational sextic, symmetroid, and jacobian surface. We examine in more detail the parameter distribution on the curves N_i described in 54 (f), (g). The involution I_1^4 cut out on $\bar{S}_2(t)$ by lines on the node p_1 contains tetrads t_3, t_4, t_5, t_6 such that the point $y(t_4 t_5 t_6)$ on N_1 has the parameter $t = t_3$. The planes $t = t_3, \cdots, t_6$ of $C_2(t)$ are the faces of a tetrahedron circumscribed to $C_2(t)$ with opposite vertices inscribed in N_1 at $t = t_3, \cdots, t_6$. Thus N_1 is an H_3 curve—a Hurwitz curve ([44] p. 315, II)—of $C_2(t)$. The axis of $C_2(t)$ on planes t_3, t_4 of $C_2(t)$ is a bisecant of N_1 on points t_5, t_6. In particular let the line on the node p_1 of $\bar{S}_2(t)$ pass through the node p_2. The tetrad $t_3, \cdots, t_6$ is then $q(2; t)^0 \cdot q(12; t)^1$. Hence the axis $q(12; t)^1$ of $C_2(t)$ is the bisecant $q(2; t)^0$ of N_1 and from symmetry is also the bisecant $q(1; t)^0$ of N_2. Due to the like behavior of the paired sextic $\bar{S}_1(\tau)$ with respect to the axes of $C_1(\tau)$ and the curves N_i, the axis $q(12; \tau)^1$ of $C_1(\tau)$ is a common bisecant of N_1 and N_2 with parameters $q(2; \tau)^0$ on N_1 and $q(1; \tau)^0$ on N_2. This axis $q(12; \tau)^1$ of $C_1(\tau)$ can not coincide with $q(12; t)^1$ of $C_2(t)$ since the ten common axes of $C_1(\tau)$, $C_2(t)$ are the axes A_i. Hence

(1) *If on each of the cubic curves $C_1(\tau)$, $C_2(t)$ there are marked the ten common axes A_i as well as respectively the 45 axes $q(ij; \tau)^1$ and the 45 axes $q(ij; t)^1$ then these axes include all of the common bisecants of any two of the curves N_i, those belonging to the pair N_i, N_j being the axis $q(ij; t)^1$ of $C_2(t)$, the axis $q(ij; \tau)^1$ of $C_1(\tau)$ and the eight common axes A_k $(k \neq i, j)$ of $C_1(\tau)$, $C_2(t)$. These ten bisecants of N_i have for parameters t on N_i respectively $q(j; t)^0$, $q(j; \tau)^0$, $q(ik; t)^1$ where parameter τ is converted into parameter t by $(l_i \tau)(\lambda_i t) = 0$.*

The two lemmas which follow are not proved.

(a) *If two cubic curves N_1, N_2 in general position and 8 of their ten common bisecants A_j $(j = 3, \cdots, 9, 0)$ are given, there is a unique quartic surface on N_1, N_2 and A_j which is the jacobian of a web of quadrics apolar to a unique* *pair of cubic curves C_1, C_2 for which the A_j are common axes and N_1, N_2 are Hurwitz curves. The pairs N_1, N_2 and C_1, C_2 are mutually but dually related.*

(b) *A planar rational sextic is uniquely determined to within a projectivity when the pairs of parameters of eight nodes are given.*

If H_1, H_2 are any two cubic curves in general position, let (H_1, H_2) denote the rational sextic whose nodal parameters are projective to those on H_1 of the ten common bisecants of H_1, H_2; and $(\bar{H}_1, \bar{H}_2)$ the rational sextic whose nodal parameters on $\bar{H}_1$ are those of the common axes of H_1, H_2. Consider then the rational sextic (N_1, N_2). Eight of the nodal pairs are $q(1k; t)^1$ $(k = 3, \cdots, 9, 0)$ (cf. (1)). The Jonquières transformation $J^5_{1,2}$ of order 5 with simple F-points at $p_3, \cdots, p_0$; 4-fold F-point at p_1 and ordinary point at p_2 transforms $\bar{S}_2(t)$ into a sextic eight of whose nodes are $q(1k; t)^1$ and this transform according to (b) must be projective to (N_1, N_2). Since p_2 is an ordinary point the transform has a node $q(2; t)^0$ as is the case according to (1) with (N_1, N_2). The transform also has the node $q(1^3 3 \cdots 9 0; t)^4$ and this according to (1) is the transform of $q(2; \tau)^0$ by $(l_1 \tau)(\lambda_j t) = 0$. Hence

(2) *The projectivity $(l_1 \tau)(\lambda_1 t) = 0$, under which $q(1j; \tau)^1$ and $q(1j; t)^1$ $(j = 2, \cdots, 9, 0)$ correspond, sends $q(j; \tau)^0$ into $q(1^3 2 \cdots, j-1, j+1, \cdots 9 0; t)^4$ and $q(1; \tau)^0$ into $q(1^8 2^3 \cdots 0^3)^{12}$ and vice versa for $q(j; t)^0$ and $q(1; t)^0$.*

The last statement will be verified in a moment. Consider now the paired sextics (N_1, N_2) and (N_2, N_1). Let (N_1, N_2) be regarded as the given sextic, the bisecants $q(13; t)^1, \cdots, q(10; t)^1$ being looked upon as nodes $p'_3, \cdots, p'_0$; the bisecant $q(2; t)^0$ as the node p'_2; and the bisecant $q(1^3 3 \cdots 9 0; t)^4$ as the node p'_1. Then $J^5_{1,2}$ transforms (N_1, N_2) back into $\bar{S}_2(t)$, the nodes p'_i of (N_1, N_2) passing into the nodes p_i of $\bar{S}_2(t)$ $(i = 1, 2, \cdots, 9, 0)$. The transformation $J^5_{2,1}$ with simple point at p_1 then transforms $\bar{S}_2(t)$ into (N_2, N_1) with nodes $p_1, p_2, p_3, \cdots, p_0$ corresponding to bisecants $q(1; t)^0, q(2^3 3 \cdots 0; t)^4, q(23; t)^1, \cdots, q(20; t)^1$. But the bisecant $q(2; t)^0$ of (N_1, N_2) is the bisecant

$q(1;t)^0$ of (N_2, N_1). Hence (N_1, N_2) is transformed into (N_2, N_1), in such a way that corresponding nodes pass into each other, by the transformation $J^5_{1,2}\, J^5_{2,1}$ followed by the transposition (12). If we set $R_1 = J^5_{1,2}\, J^5_{2,1}\,(12)$ then in the notation of 53 (1)

$$(3)\quad R_1: \begin{array}{c|ccc} & 1 & 9 & \\ \hline & 17 & -12 & -4 \\ 1 & 12 & -8 & -3 \\ 9 & 4 & -3 & 0,-1 \end{array} \;;\quad R_1\, R_2: \begin{array}{c|cccc} & 1 & 1 & 8 & \\ \hline & 65 & -40 & -24 & -16 \\ 1 & 24 & -15 & -8 & -6 \\ 1 & 40 & -24 & -15 & -10 \\ 8 & 16 & -10 & -6 & -3,-4 \end{array}\;.$$

Hence

(4) *Two paired sextics are congruent under Cremona transformation. The transformation R_1 will transform a given sextic into its paired sextic with paired nodes corresponding. The lack of symmetry in the ten nodes disappears when the integer coefficients of R_1 in $g_{10,2}$ are reduced mod. 2* (cf. 52 (4)).

Two sextics paired with a third are projective to each other. If R_1 transforms S into S' and R_2 transforms S' into S'', then S and S'', congruent under $R_1\, R_2$, must be projective. This is verified by the fact that the product $R_1\, R_2$ in (3) has coefficients which are congruent mod. 2 to the identity. The form of R_1 in (3) verifies the behavior stated in (2) of the quadratics q under transformation from $\bar{S}_2(t)$ to its paired sextic $\bar{S}_1(\tau)$.

We have observed that the sextic (N_1, N_2) has the following scheme of nodes and nodal parameters:

$$(N_1, N_2): \begin{array}{ccccc} p_1', & p_2', & p_3', & \cdots, & p_0' \\ q(1^3 3 \cdots 0;\, t)^4, & q(2;\, t)^0, & q(13;\, t)^1, & \cdots, & q(10;\, t)^1; \end{array}$$

and that (N_1, N_2) is the transform of $\bar{S}_2(t)$ by $J^5_{1,2}$. Similarly, after an additional transposition (23) of nodes, the sextic (N_1, N_3) has the scheme:

$$(N_1, N_3):$$
$$\begin{array}{cccccc} p_1'', & p_2'', & p_3'', & p_4'', & \cdots, & p_0'' \\ q(1^3 24 \cdots 0;\, t)^4, & q(3;\, t)^0, & q(12;\, t)^1, & q(14;\, t)^1, & \cdots, & q(10;\, t)^1; \end{array}$$

and (N_1, N_3) is the transform of $\bar{S}_2(t)$ by $J^5_{1,3}$ (23). Hence (N_1, N_2) is congruent to (N_1, N_3) under the quadratic transformation $J^5_{1,2} J^5_{1,3}(23) = A_{123}$. Similarly the paired sextics (N_2, N_1) and (N_3, N_1) have the following schemes which take account of the pairing of the nodes:

(N_2, N_1):

$$\begin{array}{lllll} p'_1, & p'_2, & p'_3, & p'_4, & \cdots, p'_0 \\ q(2^3 3 \cdots 0;\, t)^4, & q(1;\, t)^0, & q(23;\, t)^1, & q(24;\, t)^1, & \cdots, q(20;\, t)^1; \end{array}$$

(N_3, N_1):

$$\begin{array}{lllll} p''_1, & p''_2, & p''_3, & p''_4, & \cdots, p''_0 \\ q(3^3 24 \cdots 0;\, t)^4, & q(1;\, t)^0, & q(32;\, t)^1, & q(34;\, t)^1, & \cdots, q(30;\, t)^1. \end{array}$$

Hence (N_2, N_1) is the transform of $\bar{S}_2(t)$ by $J^5_{2,1}(12)$ and (N_3, N_1) is the transform of $\bar{S}_2(t)$ by $J^5_{3,1}(123)$. Thus (N_3, N_1) is the transform of (N_2, N_1) by $(12)\, J^5_{2,1} J^5_{3,1}(123)$. We set

$$R_{1,23} = (12)\, J^5_{2,1} J^5_{3,1}(123), \tag{5}$$

and verify by actual multiplication that

$$R_{1,90}: \quad \begin{array}{c|ccc} & & 1 & 7 \\ \hline & 10 & -6 & -3 \\ 1 & 6 & -3 & -2 \\ 7 & 3 & -2 & 0,\ -1 \end{array}\,. \tag{6}$$

Furthermore $R_{1,90}$ is the product of a Bertini involution with F-points at $p_1, \cdots, p_8$ and a Geiser involution with F-points at $p_2, \cdots, p_8$.

Since a sextic is self-congruent under a Bertini involution there follows:

(7) *If a sextic $S(t)$ is congruent to a sextic $S'(t)$ under quadratic transformation A_{123}, the sextic $S(\tau)$ paired with $S(t)$ is congruent to the sextic $S'(\tau)$ paired with $S'(t)$ under the Geiser transformation with F-points at $p_4, \cdots, p_9, p_0$.*

It is rather interesting to observe the role played in these relations by the conjugate set of generating involutions of $g_{10,2}$ (cf. **6** (4), $\cdots$, (9)). These comprise of course the transpositions and the quadratic types A_{123}, but also the quintic type $A_{1,\cdots,6} = A_{123}A_{456}A_{123}$ which appears in (14) below, the type in (6) $R_{1,90} = A_{178}A_{1\cdots6}A_{178}$, and the type in (3) $R_1 = A_{190}R_{1,90}A_{190}$.

According to **54** (h) the rational octavic curve m_1^8, cut out on Σ by the quadric M_1 on nodes $D_2, \cdots, D_0$, has nodal parameters $q(12; t)^1, \cdots, q(10; t)^1$. Projected from D_2 it yields a rational sextic with eight pairs of nodal parameters, $q(13; t)^1, \cdots, q(10; t)^1$, which therefore is the sextic (N_1, N_2) above, the transform of $\bar{S}_2(t)$ by $J_{1,2}^5$. The line $p_3' p_4'$ of (N_1, N_2) cuts (N_1, N_2) in a pair of points whose parameters are those of $\bar{S}_2(t)$ on the P-curve, $P(1^2 5 6 \cdots 0)^3$. Hence

(8) *The curve m_1^8 with parameter $t(\tau)$ is cut by the plane on its nodes $D_2 D_3 D_4$ in a further pair of points $q(1^2 5 6 \cdots 90; t)^3$ $[q(1^2 5 6 \cdots 90; \tau)^3]$.*

Let $w^{(1)}(t)$ be the web of sextics with a 5-fold point at the node p_1 of $\bar{S}_2(t)$ and simple points at the remaining nodes. This web $w^{(1)}(t)$ maps the plane on a quadric surface M_1 whose generators $g(s)$ arise from the lines (with parameter s) on p_1, and whose generators $g(\sigma)$ arise from the pencil (with parameter σ) of quintics with 4-fold point at p_1 and on the remaining nodes of $\bar{S}_2(t)$. Let $s^{(2)}, \cdots, s^{(0)}$ be the parameters s of the P-curves, $P(12)^1, \cdots, P(10)^1$; and $\sigma^{(2)}, \cdots, \sigma^{(0)}$ the parameters σ of the products of P-curves,

$$P(12)^1 \cdot P(1^3 34 \cdots 0)^4, \cdots, P(10)^1 \cdot P(1^3 23 \cdots 9)^4.$$

The exceptional points and curves of the mapping are as follows. All points on the line $p_1 p_i$ map into the point $D_i(s^{(i)}, \sigma^{(i)})$ on M_1, except p_i itself which, as a locus of directions, maps into the generator $g(s^{(i)})$ $(i = 2, \cdots, 9, 0)$. The directions at p_1 map into a quintic curve on M_1 of type $(m\sigma)(\mu s)^4 = 0$. The sextic $\bar{S}_2(t)$ maps into a rational octavic on M_1 with nodes at D_i and nodal parameters $q(1i; t)^1$. The plane on D_2, D_3, D_4 is the map of $P(12)^1 \cdot P(13)^1 \cdot P(14)^1 \cdot P(1^2 5 \cdots 0)^3$ whence this plane meets the octavic outside the nodes in $q(1^2 56 \cdots 0; t)^3$. Hence (cf. (8) and **54** (h)) the octavic is m_1^8 and

(9) *The sextic* $\bar{S}_2(t)$ *is mapped by the web* $w^{(1)}(t)\,(1^5 2 \cdots 0)^6$ *upon the rational octavic* m_1^8 *with nodes at the nodes* $D_2, \cdots, D_0$ *of the symmetroid* Σ *associated with* $\bar{S}_2(t)$.

The paired sextic $\bar{S}_1(\tau)$ is mapped by the web $w^{(1)}(\tau)$ upon the same rational octavic m_1^8. Indeed the transformation R_1 in (3) which transforms $\bar{S}_1(\tau)$ into $\bar{S}_2(t)$ transforms the web $w^{(1)}(\tau)$ into the web $w^{(1)}(t)$ but interchanges the pencils s, σ. These two pencils coincide, and M_1 has a node, if and only if the discriminant condition, $P_{\alpha\beta} = \delta(1^3 2 \cdots 0)^4 = 0$ (cf. 52 (7b)), is satisfied. If this symmetric discriminant condition vanishes the transformation R_1 becomes a collineation (cf. 9(3)) and the paired sextics are projective. If M_1 has a node the Stahl quadric K is a conic and either sextic $\bar{S}_2(t)$, $\bar{S}_1(\tau)$ has a perspective conic ([25] (k), (m), (s)). Hence

(10) *If two paired sextics,* $\bar{S}_1(\tau)$, $\bar{S}_2(t)$ *are projective then the discriminant condition which expresses that an adjoint quartic of either has a triple point at one node of the sextic is satisfied, and either has a perspective conic. Then each quadric* M_i *has a node and the Stahl quadric* K *has a double plane.*

If indeed $\bar{S}_2(t)$ has a perspective conic and nodal parameters t_1, t_2 at p_1 then tangents t_1, t_2 of the perspective conic are on p_1. Hence the pairs of nodal parameters of $\bar{S}_2(t)$, plotted with reference to the perspective conic as $K(t)$ (cf. 54 (4)), yield the nodes of $\bar{S}_1(\tau)$ in coincidence with those of $\bar{S}_2(t)$.

Conditions imposed on one part of a geometric configuration must be followed throughout the configuration with considerable caution. For example it is a single condition on $\bar{S}_2(t)$ that it have a perspective conic. Its perspective cubics then all degenerate into this conic. If in the construction of 54 (9) the perspective cubic in the plane τ_0 of $C_1(\tau)$, which is cut out by the planes of $C_2(t)$, reduces to a conic then a plane of $C_2(t)$ coincides with the plane τ_0 of $C_1(\tau)$. This is only a single condition on the two cubic curves in space but it requires that the common plane be a factor of J. The situation is rather that $\bar{S}_2(t)$ with a perspective conic determines its apolar space sextic curve $\bar{R}_2(t)$ whose catalectic point sections determine Σ as before and thereby J. The peculiarity of J however is that the web Q has, in its apolar system $\bar{Q}_1+\bar{Q}_2$, a single net of quadric envelopes with a common cubic curve, i. e. the curves, $C_1(\tau)$, and $C_2(t)$, have fallen together.

Of the $2^{13}\cdot 31\cdot 51$ rational planar sextics congruent to $\bar{S}_2(t)$ the symmetroid Σ yields two paired sextics, $\bar{S}_1(\tau)$, $\bar{S}_2(t)$; and the $2^8\cdot 51$ symmetroids congruent to Σ yield $2^8\cdot 51$ such pairs. This pairing is associated, as we have seen, with a discriminant condition, $P_{\alpha\beta}$. Those elements of the group 52 (5) which leave this condition unaltered and therefore transform a pair of rational sextics into a pair constitute a subgroup with an invariant g_2, consisting of 1 and R_1, whose factor group is the group 53 (12) of congruent symmetroids. Those further discriminant conditions on the nodal set P_{10}^2 of $\bar{S}_2(t)$ which are syzygetic to $P_{\alpha\beta}$ yield discriminant conditions on the nodal set P_{10}^3 of Σ. It is indeed clear from the mapping in (9) that

(11) *If nodes D_i, D_j of Σ coincide, then nodes p_i, p_j of $\bar{S}_2(t)$ $[\bar{S}_1(\tau)]$ coincide; if 4 nodes of Σ are coplanar then the complementary 6 nodes of $\bar{S}_2(t)$ $[\bar{S}_1(\tau)]$ are on a conic.*

Discriminant conditions azygetic to $P_{\alpha\beta}$ may be described in terms of the special behavior of the quadrics M_i with respect to Σ, and, doubtless, translated to a particular behavior of the Stahl quadric K with respect to Σ as in (10). Thus from the mapping it is at once apparent that:

(12) *If three nodes p_1, p_2, p_3 of $\bar{S}_2(t)$ are collinear then M_1 contains the line D_2D_3; or also M_4 contains the cubic curve on $D_5, \cdots, D_9, D_0$.*

The linear system (∞^5) of conics in the plane maps the plane upon a Veronese $V_2^4(t)$ in S_5, the locus of double points of a cubic spread M_4^3, the map of pairs of points of the plane. The sextic $\bar{S}_2(t)$ is mapped upon a 10-nodal rational curve $R^{12}(t)$ in S_5. If 6 of the nodes of $R^{12}(t)$ are in an S_4, then 6 of the nodes of $\bar{S}_2(t)$ are on a conic and the complementary four nodes of Σ are in a plane. Hence the set P_{10}^5 of nodes of $R^{12}(t)$ is the set associated to the nodal set P_{10}^3 of Σ. But then conics in the plane of $\bar{S}_1(\tau)$ must likewise map its plane upon a Veronese $V_2^4(\tau)$ containing the same P_{10}^5. Hence

(13) *If two Veronese surfaces,* $V_2^4(t)$, $V_2^4(\tau)$ *meet in* 10 *points,* P_{10}^5, *then this set is associated to the set,* P_{10}^3, *of nodes of a symmetroid* Σ. *The spreads* $M_4^3(t)$, $M_4^3(\tau)$, *with double* $V_2^4(t)$, $V_2^4(\tau)$ *respectively, each cut the double* V_2^4 *of the other in a rational* 12-*ic curve,* $R^{12}(\tau)$, $R^{12}(t)$, *with nodes at* P_{10}^5. *These* 12-*ic curves are the maps of the two rational sextics associated with* Σ.

A regular quintic transformation in S_5 with six F-points in P_{10}^5 converts P_{10}^5 into a congruent set Q_{10}^5 associated to a set Q_{10}^3 congruent to P_{10}^3 under regular cubic transformation with F-points at the complementary four points of P_{10}^3(16(8)). Under such transformation a V_2^4 on P_{10}^5 is converted into a V_2^4 on Q_{10}^5, their respective planes being congruent under a ternary quintic transformation $A_{1\cdots 6}$ ([21](17)). Hence

(14) *If* Σ *is congruent to* Σ' *under the cubic transformation* A_{1234} *in space, then* $\bar{S}_2(t)\,[\bar{S}_1(\tau)]$ *is congruent to* $\bar{S}_2'(t)\,[\bar{S}_1'(\tau)]$ *under the quintic transformation* A_{567890}.

The entire ternary Cremona group is generated by collineations and a single quadratic transformation, A_{123}. It contains subgroups of the same nature and degree of generality as itself; e. g. the group G generated by collineations and the quintic transformation A_{123456}. This group G is isomorphic with the regular group in S_5 which transforms a V_2^4 into itself. With congruence defined as congruence under G a planar rational sextic is congruent to $2^8 \cdot 51$ projectively distinct pairs which are in (1,1) correspondence with the $2^8 \cdot 51$ projectively distinct types of congruent symmetroids. An investigation of the types of groups G thus defined by particular types of Cremona transformations would be of decided interest.

The jacobian J of a web admits a variety of Cremona transformations (cf. [65]). Thus the two cubic involutions defined by ζ, ζ' respectively in 54 (23) each define the involution I on J and their product is a transformation for which J is a locus of fixed points. Also since Σ is a transform of J by a cubic transformation then the J, J', birationally equivalent to congruent Σ, Σ', must also be equivalent under Cremona transformation. It is nevertheless desirable to restrict somewhat the type of transformation so that a groupoid property analogous to that of congruence may appear. A suitable type is that cubic transformation with sextic F-curve which degenerates into a cubic curve N and three bisecants A. This has been studied in various connections by Fano, Tinto, Young and Morgan, and Montesano (cf. [52] pp. 203–4 and p. 217). Montesano[46] shows that for fixed N and variable triad of bisecants these transformations produce types isomorphic with the ternary types of Cremona transformation. We give without proof the theorem analogous to (14).

(c) *The jacobian J is transformed by the cubic transformation with F-curves, N_1, A_2, A_3, A_4 into a jacobian J'. The symmetroids Σ, Σ' birationally equivalent to J, J' are congruent under the regular cubic transformation A_{1234}.*

A consequence of this eventual symmetry in 1, 2, 3, 4 is that if J is transformed into J' as indicated, and into J'' with F-loci, N_2, A_1, A_3, A_4 then J' and J'' are projective. Thus the product of two such transformations and a properly chosen projectivity sends J into itself.

56. Related projective figures and projective groups. Let $\bar{S}_1(\tau)$ and $\bar{S}_2(t)$ be two paired sextics and let $K(t)$ be the covariant conic of **54** (4) in the plane of $\bar{S}_1(\tau)$. The pairs of points cut out on $\bar{S}_2(t)$ by the P-curves of the nodal P_{10}^2 of $\bar{S}_2(t)$ will be named by the quadratics determined by their parameters as before but with the parameter t deleted as in $q(1)^0$, $q(12)^1$, etc.; those similarly cut out on $\bar{S}_1(\tau)$ will have the parameter τ indicated. With reference to $K(t)$ as a norm conic, these quadratics in t determine points in the plane of $\bar{S}_1(\tau)$ which will be referred to as the "points" $q(1)^0$, $q(12)^1$, etc. According to **52** (10) et seq. if these points are marked upon the plane of $\bar{S}_1(\tau)$ there will be an initial set of 527 points (cf. **52** (3)) each of which will be the initial point of an infinite conjugate set under a ternary collineation group $\Gamma_{10,2}$ with the invariant conic $K(t)$. The 527 sets exhaust the points q. Two points q are in the same conjugate set if and only if their signatures are congruent mod. 2. The group $\Gamma_{10,2}$ is generated by a conjugate set of involutions, i. e. harmonic perspectivities, determined as follows. If $q(1^{r_1}, \cdots, 0^{r_0})^r$ and $q(1^{s_1}, \cdots, 0^{s_0})^s$ are any two points q for which

$$(1) \qquad rs - r_1 s_1 - \cdots - r_9 s_9 - r_0 s_0 = 0,$$

then the harmonic perspectivity whose axis is the line l joining the two points q and whose center is the pole of l as to $K(t)$ is a generator of $\Gamma_{10,2}$. Some of the more interesting projective properties of certains groups of these points are developed here.

We observe in the first place that the points q contain an infinite number of nodal sets P_{10}^2 of planar rational sextics. According to **54** (4) the ten points $q(1)^0, \cdots, q(9)^0, q(0)^0$ are the nodes of $\bar{S}_1(\tau)$ itself; the ten points $q(23)^1, q(31)^1, q(12)^1, q(4)^0, \cdots, q(0)^0$ whose parameters arise from $\bar{S}_2(t)$ by the quadratic transformation A_{123} are the nodes of the transform of $\bar{S}_1(\tau)$ by the Geiser involution I^7 with F-points at $q(4)^0, \cdots, q(0)^0$ (**55** (7)); and the ten points $q(1^3 3 \cdots 0)^4$, $q(2)^0, q(13)^1, \cdots, q(10)^1$ are also a nodal set (**55** (4) et seq.). In the entire configuration q there are $2^{13}\cdot 31\cdot 51$ projectively distinct nodal sets P_{10}^2 of this sort and the remaining sets arise from these by the collineations of $\Gamma_{10,2}$.

(2) *The entire configuration of points q can be obtained from a given nodal set, e. g. $q(1)^0, \cdots, q(0)^0$ by the linear process of constructing the 9th intersection of two cubics on eight points.*

If indeed C_i is the cubic curve on the nine points other than $q(i)^0$ $(i = 1, \cdots, 9, 0)$ then C_i contains the vertices of 4-lines circumscribed to $K(t)$. For, on $\bar{S}_2(t)$ the I_1^4 cut out by lines on the node p_1 will, when transferred to $K(t)$, produce a system (∞^1) of 4-lines circumscribed about $K(t)$ whose vertices are on a cubic and whose opposite vertices are in a hessian correspondence $u' \equiv u + \omega/2$ on the cubic.

Such opposite vertices are $q(i)^0$ and $q(1i)^1$ $(i=2,\cdots,9,0)$. Hence the cubic is C_1 and it contains the point $q(12)^1$ which is the intersection of C_1 and C_2. Thus the nodal set $q(23)^1$, $q(31)^1$, $q(12)^1$, $q(4)^0, \cdots, q(0)^0$ can be constructed as in (2). This belongs to the sextic paired with the transform of $\overline{S}_2(t)$ by A_{123} and the entire configuration can be obtained by successive quadratic transformations.

Another notable set of ten points q, which is not a nodal set consists of the four points $q(1)^0, \cdots, q(4)^0$ and the six points $q(ij)^1$ $(i,j=1,\cdots,4)$, say the set R. It is convenient to symmetrize the notation by setting $q(i)^0=r_{0i}$ and $q(ij)^1=r_{kl}$ $(i,j,k,l=1,\cdots,4)$. Let the six further points $q(5)^0,\cdots,q(0)^0$ be denoted by Q_6^2. Then first there are five rational sextics S_i $(i=0,\cdots,4)$ with six nodes at Q_6^2 and four nodes at r_{ij}. One of these is the original sextic $\overline{S}_1(\tau)=S_0$ with nodes at $q(1)^0,\cdots,q(4)^0$. The first of the remaining four is the sextic S_1 with nodes at $q(1)^0$, $q(34)^1$, $q(24)^1$, $q(23)^1$. The five sextics S_i are thus the sextics paired with the five which have six given pairs of nodal parameters. Secondly there are five cubic curves K_i on Q_6^2 and on those six points r_{ij} which are complementary to the four nodes of S_i. Four of these, $K_1,\cdots,K_4$ are the cubics $C_1,\cdots,C_4$ above which have the required property. The fifth, K_0, is to contain the six points $q(ij)^1$ $(i,j=1,\cdots,4)$ and $q(k)^0$ $(k=5,\cdots,9,0)$. It may be proved that K_0 with the required property exists. If then the plane is mapped upon a cubic surface by cubic curves on Q_6^2 (which isolates a sixer on the surface) the cubics K_i become an inscribed five-plane and the sextics S_i become the rational sextics cut out by the five quadrics which touch the surface at the vertices of one of the five tetrahedra of the five-plane. Conversely if a cubic surface has an inscribed five-plane these five tangent quadrics exist and, after choice of a sixer on the surface, the five cubics K_i and sextics S_i are determined.

As a consequence

(a) *A necessary and sufficient condition that* 10 *points* $p_1,\cdots,p_4$; $p_5,\cdots,p_9,p_0$ *be nodes of a rational sextic is that the four* *cubics on the last six points P_6^2 and respectively on three of the four remaining points shall meet again by pairs in six points which also lie with P_6^2 on a cubic curve.*

If, in conformity with the notation for the set R, the nodes $p_1, \cdots, p_4$ of $\bar{S}_2(t)$ have nodal parameters $r_{01}, \cdots, r_{04}$ and the line joining the nodes p_i, p_j meets $\bar{S}_2(t)$ again in r_{kl} $(i, \cdots, l = 1, \cdots, 4)$; if also the points p_i be taken as the reference and unit points respectively then the parametric equations of $\bar{S}_2(t)$ are

$$(3) \quad \begin{aligned} x_1 &= r_{02}\, r_{03}\, r_{14} & x_2 - x_3 &= r_{04}\, r_{01}\, r_{23} \\ x_2 &= r_{03}\, r_{01}\, r_{24}; & x_3 - x_1 &= r_{04}\, r_{02}\, r_{31} \\ x_3 &= r_{01}\, r_{02}\, r_{34} & x_1 - x_2 &= r_{04}\, r_{03}\, r_{12} \end{aligned} \quad (r_{ij} = -r_{ji}).$$

Then the 10 quadratics r are connected by five quadratic identities which are due to these obvious identities in x:

$$\Sigma_1^3 (x_2 - x_3) = 0; \qquad \Sigma_1^3 x_1 (x_2 - x_3) = 0;$$
$$x_3 - x_2 + (x_2 - x_3) = 0.$$

The five quadratic relations are

$$(4) \quad r_{jk}\, r_{il} + r_{ki}\, r_{jl} + r_{ij}\, r_{kl} = 0 \qquad (i, \cdots, l = 0, \cdots, 4).$$

Due to these relations one may prove that

(b) *The set of* 10 *points R can be separated in* 10 *ways, by isolating a point r_{ij}, into three points r_{lm}, r_{mk}, r_{kl}, and three pairs of points on lines $r_{ik}\, r_{jk}$, $r_{il}\, r_{jl}$, $r_{im}\, r_{jm}$, such that the triangles of three points and three lines are perspective.*

The property of the set R expressed in (b) is very similar to the perspective property of the Desargues configuration. The configuration R however has 9 absolute constants whereas the Desargues configuration has but three. It may be verified easily that eight of the points R may be chosen at random, the ninth is then any point of a unique line, which when chosen determines the tenth uniquely. This of course is on the assumption that the norm-conic $K(t)$ is not given in advance. Indeed one may prove that the five quadratic relations (4) can be replaced by the five linear relations,

$$(5)\qquad r_{i0}+r_{i1}+r_{i2}+r_{i3}+r_{i4}\equiv 0 \qquad (r_{ij}=-r_{ji};\, r_{ii}\equiv 0),$$

of which only four are independent, and by any one of the five quadratic relations. The linear relations determine the projective situation of the points R in the plane as expressed in theorem (b) and the quadratic relation then determines the location of $K(t)$.

This suggests another way of bringing in the quadratics r_{ij} or points of R. If $y_0, \cdots, y_4$ with $y_0+\cdots+y_4=0$ are the supernumerary coördinates of a point y in S_3, the line coördinates, $p_{ik}=(y_i y'_k - y_k y'_i)$, of the line yy' in S_3 satisfy the linear relations (5) as well as the quadratic relations (3). The coördinate system determines a five-plane, and a quadric, adjoined as follows, brings the number of absolute constants up to 9. Let

$$(6)\qquad y_i=(a_i\tau)(\alpha_i t),\quad \textstyle\sum_i (a_i\tau)(\alpha_i t)\equiv 0 \qquad (i=0,\cdots,4).$$

As t, τ vary the point $y(t,\tau)$ runs over a quadric with generators (t,τ). The line coördinates of a line on points $y(t,\tau)$ and $y'(t',\tau')$ are

$$(7)\qquad p_{ij}=(r_{ij}\,t)(r_{ij}\,t')\cdot(\tau\tau')+(s_{ij}\,\tau)(s_{ij}\,\tau')\cdot(tt').$$

For $t'=t$, i. e., a t-generator, the ten line coördinates are a set of quadratics $(r_{ij}\,t)^2$, and for $\tau'=\tau$, i. e. a τ-generator, the ten line coördinates are an entirely similar set of quadratics $(s_{ij}\,\tau)^2$; where

$$(8)\quad (r_{ij}t)^2=(a_i a_j)(\alpha_i t)(\alpha_j t);\qquad (s_{ij}\tau)^2=(\alpha_i\alpha_j)(a_i\tau)(a_j\tau).$$

Further details with respect to these configurations will appear in the article cited in connection with **54** (a).

Attention may be called at this point to a method suggested by the author ([22] pp. 359–61) for setting up series in the plane of $\bar{S}_2(t)$ which are formally invariant under the $i^{(2)}_{10,2}$ of $\bar{S}_2(t)$, and which are readily transformed into series invariant under $\Gamma_{10,2}$.

CHAPTER VI

THETA RELATIONS OF GENUS FOUR

The variety of geometric configurations and of algebraic forms discussed in the preceding chapter had the common property of being, in one way or another, projectively determined by the birationally general algebraic relation of genus four, $F=(a\tau)^3(\alpha t)^3=0$. In the present chapter some of these geometric figures will be deduced directly from the theta relations. The developments are due largely to Schottky who observed that the P_{10}^3 of ten nodes of a symmetroid Σ could be expressed in terms of modular functions of genus four of the abelian type defined by an algebraic curve. The bearing of this upon the problem of the determination of the tritangent planes of the canonical space sextic of genus four will be discussed at the close in the light of the similar problem presented by the double tangents of the planar quartic.

57. Derivation of certain theta relations. For $p=4$ and $2p+2$ subscripts, $i, j, \cdots = 1, 2, \cdots, 9, 0$, the odd and even theta functions of the first order comprise the 136 even functions, ϑ_i, ϑ_{ijklm}, with zero values c_i, c_{ijklm}; and the 120 odd functions, ϑ_{ijk}. The half periods are represented by the points P_{ij}, P_{ijkl} in the finite geometry mod. 2 of an S_7 referred to a null system C_4.

The projection and section of the null system C_4 by a point and its null space, say the point P_{90}, leads to a derived null system, C_3, in an S_5 (cf. 26). The points of the S_5 arise from the null lines of C_4 and are named by subscripts $i, j, \cdots = 1, \cdots, 8$. Thus the null lines P_{90}, P_{ij}, P_{ij90}; P_{90}, P_{ijkl}, P_{mnop} on P_{90} in S_7 contribute the points P_{ij} and $P_{ijkl}=P_{mnop}$ respectively in S_5. The quadrics in S_7 on P_{90} divide into 64 pairs and, with respect to these pairs, we define the products,

(1) $X = \vartheta_9\,\vartheta_0,\ X_{ijkl} = X_{mnop} = \vartheta_{ijkl9}\,\vartheta_{ijkl0};\quad X_{ij} = \vartheta_{ij9}\,\vartheta_{ij0},$

with their zeros values

$$(2)\qquad p = c_9\,c_0,\ p_{ijkl} = c_{ijkl9}\,c_{ijkl0}\,.$$

It is proved below that these products satisfy the same system of linear relations, as the theta squares for $p = 3$.

The projection and section of the null system C_3 in S_5 from one of its points is equivalent to the projection and section of the original C_4 in S_7 from one of its null lines and by the null space S_5 of the null line. If the null line in the S_7 is defined by P_{90}, P_{78}, the result is a null system C_2 in S_3 whose points are named by subscripts $i, j, \cdots = 1, \cdots, 6$. The quadrics in S_7 on the null line divide into 16 tetrads with respect to which the 16 products and 10 zero values are:

$$(3)\qquad \begin{gathered} Y_i = X_{i7}\,X_{i8},\quad Y_{ijk} = X_{ijk7}\,X_{ijk8},\quad q_{ijk} = p_{ijk7}\,p_{ijk8} \\ (i, j, k = 1, \cdots, 6). \end{gathered}$$

Similarly the projection and section of this C_2 in S_3 from one of its points and by one of its planes is a null system C_1 in S_1 which is also the projection and section of the original C_4 in S_7 from one of its null planes and by the null S_4 of the null plane. If the null plane in S_7 is that defined by P_{90}, P_{78}, P_{56}, the null system C_1 in S_1 has a basis notation with subscripts $i, j = 1, \cdots, 4$. The quadrics in S_7 on the null plane divide into 4 octads which yield 4 products and zero values, namely:

$$(4)\qquad \begin{gathered} Z = Y_5\,Y_6,\quad Z_{ij} = Y_{ij5}\,Y_{ij6},\quad r_{ij} = q_{ij5}\,q_{ij6} \\ (i, j = 1, \cdots, 4). \end{gathered}$$

Following Schottky ([63] pp. 251–2, cf. also [18] § 4) a notable set of linear relations connecting the theta squares ($p = 3$) is obtained. The eight functions, $\vartheta_{18}, \vartheta_{28}, \cdots, \vartheta_{78}, \vartheta$, are a normal fundamental set (cf. **26**). Since only eight theta squares are linearly independent, ϑ_{12}^2 must be linearly expressible in terms of the eight in the F. S. Since all except ϑ vanish for $u=0$, ϑ^2 can not appear in the relation. On adding the half period P_{18} all the functions except ϑ_{18} remain odd whence ϑ_{18}^2 can not appear in the relation and similarly ϑ_{28}^2 can not appear. Thus the relation has the form, $\sum_{\alpha=3}^{\alpha=7} A_{\alpha 8}\,\vartheta_{\alpha 8}^2 = A\,\vartheta_{12}^2$. On setting $u=P_{\alpha 8}$, $A_{\alpha 8}\,c^2 = \pm A\,c_{12\alpha 8}^2$, or

$$c^2\,\vartheta_{12}^2 = \sum\nolimits_\alpha \pm c_{12\alpha 8}^2\,\vartheta_{\alpha 8}^2 \qquad (\alpha = 3, \cdots, 7). \tag{5}$$

On replacing u by $u+P_{12}$ this becomes

$$c^2\,\vartheta^2 + \sum\nolimits_\alpha \pm c_{\alpha 128}^2\,\vartheta_{\alpha 128}^2 = 0. \tag{6}$$

These two relations contain a closed set of six azygetic functions, i. e., any three are azygetic and the six are linearly dependent (as quadrics in S_7). The general relation of this character may be described as follows:

(7) *If* $\vartheta_\alpha, \cdots, \vartheta_\zeta (\alpha, \cdots, \zeta$ *and* λ *properly chosen sets of subscripts) are a closed azygetic set of six functions then*

$$\sum\nolimits_\alpha^\zeta c_{\alpha\lambda}^2\,\vartheta_\alpha^2 = 0$$

where P_λ *is the half period (proper or zero) for which all the functions* $\vartheta_{\alpha\lambda}, \cdots, \vartheta_{\zeta\lambda}$ *are even.*

The functions X of genus four defined in (1) of the second order and characteristic P_{90} behave like the theta squares ($p=3$) in the following respects: their linear independence (cf. **20**(9)); their vanishing when $u=0$; and their permutation under addition of half periods (syzygetic with P_{90}). But this behavior was sufficient to establish (5), (6), (7). Hence

(8) *If* $X_\alpha, \cdots, X_\zeta (\alpha, \zeta, \lambda$ *properly chosen set of subscripts for* $p=3$) *are a closed asygetic set of six functions then*

$$\sum\nolimits_\alpha^\zeta p_{\alpha\lambda}\,X_\alpha = 0$$

with subscripts λ *chosen as in* (7).

One of the relations (8) is

$$\sum_{i=1}^{i=4} \pm p_{i567}\,X_{i568} \pm p_{5678}\,X_{56} \pm p\,X_{78} = 0.$$

This for $u = 0$ becomes

$$\sum_i \pm q_{i56} = 0 \qquad (i = 1, \cdots, 4), \tag{9}$$

which is one of 15 linear relations among the 10 constants q_{ijk} of the same form as the system **30** (V) for $p = 2$.

Let u_{ijk} be the linear term in the development of the odd theta function $\vartheta_{ijk}(u)$. Corresponding to (1) and (4) we set

$$v_{ij} = u_{ij9}\, u_{ij0}, \qquad w_i = v_{i7}\, v_{i8}. \tag{10}$$

The theta relations are satisfied if we set as in **45** (5)

$$\vartheta_{ijk} = k \sqrt{u_{ijk}} \sqrt{u'_{ijk}}. \tag{11}$$

The variables u are then not unrestricted but are rather proportional to algebraic functions of a variable x; and the u' to functions of x'. They are subject to two homogeneous relations of higher degree which in S_3 define the normal curve G_4^6 for which the u_{ijk} are tritangent planes. Then from (8) and (5) there follows

$$\begin{aligned} &\sum_i \pm p_{i678}\, v_{i6} \pm p\, v_{78} = 0 \qquad (i = 1, \cdots, 5), \\ &\sum_i \pm p_{i678} \sqrt{v_{i6}} \sqrt{v'_{i6}} \pm p \sqrt{v_{78}} \sqrt{v'_{78}} = 0. \end{aligned} \tag{12}$$

On G_4^6 the $\sqrt{v}$ are in the G_2^6 of contacts of contact quadrics of the system P_{90} so that any four are linearly related. Let then $\sum_j A_j \sqrt{v_{j6}} = 0$ $(j = 1, \cdots, 4)$. The lemma of **45** (4) applied to this linear relation and to the quadric relation (12) yields $\sum_j \pm A_j^2 / p_{j678}$. The constants $A_1, \cdots, A_4$ satisfy also the equations obtained from this by replacing the indices 78 by 57 and by 58. These three equations are sufficient to determine the A_j^2 to be $A_j^2 = p_{j678}\, p_{j567}\, p_{j568}$. For, the equations are satisfied due to (9) when these values are substituted in them. It is understood of course that (9) includes all the relations similar to it which can be obtained by projection from any null line. The linear relations may therefore be written as

$$\sum_j \pm (p_{j678}\, p_{j567}\, p_{j568}\, v_{j6})^{1/2} = 0 \qquad (j = 1, \cdots, 4). \tag{13}$$

If similarly $B_5\sqrt{v_{56}}+\sum_k B_k\sqrt{v_{k6}}=0$ $(k=1,2,3)$ then $B_k^2=p_{k678}\,p_{k467}\,p_{k468}$ and (applying the lemma again)

$$(14)\qquad \sum_k \pm(p_{k657}\,p_{k658}\,p_{k647}\,p_{k648})^{1/2}=\sum_k \pm(q_{k65}\,q_{k64})^{1/2}=0 \quad (k=1,2,3).$$

The three terms which appear here are the quantities r_{ij} in (4) for the projection and section from P_{90}, P_{78}, P_{45} whence (cf. 28 (11))

(15) *The three products of zero values of eight even functions defined in* (4) *satisfy the relation*

$$\sqrt{r_{14}}\pm\sqrt{r_{24}}\pm\sqrt{r_{34}}=0.$$

For each of the $255\cdot 45$ null planes (cf. 28) used for projection and section there is a relation (15). Schottky [59] has proved that they each express the single condition that the ten moduli a_{ij} of the theta functions are of the abelian type which occur in connection with the normal integrals of the first kind of the curve of genus four (cf. also Roth [57] § 9). For values of p beyond four the $p(p+1)/2-(3p-3)$ conditions of this type have not been obtained.

Schottky ([63] pp. 264–6) goes on to point out that the 28 root functions $\sqrt{v_{ij}}$, obtained by projection and section from the selected half period P_{90}, satisfy the same system of linear relations as the 28 functions u_{ij} $(p=3)$ (cf. (13) and 45 (B)), provided that the constants $\sqrt{p_{ijkl}}$ $(p=4)$ replace the constants c_{ijkl} $(p=3)$. Moreover on comparing (14) with 45 (A), and (9) with the modular relation $p=3$ of the type obtained by setting $u=P_{13}$ in (6), it is clear that both systems of constants are conditioned in the same way. Hence the 28 $\sqrt{v_{ij}}$, lying in a linear system (∞^2), are the double tangents of a planar quartic, f^4. They also are in the linear system g_2^6 of contact quadrics P_{90} and are sections of a Wirtinger sextic W related to f^4 as described in 50 (in particular (13)). As double tangents of f^4 the $\sqrt{v_{ij}}$ satisfy a system of irrational relations (45 (C)) all of which are equivalent to one relation, the rational equation of f^4 itself. These transfer to $p=4$ as follows:

$$(16)\qquad \sum_i \pm (q_{i46}\, q_{i56}\, q_{i45}\, w_i)^{1/4} = 0 \qquad (i = 1, 2, 3).$$

But the values of the $w_i = v_{i7}\, v_{i8} = u_{i79}\, u_{i70}\, u_{i89}\, u_{i80}$ are obtained from the values of $u_1, \cdots, u_4$ as functions of x on W. In (16), which is an equation of f^4, the values of $u_1, \cdots, u_4$ must be those assumed when x is one of the 24 points common to f^4 and W, i. e., the $u_1, \cdots, u_4$ and the u_{ijk} are constants. Hence

(17) *With $c_{ijklm} = \vartheta_{ijklm}(0)$ it is possible to find constants c_{ijk} for the functions ϑ_{ijk} such that the system of equations* (16) *for any isolated syzygetic half periods P_{90}, P_{78} is satisfied when $u_{ijk} = c_{ijk}$.*

It will be observed that the 24 points common to f and W are, on the canonical curve G_4^6 in space, the branch points of the two g_1^3's in the canonical involution. The values of the u_{ijk} at one of these points are independent of the particular half period P_{90} which establishes the relation between f^4 and W. The equation (16) is the basis for the connection, established in 59, of the nodes of a symmetroid with the modular functions.

58. Definition of the planar set, P_8^2, in terms of modular functions. When one of the 136 constants c, say c_0, vanishes the normal space sextic, G_4^6, of genus four is on a quadric cone A (cf. 51). The 120 tritangent planes of G_4^6 are then rationally separated by the discrete points of a set P_8^2 with which there is associated projectively a plane curve G_4^9, a birational exemplar of G_4^6. The line sections of G_4^9, a g_2^9, are contained in the complete linear series g_5^9 cut out on the normal G_4^6 by contact cubics of the system determined by $\vartheta_9(u)$. The G_4^6 is mapped by this g_5^9 upon a curve Γ_4^9 of order 9 in S_5 and G_4^9 is the projection of Γ_4^9 upon the plane of P_8^2 from a properly chosen non-secant plane. Schottky [60] has obtained a projective definition of the set P_8^2 directly from the modular functions for which $c_0 = 0$. The pertinent part of this memoir follows.

The 28 root functions defined by

$$(1)\quad F_{\alpha\beta} = \sqrt{u_{\alpha 90}}\sqrt{u_{\beta 90}}\sqrt{u_{\alpha\beta 9}}\ (\alpha, \beta = 1, \cdots, 8;\ \alpha \neq \beta)$$

determine on G_4^6 the nine contacts of three tritangent planes in the system of ϑ_9. The contacts are therefore a set of g_5^9 but, due to $c_0 = 0$, the 28 functions satisfy a system of three term relations by virtue of which only three are linearly independent. These three are line sections of the planar G_4^9. The three term relations are derived from the system 57 (13). If in 57 (13), when written in terms of the c's and u_{ijk}'s, the indices 6, 9 are interchanged, and the relation is then transformed by the period transformation I_{1238} under which the c's and u_{ijk}'s are permuted like the even and odd functions, it becomes (cf. 25 (4))

$$(2)\quad \begin{aligned} &\sum_k \pm (c_{k4580}\, c_{k4568}\, c_{k5679}\, c_{k5790}\, c_{k4780}\, c_{k4678}\, u_{k96}\, u_{k90})^{1/2} \\ &\pm (c_0\, c_6\, c_{46789}\, c_{47890}\, c_{45689}\, c_{45890}\, u_{570}\, u_{567})^{1/2} = 0\ (k = 1, 2, 3). \end{aligned}$$

If to indicate projections from points other than P_{90}, P_{78} the products p, q are written more explicitly as

$$(3)\quad p_{klmn}^{(ij)} = c_{iklmn}\, c_{jklmn}$$

then (2) takes the form

$$(4)\quad \begin{aligned} &\sum_k \pm (p_{k458}^{(60)}\, p_{k478}^{(60)}\, p_{k579}^{(60)}\, v_{k9}^{(60)})^{1/2} = \sum_k \pm (q_{k48}^{(60)\,(57)}\, p_{k579}^{(60)}\, v_{k9}^{(60)})^{1/2} = 0, \\ &c_0 = 0 \qquad (k = 1, 2, 3). \end{aligned}$$

This, multiplied by $\sqrt{u_{690}}$ and modified by (1), yields

$$(5)\quad \sum_k \pm (q_{k48}^{(60)\,(57)}\, p_{k579}^{(60)})^{1/2}\, F_{k6} = 0, \quad c_0 = 0.$$

By means of these relations any of the 28 $F_{\alpha\beta}$ can be expressed linearly in terms of a properly chosen set of three such as F_{12}, F_{13}, F_{23}. If then these three are equated to three independent linear combinations of x_0, x_1, x_2, the coördinates of the point x in S_2 are expressed in terms of the $\sqrt{u_{ijk}}$. Since the 7 functions $F_{\alpha\beta}$ with common index α can all be expressed in terms of two, the corresponding lines in S_2 all pass through a point p_α ($\alpha = 1, \cdots, 8$) and $F_{\alpha\beta}$ is the line joining the points p_α, p_β. For variation of the initial terms u with the variation of a point on the normal G_4^6, the point x runs over the curve G_4^9 with triple points at P_8^2 since the seven functions $F_{\alpha\beta}$ for fixed α vanish for the three zeros of $\sqrt{u_{\alpha 90}}$, a tritangent plane of G_4^6.

Setting for brevity

$$(6) \qquad s_\alpha = u_{\alpha 90}, \qquad P = \prod_{\alpha=1}^{\alpha=8} (\sqrt{s_\alpha}),$$

Schottky defines a number of further root functions as follows:

$$(7) \qquad \begin{aligned} G_{\alpha\beta\gamma} &= P\sqrt{u_{\alpha\beta\gamma}}/\sqrt{s_\alpha}\sqrt{s_\beta}\sqrt{s_\gamma}; \\ H_{\alpha,\beta} &= P\sqrt{u_{\alpha\beta 0}}\sqrt{s_\alpha}/\sqrt{s_\beta}; \\ J_{\alpha\beta\gamma} &= P\sqrt{s_\alpha}\sqrt{s_\beta}\sqrt{s_\gamma}\sqrt{u_{\alpha\beta\gamma}}; \\ K_{\alpha\beta} &= P^2\sqrt{u_{\alpha\beta 9}}/\sqrt{s_\alpha}\sqrt{s_\beta}; \\ L_{\alpha\beta 9} &= F_{\alpha\beta} K_{\alpha\beta}; \\ L_{\alpha\beta\gamma} &= G_{\alpha\beta\gamma} J_{\alpha\beta\gamma}; \\ L_{\alpha\beta 0} &= H_{\alpha,\beta} H_{\beta,\alpha}; \\ L_{\alpha 90} &= P^2 s_\alpha. \end{aligned}$$

These with $F_{\alpha\beta}$ are the sections of G_4^9 by the P-curves (cf. 51 (1)) of the set P_8^2 to which the improper section by $P(\alpha)^0$, i. e. $\sqrt{s_\alpha}$, should be added. The L-curves are the pairs of P-curves which in 51 (1) make up the degenerate sextics of the web with nodes at P_8^2.

The algebraic relations among these curves are consequences of the theta relations 57 (13). The linear relations among the $F_{\alpha\beta}$ have already been noted as consequences of (4). Other three and four term relations connect

$$\begin{aligned} &\text{(a)} \quad (u_{139}\,u_{249})^{1/2}, \quad (u_{239}\,u_{149})^{1/2}, \quad (u_{590}\,u_{678})^{1/2}; \\ &\text{(b)} \quad (u_{120}\,u_{890})^{1/2}, \quad (u_{194}\,u_{284})^{1/2}, \quad (u_{195}\,u_{285})^{1/2}; \\ (8)\ &\text{(c)} \quad (u_{123}\,u_{490})^{1/2}, \quad (u_{140}\,u_{239})^{1/2}, \quad (u_{240}\,u_{139})^{1/2}; \\ &\text{(d)} \quad (u_{129}\,u_{190})^{1/2}, \quad (u_{123}\,u_{130})^{1/2}, \quad (u_{124}\,u_{140})^{1/2}; \\ &\text{(e)} \quad (u_{190}\,u_{590})^{1/2}, \quad (u_{192}\,u_{592})^{1/2}, \quad (u_{193}\,u_{593})^{1/2}, \quad (u_{194}\,u_{594})^{1/2}, \end{aligned}$$

the last of these being obtained directly from **57** (13) by permutation of the indices. The relation involving the terms in (d) arises from (4) by the permutation (629134). The terms (a) arise from (4) by the transformation I_{1280} (63584); the terms (b) from (a) by I_{1670} (3582); and the terms (c) from (b) by I_{2450} (1384) (52). These transformations do not disturb the equation $c_0 = 0$.

If the terms (a), $\cdots$, (e) in (8) are multiplied respectively by $(s_1 s_2 s_3 s_4)^{1/2}$, $P(s_1)^{1/2}/(s_2 s_8)^{1/2}$, $P(s_1 s_2 s_3)^{1/2}/(s_4)^{1/2}$, $P^2/s_1 (s_2)^{1/2}$, $P^2(s_1)^{1/2}/(s_5)^{1/2}$, they yield, by comparison with (7), the terms:

$$
\begin{array}{lll}
 & \text{(a)} & F_{13} F_{24},\ F_{23} F_{14},\ G_{678}; \\
 & \text{(b)} & H_{1,2},\ F_{14} G_{284},\ F_{15} G_{285}; \\
(9) & \text{(c)} & J_{123},\ H_{1,4} F_{23},\ H_{2,4} F_{13}; \\
 & \text{(d)} & K_{12},\ G_{123} H_{3,1},\ G_{124} H_{4,1}; \\
 & \text{(e)} & L_{190},\ K_{25} F_{21},\ K_{35} F_{31},\ K_{45} F_{41}.
\end{array}
$$

Each group of terms in (9) is linearly related with coefficients which can be determined by the same transformations as were used above to produce the group of terms. Hence in (a) $G_{678} = 0$ is a conic on $p_1, \cdots, p_4$ and therefore, by virtue of the symmetry in its definition in (7), on p_5 also. In (b) $H_{1,2} = 0$ is the cubic with node at p_1, and simple points at $p_3, \cdots, p_7$ and, by symmetry, at p_8. Similarly in (c) and (d) J_{123} and K_{12} are the P-curves, $P(1^2 2^2 3^2 4 \cdots 8)^4$, $P(1 2 3^2 \cdots 8^2)^5$ respectively. In (e) the sextic curve L_{190} must have a triple point at p_1 and nodes at $p_2, p_3, p_4, p_6, p_7, p_8$ and at least a simple point p_5. Again, from the symmetry of $L_{190} = P^2 s_1$, it must have a node at p_5 and be the P-curve, $P(1^3 2^2 \cdots 8^2)^6$.

The effect of the definitions (7) is to assign definite values to the constant factors in the equations of the P-curves. The identical relations which follow from these definitions are various forms of the equation of G_4^9. Such are

$$
\begin{array}{lll}
 & \text{(a)} & H_{\alpha,\beta} H_{\beta,\gamma} H_{\gamma,\alpha} = H_{\beta,\alpha} H_{\gamma,\beta} H_{\alpha,\gamma}; \\
(10) & \text{(b)} & L_{\alpha 90} H_{\beta,\alpha} = L_{\beta 90} H_{\alpha,\beta}; \\
 & \text{(c)} & J_{\alpha\beta\gamma} K_{\alpha\beta} = F_{\alpha\beta} G_{\alpha\beta\gamma} L_{\gamma 90}.
\end{array}
$$

The concluding sections of Schottky's memoir deal with the previously known mapping of G_4^9 upon the normal G_4^6 described in 51.

We examine further the projective definition of the set P_8^2 in terms of the modular functions. The relation (5) connecting the lines F_{12}, F_{13}, F_{14} reads

$$(11) \qquad \sum_k \pm (p_{k675}^{(01)} p_{k678}^{(01)} p_{k589}^{(01)})^{1/2} F_{1k} = 0 \quad (k = 2, 3, 4).$$

This is to be compared with the projective relation

$$(12) \qquad |134| F_{12} \pm |124| F_{13} \pm |123| F_{14} = 0$$

where $|ijk|$ is the determinant of the coördinates of p_i, p_j, p_k. From a comparison of (11) and (12)

$$(13) \quad |124|/|134| = \pm (p_{3675}^{(01)} p_{3678}^{(01)} p_{3589}^{(01)})^{1/2} / (p_{2675}^{(01)} p_{2678}^{(01)} p_{2589}^{(01)})^{1/2}.$$

This value depends first upon the isolation of $c_0 = 0$, upon the isolation of the index 9 in the definition of $F_{\alpha\beta}$, and upon the isolation of 1, 2, 3, 4 in the formation of the ratio but it should be independent of permutation of 5, 6, 7, 8. This is in fact the case because of the relations 57 (15). For, if the value given in (13) be equated to that obtained by the interchange of 7, 8 then

$$(p_{3675}\, p_{3589}\, p_{2685}\, p_{2579})^{1/2} = \pm (p_{2675}\, p_{2589}\, p_{3685}\, p_{3579})^{1/2}.$$

But these are two of the terms in 57 (15) for which the third, $(p\, p_{6789}\, p_{2378}\, p_{2369})^{1/2} = 0$, since $p = c_0\, c_1 = 0$.

On replacing 4 in (13) by 5 and taking the ratio there results:

$$(14) \qquad \frac{|124|\,|135|}{|125|\,|134|} = \pm \frac{p_{2489}^{(01)}\, p_{3589}^{(01)}}{p_{2589}^{(01)}\, p_{3489}^{(01)}}.$$

This expression for a double ratio of the four lines from p_1 to $p_2, \cdots, p_5$ or, more specifically, of the double ratio of the line pair $p_1 p_2$, $p_1 p_3$ with respect to the point pair, p_4, p_5, in terms of modular functions is the desired projective definition of P_8^2. As before its value is unaltered if the index 8 is replaced by 7 or 6.

Under quadratic transformation A_{123}, P_8^2 is congruent to $P_8'^2$ for which $p_1', p_2', p_3', p_4', p_5'$ is projective to p_1, p_2, p_3, p_5, p_4. Under period transformation I_{1230}, $p_{2489}^{(01)} = c_{02489}\, c_{12489}$ is converted into $c_{13489}\, c_{03489} = p_{3489}^{(01)}$. Thus the effect on the left of (14) is to interchange 4, 5, and on the right to interchange 2, 3; in either case the ratio is inverted. Hence

(15) *If, in the definition* (14) *of the set P_8^2 in terms of modular functions, the moduli are subjected to a period transformation, the set P_8^2 is transformed into a set $P_8'^2$ congruent to P_8^2 under Cremona transformation.*

59. Definition of the nodal set, P_{10}^3, of a symmetroid in terms of modular functions. The theorem **57** (17) states that the equations **57** (16) can be satisfied by certain values c_{ijk} of the odd functions u_{ijk}. The equations then read

$$(1) \qquad \sum_i \pm (q_{i46}^{(90)(78)}\, q_{i56}^{(90)(78)}\, q_{i45}^{(90)(78)}\, q_i^{(90)(78)})^{1/4} = 0 \qquad (i = 1, 2, 3).$$

An equation of this system is determined when the null line, P_{90}, P_{78}, P_{7890}, from which the projection is made, is chosen in one of $255 \cdot 21$ ways (cf. **28**) and when thereafter in the projected space ($p = 2$) any three of the six odd functions are selected. The number of such equations is therefore $255 \cdot 21 \cdot 20$.

These equations take more simple forms when the constants c are replaced by constants e from the equations of Schottky (**28**(1)) in which

$$(2) \qquad f_i = c_i^4, \qquad f_{ijklm} = c_{ijklm}^4, \qquad f_{ijk} = c_{ijk}^4.$$

The conversion is accomplished by the method explained in **45** (11) et seq. The space G allied with each of the terms in (1) is the S_3 determined by $P_{90}, P_{78}, P_{45}, P_{56}$. The space G' syzygetic to G is the S_3 determined by $P_{90}, P_{78}, P_{12}, P_{13}$. The first factor of the first term, c_{14679}, corresponds to the quadric Q_{14679} which is not on the four points $P_{23}, P_{2378}, P_{2390}, P_{1456}$ of G'. Hence $e_{23}, \cdots$ occur in each of the 16 factors in the first term of (1). On extracting the roots indicated by $(c_{14679})^{1/4} = (f_{14679})^{1/16}$ the relation (1) becomes

$$\text{(3a)} \qquad \Sigma_1^3 \pm e_{23}\, e_{2378}\, e_{2390}\, e_{1456} = 0.$$

Recalling that the e's are identified with the points in the finite geometry in S_7, it is apparent that each of the products in (3a) is identified with four points which with the null line e_{78}, e_{90}, e_{7890} make up a null plane. The various types of such products may be obtained by projection from the null line, and selection in the resulting $S_3\,(p=2)$ of one of the 20 ordinary lines. They may also be obtained by transformation of (3a) by the involutions I_{ijkl}. A complete set includes, in addition to (3a), the following (cf. Schottky[63] pp. 280–4):

$$\text{(3b)} \qquad \Sigma_1^3 \pm e_{23}\, e_{14}\, e_{2390}\, e_{1490} = 0;$$

$$\text{(3c)} \qquad \pm e_{12}\, e_{34}\, e_{1290}\, e_{3490} = \begin{vmatrix} e_{1456}\, e_{1478} & e_{1356}\, e_{1378} \\ e_{2456}\, e_{2478} & e_{2356}\, e_{2378} \end{vmatrix};$$

$$\text{(3d)} \qquad \Sigma_1^3 \pm e_{23}\, e_{1456}\, e_{1478}\, e_{1490} = 0;$$

$$\text{(3e)} \qquad e_{12}\, e_{3456}\, e_{3478}\, e_{3490} = \Sigma_1^2 \pm e_{1679}\, e_{1589}\, e_{1570}\, e_{1680};$$

$$\text{(3f)} \qquad \pm e_{56}\, e_{78}\, e_{90}\, e_{1234} = \begin{vmatrix} e_{1579}\, e_{1580} & e_{1589}\, e_{1570} \\ e_{1679}\, e_{1680} & e_{1689}\, e_{1670} \end{vmatrix}.$$

If we set

$$\text{(4a)} \qquad D_{\alpha\beta\gamma\delta} = e_{\alpha\beta\gamma\delta}\, e_{\alpha\beta}\, e_{\alpha\gamma}\, e_{\alpha\delta}\, e_{\beta\gamma}\, e_{\beta\delta}\, e_{\gamma\delta},$$

then on multiplying (3b) by $e_{19}\, e_{10}\, e_{29}\, e_{20}\, e_{39}\, e_{30}\, e_{49}\, e_{40}\, e_{90}^2$ it becomes

$$\text{(3b}'\text{)} \qquad D_{2390}\, D_{1490} \pm D_{3190}\, D_{2490} \pm D_{1290}\, D_{3490} = 0.$$

The system of relations of this type shows that the D_{ijkl} are the determinants formed from the coördinates of four points p_i, p_j, p_k, p_l of a set P_{10}^3 in space. In (4a) the significant factor of $D_{\alpha\beta\gamma\delta}$ is $e_{\alpha\beta\gamma\delta}$—the coplanar condition—and the other factors correspond to coincidences among the points of P_{10}^3. These latter factors are more easily handled in certain combinations introduced by Schottky as follows:

$$\text{(4b)} \quad e = \prod (e_{\alpha\beta}); \quad e_\alpha = \prod_\beta (e_{\alpha\beta}) \quad (\alpha, \beta = 1, 2, \cdots, 9, 0);$$

$$\text{(4c)} \qquad \xi_\alpha = 1/e_\alpha^2; \quad f_{\alpha\beta} = e_\alpha^3\, e_\beta^3 / e\, e_{\alpha\beta}^2.$$

The relations, (3a), $\cdots$, (3f) then become

$$(3a') \qquad \Sigma_1^3 \pm \xi_1^3 f_{14} f_{15} f_{16} D_{2378} D_{2390} D_{1456} = 0;$$

$$(3b') \qquad \Sigma_1^3 \pm D_{2390} D_{1490} = 0;$$

$$(3c') \qquad \pm f_{57} f_{58} f_{67} f_{68} D_{1290} D_{3490} = \frac{\xi_1^2 \xi_2^2 \xi_3^2 \xi_4^2}{\xi_5^2 \xi_6^2 \xi_7^2 \xi_8^2} \begin{vmatrix} f_{14} D_{1456} D_{1478} & f_{13} D_{1356} D_{1378} \\ f_{24} D_{2456} D_{2478} & f_{23} D_{2356} D_{2378} \end{vmatrix};$$

$$(3d') \qquad \Sigma_1^3 \pm \xi_1^2 f_{14} D_{1456} D_{1478} D_{1490} = 0;$$

$$(3f') \qquad \pm \frac{\xi_1^3 \xi_2 \xi_3 \xi_4}{\xi_5 \xi_6 \xi_7 \xi_8 \xi_9 \xi_0} f_{12} f_{13} f_{14} D_{1234} = \begin{vmatrix} D_{1579} D_{1580} & D_{1589} D_{1570} \\ D_{1679} D_{1680} & D_{1689} D_{1670} \end{vmatrix}.$$

The rather complicated transcription of (3e) is omitted.

The right member of (3f′) equated to zero is the condition that, of the points P_{10}^3 defined by (3b′), $p_5, p_6, p_7, p_8, p_9, p_0$ are on a quadric cone with vertex at p_1. If this member is designated by $g_{234,1}$ there follows from the values in the left members of (3f′) that

$$(4) \qquad g_{145,2}\, g_{245,3}\, g_{345,1} = g_{245,1}\, g_{345,2}\, g_{145,3}.$$

This equation does not contain the coördinates of p_4, p_5 and is of degree 6 in each of the other eight points. By comparison with **47** (3a) it is the condition that the eight points may be the nodes of an azygetic 8-nodal surface whence P_{10}^3 is the set of ten nodes of an azygetic 10-nodal quartic surface.

In the terms of (3d′) replace the point p_1 with index 1 by variable x. The terms then are of the form

$$(5) \qquad c_{ij,kl,mn} = D_{x4ij} D_{x4kl} D_{x4mn} \quad (i, \cdots, n = 5, \cdots, 9, 0).$$

This is a cubic cone with vertex at p_4 and containing the lines from p_4 to $p_5, \cdots, p_9, p_0$. A set of four linearly independent cones of this character can be obtained by making four proper selections of ij, kl, mn from the indices $5, \cdots, 9, 0$. Let $i_1, \cdots, n_1; \cdots; i_4, \cdots, n_4$ be four proper selections. If

$F(x)$ is any cubic cone with vertex at p_4 and on $p_5, \cdots, p_9, p_0$ as well as on p_2, p_3, then $F(x)$ will have an equation of the form,

$$F(x) = k_1 c_{i_1 j_1, k_1 l_1, m_1 n_1}(x) + \cdots + k_4 c_{i_4 j_4, k_4 l_4, m_4 n_4}(x) = 0.$$

Then

$$\xi_1^2 f_{14} F(p_1) \pm \xi_2^2 f_{24} F(p_2) \pm \xi_3^2 f_{34} F(p_3) = 0$$

since the coefficient of k_j $(j = 1, \cdots, 4)$ vanishes due to (3d′). But the k's were so chosen that $F(p_2) = 0$ and $F(p_3) = 0$ whence also $F(p_1) = 0$. Thus the cubic cones with vertex at p_4 and on $p_2, p_3, p_5, \cdots, p_9, p_0$ are on p_1 also, or P_{10}^3 has the property that the nine lines from one point to the remaining points are the base lines of a pencil of cubic cones.

We have seen (cf. **53**) that the two geometric properties of P_{10}^3 thus deduced by Schottky ([63] pp. 286–87) imply that P_{10}^3 is the nodal set of a Cayley symmetroid. There follows also from (4a) that the $e_{\alpha\beta}$ and $e_{\alpha\beta\gamma\delta}$ constitute the 255 discriminant factors of the symmetroid (**53** (12)). The identity of further discriminant conditions on P_{10}^3 with these (as expressed in **53** (14)) is in the first case an immediate consequence of the cubic cone property of P_{10}^3; and in the second case is read off at once from (3f′). Hence

(6) *The nodal set P_{10}^3 of the symmetroid, whose double ratios are defined by the theta modular functions*

$$D_{1235} D_{1246}/D_{1236} D_{1245} = e_{1235} e_{1246} e_{35} e_{46}/e_{1236} e_{1245} e_{36} e_{45},$$

is transformed into a set $P_{10}'^3$ congruent to P_{10}^3 under regular Cremona transformation when the moduli are subjected to a period transformation.

For the argument used in connection with **57** (15) can be applied in precisely the same way to show that the effect of the cubic transformation A_{1234} and the period transformation I_{1234} upon these double ratios is the same.

60. The tritangent planes of the space sextic of genus four. We may regard the 120 tritangent planes of G_4^6 as satisfactorily determined if a geometric configuration can be given in terms of whose elements each of the planes is rationally known. Thus the eight points of P_8^2 when individually given serve as a basis for the rational representation of G_4^6 on a quadric cone A and of the individual tritangent planes of G_4^6 (cf. **51**, **58**). The various contact systems defined by groups of tritangent planes are then also rationally known.

The nodal set P_{10}^3 of a symmetroid should play a similar part for the general G_4^6. Let us compare the behavior of this set with that of the analogous set P_8^3 of base points of a net of quadrics in connection with the determination of the bitangents of a quartic curve. Both P_8^3 and P_{10}^3 are *special* sets with projective peculiarities which cause certain of the points to remain fixed under Cremona transformation defined by the others. Each set is congruent to only a finite number (36 and $2^8 \cdot 51$ respectively) of projectively distinct sets of similar character under regular Cremona transformation. Under such transformation the projectively distinct types with ordered points are permuted according to a finite group g_p isomorphic with the modular group ($p = 3, 4$). The discriminant conditions of each set are finite in number and are permuted like the half periods under g_p. In each case the number of projectively distinct types congruent to each other is the number of basis configurations of half periods. Most significant of all is the fact that the coördinates of the points of each set can be expressed in terms of modular functions and that from one set thus determined the sets congruent to it under Cremona transformation arise by period transformation of the moduli.

The analogy fails in one important respect. When the base points, P_8^3, of the net are given there is determined in space a curve, G_3^6, of genus three, the locus of nodes of the net, in birational correspondence with the normal planar quartic, such that any pair of points of P_8^3 is on a bisecant of G_6^3 which cuts G_6^3 in the pair of contacts of a bitangent of the quartic. The discriminant conditions of the set P_8^3 are precisely the discriminant factors of the quartic curve. Thus far the geometric investigation of the symmetroid and its nodal set P_{10}^3 has failed to disclose an algebraic curve of genus four, G_4^t, with the property that a triad of P_{10}^3 will rationally isolate on G_4^t a triad of points which correspond on the normal G_4^6 to the triad of contacts of a tritangent plane; or with the equivalent property that if a discriminant condition on P_{10}^3 is satisfied the genus of G_4^t will be reduced to three. In other words Schottky's derivation of the configuration P_{10}^3 of the symmetroid from the abelian modular functions is not supplemented by an exemplar of the algebraic curve which defines the functions.

That a curve G_4^t, projectively related to the symmetroid, exists is not to be doubted. It should be symmetrically related either to P_{10}^3, or to any P_9^3 contained in P_{10}^3, since P_9^3 uniquely determines the tenth node. It is not necessary that the G_4^t attached to P_{10}^3 should be transformed by regular Cremona transformation into the $G_4'^t$ attached to a set $P_{10}^{3'}$ congruent to P_{10}^3 under such transformation. The $G_4'^t$ is merely to be birationally equivalent to G_4^t. This for example is the behavior of the envelope E^4 defined by its Aronhold set of seven nodes, P_7^2.

One of the main purposes of this book has been to indicate the richness and variety of the analytic, algebraic, and geometric domain within which such a curve G_4^t may be found. Another desideratum of like character is the algebraic curve of genus five with isolated even theta characteristic whose modular group appears in connection with the nodal P_{10}^2 of a planar rational sextic.

From the applications made herein of congruence under Cremona transformation it might be inferred that such applications are restricted in the plane to curves of order $3r$ with r-fold points, in space to surfaces of order $2r$ with r-fold points, etc. If this were true the number of sets with interesting connections would be small. It may well be, however, that the method cited at the close of 56 will provide series which behave like curves of order $3r$ with as many r-fold points as we please.

REFERENCES

[1] Autonne, L. — ... groupes d'ordre fini contenus dans le groupe Cremona. Jour. de Math. (4), 1 (1885), 431–454. [8.

[2] Baker, H. F. — Multiply periodic functions. Cambridge (1907). [41.

[3] Bateman, H. — A type of hyperelliptic curve and the transformations connected with it. Quart. Jour., 37 (1906), 277–286. [38.

[4] Bertini, E. — ... sulle trasformazioni univoche involutorie nel piano. Ann. di Mat. (2), 8 (1877), 244–286. [51.

[5] — Sulle curve razionali per le quali si possono assegnare arbitrariamente i punti multipli. Gior. di Mat. (1), 15 (1877), 329–335. [9.

[6] — Sui sistemi lineari. Ist. Lomb. Rend. (2), 15 (1882), 24–28. [1.

[6.1] Bolza, O. — Darstellung der rationalen ganzen Invarianten der Binärform sechsten Grades durch die Nullwerthe der zugehörigen ϑ-Functionen. Math. Ann., 30 (1887), 478–495. [30.

[7] Brahana, H. R., and Coble, A. B. — Maps of twelve countries with five sides with a group of order 120 .... Amer. Jour., 48 (1926), 1–20. [9.

[8] Caporali, E. — Sulla teoria delle curve piane del quarto ordine. Memorie di Geometria. Naples (1888). [50.

[9] Castelnuovo, G. — Sulla razionalità delle involuzioni piane. Math. Ann., 44 (1894), 125–154. [2.

[10] Cayley, A. — Sur les courbes du troisième ordre. Jour. de Math., 9 (1844), 285–293. [50.

[11] — A memoir on quartic surfaces. Proc. Lond. Math. Soc., 3 (1869), I: 19–69, II: 198–202, III: 234–266. [44, 53.

[12] Clebsch, A. — Über die Anwendung der Abelschen Functionen in der Geometrie. Jour. für Math., 63 (1864), 189–243. [50.

[13] Coble, A. B. — ... form-problems associated with certain Cremona groups. ... solution of equations of higher degree. Trans. Amer. Math. Soc., 9 (1908), 396–424. [8.

[14] — Symmetric binary forms and involutions. Am. Jour. I: 31 (1909), 183–212; II: 31 (1909), 355–364; III: 32 (1910), 333–364. [17.

[15] — ... Moore's cross-ratio group ... solution of the sextic equation. Trans. Amer. Math. Soc., 12 (1911), 311–325. [8, 35, 36, 42.

[16] — ... finite geometry ... characteristic theory of the odd and even theta functions. Trans. Amer. Math. Soc., 14 (1913), 241–276. [22, 23, 24, 26.

[17] — Point sets and allied Cremona groups. Trans. Amer. Math. Soc., I: 16 (1915), 155–198; II: 17 (1916), 345–385; III: 18 (1917) 331–372. [1, 7, 15, 16, 30, 35, 36, 40, 42, 43, 44, 49, 51, 52, 53.

[18] Coble, A. B. — ... An isomorphism between theta characteristics and the $(2p+2)$-point. Annals of Math. (2), 17 (1916), 101–112. [23, 25, 29, 57.

[19] — Theta modular groups determined by point sets. Amer. Jour. 40 (1918), 317–340. [26, 27, 52, 53.

[20] — The ten nodes of the rational sextic and of the Cayley symmetroid. Amer. Jour. 41 (1919), 243–265. [51, 52, 53.

[21] — Associated sets of points. Trans. Amer. Math. Soc., 24 (1922), 1–20. [16, 55.

[22] — Cremona transformations and applications .... Bull. Amer. Math. Soc., 28 (1922), 329–364. [56.

[23] — The equation of the eighth degree. Bull. Amer. Math. Soc., 30 (1924), 301–313. [39.

[24] — Geometric aspects of the abelian modular functions of genus four. Amer. Jour., 46 (1924), 143–192. [17, 54.

[25] Conner, J. R. — The rational sextic curve and the Cayley symmetroid. Amer. Jour., 37 (1915), 29–42. [54.

[26] — Correspondences determined by the bitangents of a quartic. Amer. Jour., 38 (1916), 155–176. [43, 44, 47.

[27] Dickson, L. E., — Linear groups. Leipzig (1901). Teubner. [24.

[28] — Determination of all general homogeneous polynomials expressible as determinants with linear elements. Trans. Amer. Math. Soc., 22 (1921), 167–179. [14.

[29] Everett, H. S. — Determination of all general homogeneous polynomials expressible as determinants .... Trans. Amer. Math. Soc., 24 (1923), 185–194. [14.

[30] Frobenius, G. — ... Beziehungen zwischen den 28 Doppeltangenten einer ebenen Curve vierter Ordnung. Jour. für Math., 99 (1886), 275–314. [44.

[31] Gordan, P. — Invariantentheorie. Leipzig (1885). Teubner. [41.

[32] Grassmann, H. — Die Ausdehnungslehre .... Berlin (1862). A. Enslin. [1.

[33] Hesse, O. — Über die Doppeltangenten der Curven vierter Ordnung. Jour. für Math., 49 (1855), 279–332. [14.

[34] Hudson, Hilda P. — The Cremona transformation of a certain plane sextic. Proc. Lond. Math. Soc. (2), 15 (1916–7), 385–400. [52.

[35] — Cremona transformations in plane and space. Cambridge (1927). [38, 43.

[36] Hudson, R. W. H. T. — Kummer's quartic surface. Cambridge (1905). [32, 41.

[37] Kantor, S. — Premiers fondements ... transformations périodiques univoques. Naples (1891). [4.

[38] — ... periodischen kubischen Transformationen im Raume R_3. Amer. Jour., 19 (1897), 1–59; 382. [47.

[39] — ... Transformationen im R_3, welche keine Fundamentalcurven erster Art besitzen. Acta Math., 21 (1897), 1–78. [8, 15.

[40] Kantor, S. — ... eine Frage über die birationalen Transformationen. Wien. Ber., 112_{2a} (1903), 667–754. [5.

[41] Krazer, A. — Lehrbuch der Thetafunktionen. Leipzig (1903). Teubner. [18, 19, 21, 22, 26, 31, 32, 34, 39.

[42] Krazer, A.; Wirtinger, W. — Abel'sche Funktionen und allgemeine Thetafunctionen. Encyklopädie II B 7, 604–873. [18.

[43] Marletta, G. — Sulla identità cremoniana di due curve piane. Palermo Rend., 24 (1907), 229–242. [13.

[44] Meyer, W. F. — Apolarität und rationale Curven. Tübingen (1883). Fues. [17, 54, 55.

[44.1] Miller, G. A.; Blichfeldt, H. F.; Dickson, L. E. — Theory and applications of finite groups. New York (1916). Wiley and Sons. [6.

[45] Milne, W. P. — Sexactic cones and tritangent planes of the same system of a quadri-cubic curve. Proc. Lond. Math. Soc. (2), 21 (1922–3), 373–380. [50.

[46] Montesano, D. — Su alcune tipi di corrispondenze cremoniane spaziali collegati alle corrispondenze birazionali piane di ordine n. Napoli Rend. (3), 27 (1921), 164–175. [55.

[47] Moore, E. H. — The cross-ratio group of $n!$ Cremona transformations of order $n-3$ in flat space of $n-3$ dimensions. Amer. Jour. 22 (1900), 279–291. [8. 36.

[48] Morley, F. — On the Lüroth quartic curve. Amer. Jour., 41 (1919), 279–282. [49.

[49] Musselman, J. R. — The set of eight self-associated points in space Amer. Jour., 40 (1918), 69–86. [16.

[50] Pascal, E.; Timerding, H. — Repertorium der höheren Mathematik. Leipzig: II 1 (1910), II 2 (1922). Teubner. [14, 43.

[51] Picard, E., and Simart, G. — Théorie des fonctions algébriques de deux variables indépendantes. Vol. II. Paris (1906). Gauthier-Villars. [1.

[52] Report: — Selected topics in algebraic geometry. Bulletin, National Research Council, Washington, No. 63 (April 1928). [4, 5, 6, 10, 32, 36, 38, 43, 55.

[53] Reye, T. — Über lineare Systeme und Gewebe von Flächen zweiten Grades. Jour. für Math., 82 (1877), 54–83. [54.

[54] Rohn, K. — Über Flächen 4. Ordnung mit acht bis sechzehn Knotenpuncten. Leipzig Berichte (1884), 52–60. [53.

[55] — Die Flächen vierter Ordnung .... Jablonowski'sche Preisschrift. Leipzig (1886). Hirzel. [53.

[56] — Die Flächen vierter Ordnung hinsichtlich ihrer Knotenpunkte und ihrer Gestaltung. Math. Ann., 29 (1887), 81–96. [53.

[57] Roth, P. — Über Beziehungen zwischen algebraischen Gebilden vom Geschlechte drei und vier. Monatshefte, 22 (1911), 64–88. [50, 57.

[58] Schottky, F. — Abriss einer Theorie der Abelschen Functionen von drei Variabeln. Leipzig (1880). Teubner. [46.

[59] Schottky, F. — Zur Theorie der Abelschen Functionen von vier Variabeln. Jour. für Math., 102 (1888), 304–352. [28, 57.

[60] — Über specielle Abelsche Functionen vierten Ranges. Jour. für Math., 103 (1888), 185–203. [51, 58.

[61] — ... Beziehungen zwischen den sechzehn Thetafunctionen von zwei Variabeln. Jour. für Math., 105 (1889), 233–249. [37, 41.

[62] — Eine algebraische Untersuchung über Thetafunctionen von drei Argumenten. Jour. für Math., 105 (1889), 269–297. [47.

[63] — Über die Moduln der Thetafunctionen. Acta Mat., 27 (1903), 235–288. [28, 30, 45, 46, 57, 59.

[64] Severi, F. (Löffler). — Vorlesungen über algebraische Geometrie. Leipzig (1921). Teubner. [1, 11, 14.

[65] Snyder, V. and Sharpe, F. R. — Space involutions defined by a web of quadrics. Trans. Amer. Math. Soc., 19 (1918), 275–290. [55.

[66] Speiser, A. — Die Theorie der Gruppen von endlicher Ordnung. Berlin (1927). [6.

[67] Stahl, H. — Theorie der Abel'schen Functionen. Leipzig (1896). Teubner. [18, 21, 34, 45, 47.

[68] Stahl, W. — Über die Fundamentalinvolutionen auf rationalen Curven. Jour. für Math., 104 (1889), 38–61. [54.

[69] — Zur Erzeugung der ebenen rationalen Curven. Math. Ann., 38 (1891), 561–585. [49.

[70] Veronese, G. — ... projectivischen Verhältnisse der Räume von verschiedenen Dimensionen .... Math. Ann., 19 (1882), 161–234. [48.

[71] — La superficie omaloide normale a due dimensioni e del quarto ordine dello spazio a cinque dimensioni .... Rom. Acc. L. Mem. (3), 19 (1884). [48.

[72] Weber, H. — Über einige Ausnahmefälle in der Theorie der Abelschen Functionen. Math. Ann., 13 (1878), 35–48. [51.

[73] Wiman, A. — Zur Theorie ... birationalen Transformationen in der Ebene. Math. Ann., 48 (1896–7), 195–240. [51, 58.

[74] Wirtinger, W. — Über eine Verallgemeinerung der Theorie der Kummer'schen Fläche und ihrer Beziehungen zu den Thetafunctionen zweier Variabeln. Monatshefte, 1 (1890), 113–128. [32.

[75] — Untersuchung über Abel'sche Functionen vom Geschlechte drei. Math. Ann., 40 (1892), 261–312. [50.

[76] — Untersuchung über Thetafunctionen. Leipzig (1895). Teubner. [33, 50.

[77] Zariski, O. — On hyperelliptic ϑ-functions with rational characteristics. Amer. Jour., 50 (1928), 315–344. [51.